Lecture Notes in Computer Science

Lecture Notes in Artificial Intelligence 15154

Founding Editor

Jörg Siekmann

Series Editors

The series Lecture Notes in Artificial Intelligence (LNAI) was established in 1988 as a topical subseries of LNCS devoted to artificial intelligence.

The series publishes state-of-the-art research results at a high level. As with the LNCS mother series, the mission of the series is to serve the international R & D community by providing an invaluable service, mainly focused on the publication of conference and workshop proceedings and postproceedings.

Ching Yee Suen · Adam Krzyzak ·
Mirco Ravanelli · Edmondo Trentin ·
Cem Subakan · Nicola Nobile
Editors

Artificial Neural Networks in Pattern Recognition

11th IAPR TC3 Workshop, ANNPR 2024
Montreal, QC, Canada, October 10–12, 2024
Proceedings

Springer

Editors
Ching Yee Suen
Concordia University
Montreal, QC, Canada

Adam Krzyzak
Concordia University
Montreal, QC, Canada

Mirco Ravanelli
Concordia University
Montreal, QC, Canada

Edmondo Trentin
Università degli Studi di Siena
Siena, Italy

Cem Subakan
Université Laval
Quebec, QC, Canada

Nicola Nobile
Concordia University
Montreal, QC, Canada

ISSN 0302-9743 ISSN 1611-3349 (electronic)
Lecture Notes in Artificial Intelligence
ISBN 978-3-031-71601-0 ISBN 978-3-031-71602-7 (eBook)
https://doi.org/10.1007/978-3-031-71602-7

LNCS Sublibrary: SL7 – Artificial Intelligence

This Springer imprint is published by the registered company Springer Nature Switzerland AG
The registered company address is: Gewerbestrasse 11, 6330 Cham, Switzerland

Preface

This volume contains the papers presented at the 11th IAPR TC3 workshop on *Artificial Neural Networks in Pattern Recognition (ANNPR 2024)*, held at Concordia University, Montréal, Québec, Canada on October 10-12, 2024. ANNPR 2024 follows the success of the ANNPR workshops of 2003 (Florence), 2006 (Ulm), 2008 (Paris), 2010 (Cairo), 2012 (Trento), 2014 (Montréal), 2016 (Ulm), 2018 (Siena), 2020 (Winterthur), and 2022 (Dubai). The series of ANNPR workshops has acted as a major forum for international researchers and practitioners from the communities of pattern recognition and machine learning based on artificial neural networks.

Among the 46 submitted manuscripts, the program committee selected 27 papers after a single-blind peer review. The scientific program was organized in regular oral presentations and poster sessions. The workshop was enriched by two *keynotes*: *Responsible AI: AI Ethics and Safety* given by Jay Liebowitz, Harrisburg University of Science and Technology, USA and *A Unified Framework of Data and Knowledge Environment of Machine Learning* given by Witold Pedrycz, University of Alberta, Canada. We also hosted two *invited talks*: *Generative models as queryable world models* given by Adriana Romero Soriano, Research Scientist at Meta AI (FAIR), Canada and *Neural Speech and Audio Coding* given by Minje Kim, University of Illinois at Urbana-Champaign, USA.

The workshop was made possible by the support of numerous individuals and organizations. First of all, we are grateful to all the authors and the members of the program committee. Special thanks to the Gina Cody School of Engineering and Computer Science of Concordia University for all the arrangements and support for hosting ANNPR 2024. ANNPR 2024 received support from the International Association for Pattern Recognition (IAPR) and its Technical Committee on Neural Networks. Additional support came from the Applied AI Institute and the Centre for Pattern Recognition and Machine Intelligence (CEMPARMI) at Concordia University, the Institute of Intelligence and Data at Laval University, the Centre for Research on Brain, Language, and Music (CRBLM), and Nexadeeds company. Finally, we wish to express our gratitude to Springer for publishing our workshop proceedings within their LNCS/LNAI series. We hope that readers of this volume will be inspired by its contributions.

Ching Yee Suen
Adam Krzyzak
Mirco Ravanelli
Edmondo Trentin
Cem Subakan
Nicola Nobile

Organization

Conference Chairs

Ching Y. Suen	Concordia University, Canada
Adam Krzyzak	Concordia University, Canada
Mirco Ravanelli	Concordia University, Canada
Edmondo Trentin	Università di Siena, Italy
Cem Subakan	Université Laval, Canada
Nicola Nobile	Concordia University, Canada

Program Committee

Alessandro Betti	Institute for Advanced Studies Lucca, Italy
Alessio Brutti	Fondazione Bruno Kessler, Italy
Arsenii Gorin	Ubenwa Health, Canada
Artem Ploujnikov	Université de Montréal, Canada
Benjamin Leblanc	Laval University, Canada
Davide Borra	University of Bologna, Italy
Dimitrios Bralios	University of Illinois at Urbana-Champaign, USA
Eleonora Mancini	University of Bologna, Italy
Erwin M. Bakker	Leiden University, Netherlands
Fırat Öncel	Concordia University, Canada
Francesco Paissan	Fondazione Bruno Kessler, Italy
François Grondin	Université de Sherbrooke, Canada
Friedhelm Schwenker	Ulm University, Germany
Haici Yang	Indiana University, USA
Heitor Guimarães	Institut national de la recherche scientifique, Canada
Jayneel Parekh	Télécom Paris, France
Julien Laumonier	Université Laval, Canada
Junhao Wang	Ubenwa Health, Canada
Kota Shamanth Ramanath Nayak	Concordia University, Canada
Krishna Subramani	University of Illinois at Urbana-Champaign, USA
Luca Della Libera	Concordia University, Canada
Mahdi S. Hosseini	Concordia University, Canada
Marco Maggini	University of Siena, Italy
Mathieu Bazinet	Université Laval, Canada

Maxime Heuillet	Université Laval, Canada
Mohammad M. Hamdan	Yarmouk University, Jordan
Monica Bianchini	University of Siena, Italy
Neamat El Gayar	Heriot-Watt University, United Arab Emirates
Pooneh Mousavi	Concordia University, Canada
Renato De Mori	McGill University, Canada
Samuele Cornell	Università Politecnica delle Marche, Italy
Sara Karami	Université Laval, Canada
Seyed Mahed Mousavi	University of Trento, Italy
Shigeo Abe	Kobe University, Japan
Shubham Gupta	Université Laval, Canada
Thilo Stadelmann	ZHAW School of Engineering, Switzerland
Umberto Cappellazzo	University of Trento, Italy
Wei Pang	Heriot-Watt University, UK
Yannick Estève	University of Avignon, France
Yiming Xiao	Concordia University, Canada
Zeyu Zhao	University of Edinburgh, UK
Zhepei Wang	Amazon, USA

Sponsoring Institutions

International Association for Pattern Recognition (IAPR)
Technical Committee 3 (TC 3) of the IAPR
Gina Cody School of Engineering and Computer Science, Concordia University, Montréal, Canada
Applied AI Institute, Concordia University, Montréal, Canada
Centre for Pattern Recognition and Machine Intelligence (CEMPARMI), Concordia University, Montréal, Canada
Institute of Intelligence and Data, Laval University, Montréal, Québec City, Canada
Centre for Research on Brain, Language and Music (CRBLM), Montréal, Canada
Nexadeeds, Montréal, Canada

Contents

Applications in NLP, Speech, and Music

Applications in Environmental and Biological Sciences

Learning Algorithms and Architectures

Learning Graph Matching with Graph Neural Networks

Kalvin Dobler(✉) and Kaspar Riesen

Institute of Computer Science, University of Bern,
Neubrückstrasse 10, 3012 Bern, Switzerland
{kalvin.dobler,kaspar.riesen}@unibe.ch

Abstract. Graph matching aims at evaluating the dissimilarity of two graphs by defining a constrained correspondence between their nodes and edges. *Error-tolerant graph matching*, for instance, introduces the concept of a cost for penalizing structural differences in the matching. A popular method for this approach is *graph edit distance*, which is based on the cost of the minimal sequence of edit operations to transform a source graph into a target graph. One of the main problems of graph edit distance is the computational complexity, which is exponential in its exact form. In recent years, several approximation methods for graph edit distance have been presented which offer polynomial runtimes. In this paper, we approach the graph edit distance problem in a fundamentally different way. In particular, we propose to learn graph edit distance by means of graph neural networks. In a comprehensive experimental evaluation on six data sets, we verify that our approach not only provides comparable classification performance but also substantially reduces the runtime compared to a prominent algorithm for approximate graph edit distance computation.

Keywords: Structural Pattern Recognition · Graph Matching · Graph Edit Distance · Graph Representation Learning

1 Introduction

Pattern recognition can be categorized into two main approaches depending on the formal representation of the underlying data. *Statistical pattern recognition* employs n-dimensional feature vectors for data representation, whereas *structural pattern recognition* employs trees or graphs for the same task. On one hand, we observe an increased algorithmic complexity and lack of mathematical operations in the graph domain. On the other hand, graphs inherently capture richer information than ordered and fixed-size vectors of real numbers. Due to their flexibility and power, graph structures arise in diverse real-world pattern recognition applications, such as malware classification [1] or detection of Alzheimer's Disease [2], to name just two examples.

Research on structural pattern recognition can be roughly subdivided into three eras of research. The first era is primarily resolved around the graph matching paradigm [3], while the paradigm of the second era relies on adapting the

C. Y. Suen et al. (Eds.): ANNPR 2024, LNAI 15154, pp. 3–12, 2024.
https://doi.org/10.1007/978-3-031-71602-7_1

kernel theory, originally developed for vectorial representations, to graphs (e.g., by means of Random Walk Kernels [4], or similar). In parallel to the advent of neural networks, there was a surge in adapting and generalizing deep neural networks to non-Euclidean spaces. This transformative phase saw the maturation of graph neural networks in the third era of structural pattern recognition (e.g., by means of Graph Convolution Networks [5], or similar).

The main contribution of the present paper is to bridge the gap between the first and third era of structural pattern recognition. In particular, we connect *graph neural networks* (GNNs) with *graph edit distance* (GED) [6]. Although introduced about 40 years ago, GED is still considered one of the most popular error-tolerant graph matching methods. GED computes the minimum cost edit path to transform a source graph into a target graph. As the exact computation of GED has exponential time complexity w.r.t. the number of nodes, many approximation algorithms have been proposed for this task (e.g. BP-GED [7] or Hausdorff-GED [8]). In the present paper, we propose a novel approach to speed up GED computation - in particular, we propose a GNN-based approach to derive the edit operations for GED.

Our work is somewhat related to previous works that leverage deep learning for graph similarity computation. For instance, both GMN [9] and SimGNN [10] use GNNs as encoders and compute graph similarities based on the embedded graphs. A similar line of research relies on unsupervised graph similarity learning using contrastive learning, e.g. CGMN [11]. The approach proposed in the present paper differs to these approaches as follows. Rather than relying on the embedding vector resulting from the GNN model, we intend to use the node embeddings in a more direct way to define explicit node-to-node mappings. Moreover, unlike vector similarity computation, we aim to compute interpretable graph dissimilarities. That is, the resulting values directly relate to edit operations to transform both nodes and edges of the underlying graphs.

The remainder of this paper is organized as follows. First, in Sect. 2, we provide the necessary background to comprehend our novel approach. Then, in Sect. 3, we describe the details of the proposed graph matching procedure. In Sect. 4, we present the experimental setup and provide the results of our experimental evaluation. Finally, in Sect. 5, we derive some conclusions and discuss future research avenues.

2 Background

In order to keep the paper self-contained, we briefly review some concepts, viz. Graph Neural Networks and the graph edit distance framework BP-GED.

2.1 Graph Theory Basics

In the present paper, we denote a *graph* by $G = (\mathcal{V}, \mathcal{E}, \mathbf{X})$, where $\mathcal{V}$ is a set of n *nodes*, $\mathcal{E}$ is a set of *edges*, i.e. unordered pairs of nodes ($\mathcal{E} \subseteq \mathcal{V} \times \mathcal{V}$), and $\mathbf{X} \in \mathbb{R}^{n \times d}$ is an $n \times d$ feature matrix, where d is the dimension of the node features (in our scenario, we consider node labels only). We encode the structure of a graph by means of the square adjacency matrix $\mathbf{A} := \{0, 1\} \in \mathbb{R}^{n \times n}$, where $\mathbf{A}_{i,j} = 1$ if $(v_i, v_j) \in \mathcal{E}$, and 0 otherwise.

2.2 Graph Neural Networks (GNNs)

GNNs generalize the set of methods of deep neural networks to graph-structured data. At a high level, they learn effective representations for each node within a graph by propagating information from their local neighborhood, which leverages the overall structure as well as local patterns. The architecture of a typical GNN model consists of several layers, each of them contains a message-passing mechanism, aggregation functions, and learnable parameters, which can be expressed as follows:

$$h_v^{t+1} = \text{UPDATE}\left(h_v^{(t)}, \text{AGGREGATE}\left(\{h_u^{(t)}, \forall u \in \mathcal{N}(v)\}\right)\right),$$

where $h_v^{(t)}$ corresponds to the representation of node v at the t-th layer. During each message-passing step, the AGGREGATE function takes as input the set of representations of the neighbors of node v to generate a new message based on this aggregated neighborhood information. The UPDATE function combines this message with the current node representation of v to generate the updated representation. The specific form of the UPDATE function depends on the architecture and design of the GNN. It is usually a transformation, which may contain parameters such as weights and biases that are learned during the learning process. This new representation is then passed on to the next GNN layer or finally used to make predictions, either on the node or on the graph level.

2.3 Bipartite Graph Edit Distance (BP-GED)

Bipartite Graph Edit Distance (BP-GED) [7] is an approximate algorithm for the computation of the GED in a substantially faster way than traditional methods. In this approach, the *quadratic assignment problem* (QAP) of GED computation is reduced to an instance of a *linear sum assignment problem* (LSAP). To this end, the quadratic term of the QAP is omitted, neglecting the structural relationships between the nodes in a global way. However, in order to consider knowledge about the local graph structure, the minimum sum of edge edit operation costs, implied by the corresponding node operation, is integrated in the LSAP solving problem. BP-GED can be used to derive two distance values from the LSAP solution, namely a lower and an upper bound on the true GED [12].

3 Graph Matching Through Node Embedding

As stated in the introduction, we aim to bridge the gap between the first and third era of structural pattern recognition. More specifically, we are interested in learning explicit matchings of graphs in the context of GED with the aid of

GNNs. To this end, we first use GNNs for learning effective node embeddings. Node embedding involves encoding nodes into real-valued vectors in a lower-dimensional space that capture both their position in the graph and the structure of their local neighborhood.

In the present paper, we adopt a graph-focused approach. That is, we train a *Graph Isomorphism Network* (GIN) [13] in an end-to-end fashion for graph classification (other GNN architectures could be used as well). Next, however, we do not rely on this final classification. Rather, we extract the node embeddings of the final message passing layer of the trained GIN and define an optimal mapping between these node embeddings using a suitable metric for real feature vectors. Finally, we feed the found node mapping back into the graph domain to compute the corresponding GED.

Formally, we assume two graphs, viz. a source graph $G = (\mathcal{V}, \mathcal{E}, \mathbf{X})$ and a target graph $G' = (\mathcal{V}', \mathcal{E}', \mathbf{X}')$, with n and n' nodes, respectively. In our procedure, we ensure that the graph with the smaller number of nodes is consistently regarded as the source graph (i.e., we can assume $n \leq n'$ from now on). Using the trained GIN model, we obtain n and n' embeddings $\mathbf{x}_1, \ldots, \mathbf{x}_n \in \mathbb{R}^m$ and $\mathbf{x}_1, \ldots, \mathbf{x}_{n'} \in \mathbb{R}^m$, which describe the n and n' nodes of G and G', respectively, with real-valued vectors of dimension m (m results from the dimension of the last layer of the GIN).

Next, using the Euclidean distance, we compute all pairwise distances between each of the n node embeddings of the source graph G to each of the n' node embeddings of the target graph G', resulting in a $n \times n'$ distance matrix $\mathbf{D} = (d_{ij})_{n \times n'}$. Given matrix $\mathbf{D} = (d_{ij})_{n \times n'}$, we now solve the optimization problem to find a permutation $(\pi_1, \ldots, \pi_n)$ of the integers $(1, \ldots, n)$ such that the sum

$$\sum_{i=1}^{n} d_{i\pi_i}$$

becomes minimal (which is an instance of an LSAP). Depending on whether $n < n'$ or $n = n'$, the optimal permutation $(\pi_1, \ldots, \pi_n)$ corresponds to an injective or a bijective assignment of the underlying nodes, respectively. Formally, we have an assignment

$$\lambda = \{u_1 \rightarrow v_{\pi_1}, u_2 \rightarrow v_{\pi_2}, \ldots, u_n \rightarrow v_{\pi_n}\}$$

that includes the substitutions of all n nodes of G to either all nodes of G', or a subset of nodes of G'. In the injective case with $n < n'$, we complete λ with the corresponding insertions. For example, assume that $n' - n = k > 0$ and that the k nodes of G' not used in λ are $v_{(1)}, \ldots, v_{(k)}$, then we complete λ with the k edit operations

$$\{\emptyset \rightarrow v_{(1)}, \ldots, \emptyset \rightarrow v_{(k)}\},$$

where $\emptyset$ refers to the null node. Similar to the algorithm BP-GED [7], we define a graph dissimilarity based on the found permutation λ. This dissimilarity corresponds to the overall cost of editing the source graph G into the target graph G'. Formally, we subdivide the overall cost into the sum of costs of all node edit operations $C_\lambda^{\langle V\rangle}$ and the sum of all edge edit operations $C_\lambda^{\langle E\rangle}$ implied by the node-to-node assignment λ. Formally,

$$\begin{aligned} C_\lambda^{\langle V\rangle} &= \sum_{i=1}^{n} c(u_i \rightarrow v_{\pi_i}) + \sum_{i=1}^{n'} \mathbb{1} c(\emptyset \rightarrow v_i) \\ C_\lambda^{\langle E\rangle} &= \sum_{i=1}^{n} \sum_{j=i+1}^{n} c(\mathbf{A}_{ij} \rightarrow \mathbf{A}'_{\pi_i\pi_j}) + \sum_{i=1}^{n'} \sum_{j=1}^{n'} \mathbb{1} c(\emptyset \rightarrow \mathbf{A}'_{ij}) \end{aligned} \tag{1}$$

where $\mathbb{1}$ returns 1 if index i is not present in the original assignment λ, and $\mathbf{A}$ and $\mathbf{A}'$ refer to the adjacency matrices of G and G', respectively. The cost function $c(\cdot \rightarrow \cdot)$ in Eq. (1) measures the magnitude of the corresponding edit operation on both nodes and edges. The cost function is often optimized for each application. Yet, in our scenario, we choose unit cost for insertions of nodes and edges as well as for substitutions of nodes with unequal labels.

By means of the adaptation of the node assignment λ into the model of Eq. (1), we ensure that the structural relationships between the nodes are taken into consideration in a global and consistent way. That is, the defined distance corresponds to a complete and admissible edit path between G and G'. From now on, we term the GED computed by this procedure as GNN-GED.

Although our new model GNN-GED and BP-GED [7] have similarities, there are two major advantages of our novel approach. In contrast to BP-GED, which solves the assignment problem based on a cost matrix with limited structural knowledge, GNN-GED solves the assignment based on node embeddings which implicitly encode wide structural knowledge by means of the message passing mechanism of GNNs. Furthermore, it can be assumed that GNN-GED is faster than BP-GED since, in contrast to BP-GED, no optimal assignment of the local edge structure has to be found during the LSAP solving step.

4 Experimental Evaluation

The goal of the experimental evaluation is twofold. In the first part, we aim at comparing the classification performance and runtime of our approach GNN-GED and BP-GED [7]. In the second part, we provide scatter plots to visually inspect the correlation of the obtained values between both methods.

4.1 Data Sets

Our evaluation is conducted on six data sets from the TUDataset graph repository [14]. We use three molecules data sets that consist of small molecules (MUTAG, AIDS and NCI1). The MUTAG data set consists of *mutagenic* and *non-mutagenic* compounds, the AIDS data set consists of *confirmed active* and *confirmed inactive* molecules, depending on their effect in protecting human cells from HIV, and the NCI1 data set consists of *active* and *inactive* molecules, depending on their activity for growth inhibition of non-small cell lung cancer. We also use three bioinformatics data sets that consist of macromolecules (DD and ENZYMES) and brain networks (OHSU). The DD data set consists of *enzymes* and *non-enzymes* proteins, and the ENZYMES data set consists of enzymes according to the 6 EC-top level classes. The OHSU data set consists of functional brain networks based on hyperactive-impulsive regions.

4.2 Experimental Setup

For each data set, we first train a GIN model [13] in an end-to-end fashion to classify the corresponding graphs. The model consists of GIN layers (implemented as linear layer, batch normalization, ReLU, linear layer, and ReLU), graph embedding concatenation, and a final multilayer perceptron (MLP) (implemented as two linear layers). In our approach, we concatenate the embeddings instead of adding them in order to preserve the information from previous layers.

We consider as hyperparameters the number of GIN layers $\in \{3, 4, 5\}$, the hidden node feature dimension $\in \{32, 64, 128\}$, and the learning rate $\in \{0.0005, 0.0006, \ldots, 0.01\}$. To finetune these parameters, we perform a 10-fold cross-validation on each data set. That is, we produce stratified splits of each shuffled data set into train and test sets using a $80\% - 20\%$ split size. The former split is used for 10-fold cross-validation and subsequent training of the models, whereas the latter split is left out from the beginning and only considered for testing. In Table 1 we show the best performing parameters and the average validation accuracy of the GNN for each data set.

Table 1. Best performing hyperparameters and the average validation accuracies of the GNN in a 10-fold cross-validation for all data sets.

Dataset	# GIN layers	Node feature dimension	Learning rate	Acc. (%)
MUTAG	3	64	$3e-4$	85.3 ± 8.8
AIDS	3	64	$1e-4$	98.4 ± 0.8
NCI1	3	64	$5e-4$	73.8 ± 2.1
ENZYMES	5	64	$5e-4$	41.6 ± 4.1
OHSU	5	64	$5e-4$	54.1 ± 18.9
DD	5	64	$1e-3$	73.7 ± 6.2

As a reminder, we only use the GIN as an auxiliary encoder to generate embeddings from the n nodes of a graph. These embeddings are then optimally mapped to each other in a second step to finally compute the novel GNN-GED (as thoroughly described in Sect. 3). Given the GNN-GED, any distance-based classifier can now be used for the final classification. In the present study, we employ a k-nearest neighbor classifier (k-NN) with $k \in \{3, 5, 7, 11\}$.

4.3 Classification Accuracy and Runtime Computation

First, we investigate the differences of the accuracies of a k-NN classifier that either operates on the reference distances (calculated by means of BP-GED [7]) or on our novel distances obtained by GNN-GED. In Table 2 we report the F1 measures for both variants and for all values of k on all datasets. We observe that both systems, BP-GED and GNN-GED, achieve quite similar recognition accuracies. Our novel distance measure generally achieves better results than BP-GED on the three small-molecules as well as on the ENZYMES data set, while the reference system BP-GED outperforms our novel dissimilarity function on OHSU and DD.

Table 2. F1-scores (%) for all data sets with the associated number of neighbors k. The best result per data set is shown in boldface.

Dataset	GNN-GED				BP-GED [7]			
	$k = 3$	$k = 5$	$k = 7$	$k = 11$	$k = 3$	$k = 5$	$k = 7$	$k = 11$
MUTAG	89.8	89.8	87.5	**91.6**	90.5	**91.6**	85.1	91.6
AIDS	**100.0**	100.0	100.0	100.0	99.8	99.7	99.5	98.4
NCI1	**80.6**	79.8	77.5	76.4	77.2	74.5	74.8	73.4
ENZYMES	**41.6**	37.5	36.6	33.3	36.6	30.1	29.1	23.3
OHSU	50.0	63.2	52.6	63.6	63.1	**69.5**	66.6	66.6
DD	25.0	33.3	33.3	33.3	**53.8**	53.8	53.8	53.8

Next, in Table 3, we compare the runtime for computing all pairwise dissimilarities on one specific dataset, as well as the runtime per individual matching (averaged over five runs). Overall, the runtime computations of our novel approach GNN-GED is consistently lower than the runtime of BP-GED. We observe speed-up factors ranging from at least 3 to at most 11. The most significant runtime improvements are observed with OHSU and DD, which represent data sets with a high average number of nodes and edges. This trend may indicate that our approach is notably more efficient for more complex graphs.

4.4 Qualitative Analysis

For a qualitative analysis, we present scatter plots illustrating the correlations between the distances derived from GNN-GED and BP-GED. Each point in Fig. 1 represents a dissimilarity computed via GNN-GED on the x-axis and the corresponding distance calculated using BP-GED on the y-axis. Dark and light gray points correspond to the upper and lower bound values derived from BP-GED, respectively. We show the results on MUTAG only, but similar plots are obtained on the other data sets. The values are normalized to the range $[0, 1]$. In general, the values obtained with our novel approach underestimate and overestimate the upper and lower bounds distance values of BP-GED, respectively.

Table 3. Total computation time and time per graph matching in minutes and milliseconds, respectively (averaged over five runs).

Dataset	Computation time (min)		Matching time (ms)		Speed-up
	GNN-GED	BP-GED	GNN-GED	BP-GED	
MUTAG	0.02	0.07	0.21	0.72	3.5
AIDS	1.80	7.30	0.17	0.69	4.1
NCI1	20.20	84.50	0.45	1.88	4.2
ENZYMES	0.40	2.10	0.43	2.16	5.3
OHSU	0.03	0.20	2.40	13.05	6.7
DD	67.10	739.90	18.11	199.69	11.0

One can thus conclude that our approach tends to provide a particularly accurate approximation of the true GED.

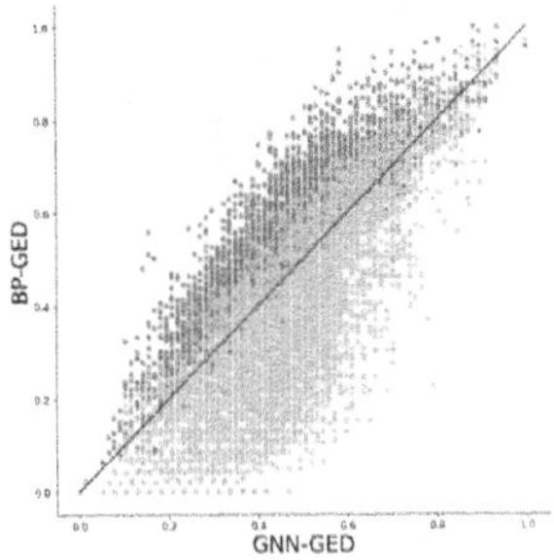

Fig. 1. GNN-GED (x-axis) versus BP-GED (y-axis). Dark and light gray points correspond to the upper and lower bound values derived from BP-GED, respectively (Color figure online)

5 Conclusions and Future Work

In the present paper, we propose and research a novel graph matching framework using GNNs. In particular, we first train a GNN in an end-to-end fashion in order to learn node representations. The node representations are then used to obtain an optimal assignment of every node in the source graph to every node in the target graph. Similarly to the BP-GED framework [7], we then define the overall cost of editing the source graph into the target graph based on the found assignment. In an experimental evaluation, we observe that our approach substantially improves the runtime of GED computation, while achieving comparable classification accuracies as BP-GED. Moreover, by means of a qualitative analysis we observe that the dissimilarity values of our novel approach tend to approximate the true GED quite well.

We see several promising avenues to be pursued in future work. First, other neural network architectures rather than GIN may be employed. Moreover, the graph-focused task for classification serves as a proxy to extract node representations. We plan to develop training losses that align more closely with our original objective. Another interesting idea would be to investigate different approaches that solve the assignment problem on the node embeddings (rather than LSAP solvers).

Acknowledgements. This study is funded by the Swiss National Science Foundation (SNSF) Project Nr 200020_219388.

Disclosure of Interests. The authors have no competing interests to declare that are relevant to the content of this article.

References

1. Kinable, J., Kostakis, O.: Malware classification based on call graph clustering. J. Comput. Virol. **7**(4), 233–245 (2011). https://doi.org/10.1007/S11416-011-0151-Y
2. Curado, M., et al.: Early detection of Alzheimer's disease: detecting asymmetries with a return random walk link predictor. Entropy **22**(4), 465 (2020). https://doi.org/10.3390/E22040465
3. Conte, D., et al.: Thirty years of graph matching in pattern recognition. Int. J. Pattern Recogn. Artif. Intell. **18**(3), 265–298 (2004). https://doi.org/10.1142/S0218001404003228
4. Kashima, H., Tsuda, K., Inokuchi, A.: Marginalized kernels between labeled graphs. In: Machine Learning, Proceedings of the Twentieth International Conference (ICML 2003), 21–24 August 2003, Washington, DC, USA. Ed. by Tom Fawcett and Nina Mishra. AAAI Press, pp. 321–328 (2003). http://www.aaai.org/Library/ICML/2003/icml03-044.php
5. Kipf, T.N., Welling, M.: Semi-supervised classification with graph convolutional networks. CoRR abs/1609.02907 (2016). arXiv:1609.02907
6. Bunke, H., Allermann, G.: Inexact graph matching for structural pattern recognition. Pattern Recogn. Lett. **1**(4), 245–253 (1983). https://doi.org/10.1016/0167-8655(83)90033-8
7. Riesen, k., Bunke, H.: Approximate graph edit distance computation by means of bipartite graph matching. Image Vis. Comput. **27**(7), pp. 950–959 (2009). https://doi.org/10.1016/J.IMAVIS.2008.04.004
8. Fischer, A., et al.: Approximation of graph edit distance based on Hausdorff matching. Pattern Recogn. **48**(2), 331–343 (2015). https://doi.org/10.1016/J.PATCOG.2014.07.015
9. Li, Y., et al.: Graph matching networks for learning the similarity of graph structured objects. In: Proceedings of the 36th International Conference on Machine Learning, ICML 2019, 9–15 June 2019, Long Beach, California, USA. Ed. by Kamalika Chaudhuri and Ruslan Salakhutdinov. vol. 97. Proceedings of Machine Learning Research. PMLR, pp. 3835–3845 (2019). http://proceedings.mlr.press/v97/li19d.html

10. Bai, Y., et al.: SimGNN: a neural network approach to fast graph similarity computation. In: Proceedings of the Twelfth ACM International Conference on Web Search and Data Mining, WSDM 2019, Melbourne, VIC, Australia, 11–15 February 2019. Ed. by J. Shane Culpepper et al. ACM, 2019, pp. 384–392. https://doi.org/10.1145/3289600.3290967
11. Di Jin, et al.: CGMN: a contrastive graph matching network for self-supervised graph similarity learning. In: Proceedings of the Thirty-First International Joint Conference on Artificial Intelligence, IJCAI 2022, Vienna, Austria, 23–29 July 2022. Ed. by Luc De Raedt. ijcai.org, pp. 2101–2107 (2022). https://doi.org/10.24963/IJCAI.2022/292
12. Riesen, K.: Structural Pattern Recognition with Graph Edit Distance. ACVPR, Springer, Cham (2015). https://doi.org/10.1007/978-3-319-27252-8
13. Xu, K., et al.: How powerful are graph neural networks? In: 7th International Conference on Learning Representations, ICLR 2019, New Orleans, LA, USA, 6–9 May 2019. OpenReview.net (2019). https://openreview.net/forum?id=ryGs6iA5Km
14. Morris, C., et al.: TUDataset: a collection of benchmark datasets for learning with graphs. In: CoRR abs/2007.08663 (2020). arXiv: 2007.08663, https://arxiv.org/abs/2007.08663

Gaussian-Mixture Neural Networks

Duccio Meconcelli and Edmondo Trentin(✉)

DIISM, Università di Siena, Siena, Italy
duccio.meconcelli@student.unisi.it, trentin@dii.unisi.it

Abstract. Density estimation is crucial to statistical pattern recognition, in both the supervised and unsupervised frameworks. It is still an open problem, due to its intrinsic difficulties and to the many shortcomings of statistical parametric and non-parametric techniques. Artificial neural networks (ANN) have long been applied to the estimation of posterior probabilities for pattern classification (a simple supervised learning task), yet only a few attempts have been made to devise ANN-based density estimation algorithms. The paper proposes a novel algorithm for training an ANN from an unlabeled data sample of patterns randomly drawn from an underlying probability density function (PDF), say $p(\cdot)$, such that the ANN learns a robust non-parametric model of $p(\cdot)$. The algorithm leverages both the generalization capabilities of ANNs and the generality of the maximum-likelihood estimates of the parameters of Gaussian mixture models. Therefore, the proposed machine is termed Gaussian-mixture Neural Network (GNN). The best selling points of the GNN lie in its simplicity and effectiveness. Preliminary experimental results are reported and analyzed that involve data samples of variable size randomly drawn from PDFs of known form, either unimodal or multimodal. The GNN favorably compares with established statistical and ANN-based estimators, scoring generally higher than its competitors over a range of evaluation metrics. The code used in the experiments is made publicly available online on GitHub.

Keywords: Density estimation · Gaussian mixture model · artificial neural network · non-parametric estimation · deep learning

1 Introduction

The estimation of probability density functions (PDF) from unlabeled data is a fundamental problem in machine learning and plays a crucial role in various domains, including pattern recognition, anomaly detection, and clustering [4,12]. In pattern recognition, estimating a PDF is relevant to both the supervised setup (where the evidence $p(\mathbf{x})$ and the class-conditional PDFs $p(\mathbf{x}|\omega_i)$ for each class ω_i have to be substituted into the tight-hand side of Bayes theorem in order to apply Bayes decision rule [4]) and the unsupervised setup (where the probabilistic law underlying the distribution of an unlabeled collection of observations has to be discovered [4]). Even regression problems, which involve modeling the

C. Y. Suen et al. (Eds.): ANNPR 2024, LNAI 15154, pp. 13–24, 2024.
https://doi.org/10.1007/978-3-031-71602-7_2

dependences of a random vector $\mathbf{y}$ on an independent random vector $\mathbf{x}$ can be reformulated in terms of estimating a PDF, that is the conditional PDF $p(\mathbf{y}|\mathbf{x})$.

Traditional statistical approaches [4], either parametric or non-parametric have long been used by practitioners. While parametric techniques rely on the strong assumption of knowing in advance the form of the PDF to be estimated, non-parametric estimators overcome this issue but do present a number of shortcomings, mostly due to their poor generalization capabilities (since they are memory-based), their excessive burden in terms of memory requirements and computational cost over large dataset, and (to the contrary) their lack of robustness over small datasets. See [12] for a detailed analysis of the drawbacks of statistical PDF estimators.

On the other hand, artificial neural networks (ANN) excel at learning intricate nonlinear relationships but they generally lack the statistical robustness of traditional models, and may not be suited for the PDF estimation task, which is intrinsically unsupervised and poses quite a few issues to any perspective ANN-based estimator [9]. Major exceptions are represented by [5,6]. The former applies a maximum-likelihood (ML) learning rule for ANNs, but the algorithm relies on a non-specified numerical integration technique for the ANN and does not apply over multidimensional feature spaces. The latter trains the ANN to mimic an empirical estimate of the cumulative distribution function (CDF), retrieving the PDF by taking the derivative of the approximated CDF computed by the ANN. Unfortunately, a good estimate of the CDF does not entail a good estimate of the PDF upon differentiation (e.g., even negative outcomes may emerge). Besides, the algorithm does not apply to multivariate scenarios, as well. More recently, [9,10,12] introduced robust ANN-based PDF estimators whose training algorithms exploit non-parametric models suitable to multi-dimensional feature spaces. In [12] the Parzen neural network (PNN) is proposed, along with a study of its convergence and modeling capabilities, and empirical evidence is reported that proves the PNN outperforms the established approaches. In [9] a novel multi-dimensional, efficient technique is devised for computing the numerical integral of the ANN during training, such that the ANN realizes a proper PDF (i.e., having unit integral over its definition domain). The reader is referred to [11,12] for an exhaustive survey of the literature.

This paper contributes a novel approach to unsupervised PDF estimation via ANNs, termed Gaussian-mixture neural network (GNN), whose training algorithm (presented in Sect. 2) exploits the ML estimation of the parameters of an underlying Gaussian mixture model (GMM) [4], benefitting from both the GMM and the ANN capabilities and overcoming the respective limitations. The GNN can be seen as a novel instance of the teacher-student learning paradigm. Note that it has nothing to do with the so-called deep neural mixture model [11]. A preliminary experimental investigation is reported on in Sect. 3, relying on random data samples of various size drawn from PDFs of known forms (either unimodal or multimodal). The GNN is compared with a number of established techniques, confirming the effectiveness of the approach. In fact, being simple and effective are the best selling points of the GNN. The code used in the experiments can be downloaded from GitHub [1]. Conclusions are drawn in Sect. 4, that pinpoints also the directions of the on-going research on GNNs.

2 The Training Algorithm

In this section we first provide a review of the required fundamentals of GMMs (Sect. 2.1), then we present the GNN training algorithm (Sect. 2.2).

2.1 Re-Estimation Equations for the Parameters of a GMM

A GMM is a parametric model that realizes a particular multivariate mixture density function having the following form:

$$p(\boldsymbol{x}|\boldsymbol{\mu}, \Sigma) = \sum_{k=1}^{K} P(w_k)\mathcal{N}(\boldsymbol{x}|\boldsymbol{\mu}_k, \Sigma_k) \tag{1}$$

where $p(\boldsymbol{x}|\boldsymbol{\mu}, \Sigma)$ is the mixture density evaluated over the column vector $\boldsymbol{x}$, K is the number of multivariate Gaussian component densities in the mixture, $P(w_k)$ is the k-th mixing parameter (that is, the prior probability of k-th component PDF) under the assumption that $\sum_{k=1}^{K} P(w_k) = 1$, $\boldsymbol{\mu}_k$ is the (column) mean vector of the k-th Gaussian component and Σ_k is the corresponding covariance matrix. The quantities $\boldsymbol{\mu}$ and Σ represent the set of all the component-specific parameters, namely $\boldsymbol{\mu} = (\boldsymbol{\mu}_1, \ldots, \boldsymbol{\mu}_K)$ and $\Sigma = (\Sigma_1, \ldots \Sigma_K)$, respectively. The generic k-th multivariate Gaussian component $\mathcal{N}(\boldsymbol{x}|\boldsymbol{\mu}_k, \Sigma_k)$ is given by the usual equation

$$\mathcal{N}(\boldsymbol{x}|\boldsymbol{\mu}_k, \Sigma_k) = \frac{1}{(2\pi)^{\frac{D}{2}}|\Sigma_k|^{\frac{1}{2}}} \exp\left(-\frac{1}{2}(\boldsymbol{x} - \boldsymbol{\mu}_k)^T \Sigma_k^{-1}(\boldsymbol{x} - \boldsymbol{\mu}_k)\right) \tag{2}$$

where D is the dimensionality of the feature space, $|\Sigma_k|$ is the determinant of the k-th covariance matrix, Σ_k^{-1} is its inverse, and the writing $\boldsymbol{x}^T$ represents the transpose of the generic vector $\boldsymbol{x}$. The parameters of the GMM, namely $P(w_1), \ldots, P(w_K)$, $\boldsymbol{\mu}$, and Σ, are estimated from a sample $\mathcal{T} = \{\boldsymbol{x}_1, \ldots, \boldsymbol{x}_N\}$ of N independent and identically distributed feature vectors drawn from an underlying, unknown PDF $p(\boldsymbol{x})$ that the GMM is expected to estimate relying on the ML criterion by means of the following iterative steepest ascent algorithm (an instance of the expectation-maximization strategy [3]):

1. At time $t = 0$, initialize the GMM parameters, usually via k-Means clustering [4,7]
2. For $t = 1, 2, \ldots$ do
   (a) Update the mixing parameters:

$$P^{(t+1)}(w_k) = \frac{\sum_{i=1}^{N} P(w_k|\boldsymbol{x}_i, \boldsymbol{\mu}^{(t)}, \Sigma^{(t)})}{N} \tag{3}$$

   (b) Update the mean vectors:

$$\boldsymbol{\mu}_k^{(t+1)} = \frac{\sum_{i=1}^{N} P(w_k|\boldsymbol{x}_i, \boldsymbol{\mu}^{(t)}, \Sigma^{(t)})\mathbf{x}_i}{\sum_{i=1}^{N} P(w_k|\boldsymbol{x}_i, \boldsymbol{\mu}^{(t)}, \Sigma^{(t)})} \tag{4}$$

(c) Update the covariance matrices:

$$\Sigma_k^{(t+1)} = \frac{\sum_{i=1}^{N} P(w_k|\boldsymbol{x}_i, \boldsymbol{\mu}^{(t+1)}, \Sigma^{(t)})(\mathbf{x}_i - \boldsymbol{\mu}_k^{(t+1)})(\mathbf{x}_i - \boldsymbol{\mu}_k^{(t+1)})^T}{\sum_{i=1}^{N} P(w_k|\boldsymbol{x}_i, \boldsymbol{\mu}^{(t+1)}, \Sigma^{(t)})} \tag{5}$$

At any given iteration t, the posterior probability $P(w_k|\boldsymbol{x}, \boldsymbol{\mu}^{(t)}, \Sigma^{(t)})$ of the component density w_k given the observation $\boldsymbol{x}$ (and, given the estimate of the GMM parameters at time t) is computed via Bayes' theorem [4].

Algorithm 1. GNN_Training.

Input: randomly initialized network *net* to be trained, number K of Gaussian components, number T_{EM} of EM iterations, unlabeled training sample $\mathcal{T} = \{\mathbf{x}_1, \ldots, \mathbf{x}_N\}$ drawn from an unknown pdf $p(\cdot)$, dimensionality D of the feature space.
Output: the neural estimate $\phi(\cdot)$ of the pdf $p(\cdot)$.

1 Appply k-Means to partition $\mathcal{T}$ into K clusters $C_1, \ldots, C_K$

2 for(k = 1 to K)
2.1 Use the statistics of C_k to initialize $P^{(0)}(w_k), \boldsymbol{\mu}_k^{(0)}, \Sigma_k^{(0)}$

3 $\boldsymbol{\mu}^{(0)} = (\boldsymbol{\mu}_1^{(0)}, \ldots, \boldsymbol{\mu}_K^{(0)})$
4 $\Sigma^{(0)} = (\Sigma_1^{(0)}, \ldots, \Sigma_K^{(0)})$

5 for(t = 1 to T_{EM})
5.1 for(k = 1 to K)

5.1.1 $P^{(t)}(w_k) = \frac{\sum_{i=1}^{N} P(w_k|\boldsymbol{x}_i, \boldsymbol{\mu}^{(t-1)}, \Sigma^{(t-1)})}{N}$

5.1.2 $\boldsymbol{\mu}_k^{(t)} = \frac{\sum_{i=1}^{N} P(w_k|\boldsymbol{x}_i, \boldsymbol{\mu}^{(t-1)}, \Sigma^{(t-1)})\mathbf{x}_i}{\sum_{i=1}^{N} P(w_k|\boldsymbol{x}_i, \boldsymbol{\mu}^{(t-1)}, \Sigma^{(t-1)})}$

5.1.3 $\Sigma_k^{(t)} = \frac{\sum_{i=1}^{N} P(w_k|\boldsymbol{x}_i, \boldsymbol{\mu}^{(t)}, \Sigma^{(t-1)})(\mathbf{x}_i - \boldsymbol{\mu}_k^{(t)})(\mathbf{x}_i - \boldsymbol{\mu}_k^{(t)})^T}{\sum_{i=1}^{N} P(w_k|\boldsymbol{x}_i, \boldsymbol{\mu}^{(t)}, \Sigma^{(t-1)})}$

5.2 $\boldsymbol{\mu}^{(t)} = (\boldsymbol{\mu}_1^{(t)}, \ldots, \boldsymbol{\mu}_K^{(t)})$
5.3 $\Sigma^{(t)} = (\Sigma_1^{(t)}, \ldots, \Sigma_K^{(t)})$

6 for(i = 1 to N)
6.1 $y_i = \sum_{k=1}^{K} P^{(t)}(w_k)\mathcal{N}(\boldsymbol{x}_i|\boldsymbol{\mu}_k^{(t)}, \Sigma_k^{(t)})$

7 $\mathcal{S} = \{(\mathbf{x}_i, y_i) \mid i = 1, \ldots, N\}$;

8 BP_Training(*net*, $\mathcal{S}$);
9 $\phi(\cdot)$ = Function_Computed_By(*net*);
10 return $\phi(\cdot)$

2.2 GNN Training Algorithm

Algorithm 1 presents the pseudo-code of `GNN_Training`, the proposed unsupervised training procedure for the GNN. The architecture of the GNN, say *net*, may be of any ANN form suitable to learn a mapping $\phi : \mathbb{R}^D \rightarrow \mathbb{R}$, where D is the dimensionality of the feature space. Any feed-forward ANN (either deep or shallow) may be used to this aim. In this paper we apply ANNs that can be described as instances of the broad family of the multilayer perceptrons, such that the backpropagation (BP) algorithm can be exploited within Algorithm 1. Upon learning from the unlabeled training sample $\mathcal{T} = \{\mathbf{x}_1, \ldots, \mathbf{x}_N\}$ of independent feature vectors drawn from an unknown PDF $p(\cdot)$, the function $\phi(\cdot)$ computed by the GNN is expected to be an estimate of $p(\cdot)$. Formally, according to the statistical meaning of the term, this is a non-parametric estimation technique, insofar that we do not make any prior assumption on the form of $p(\cdot)$ [4]. While any kind of activation functions may be used in the hidden layers of the GNN (at least in principle), e.g. logistic sigmoids or ReLUs, there are some constraints on the output activation function. In fact, the latter shall yield the value $\phi(\mathbf{x})$ of a proper PDF evaluated over current input pattern $\mathbf{x}$, i.e. it must be non-negative and it may possibly range over the $[0, +\infty)$ counterdomain. This is readily achieved by resorting to a ReLU output, to a softmax function, or to a sigmoid with adaptive amplitude [8]. The latter can spontaneously extend over the range of values that naturally span the actual counterdomain of $p(\cdot)$. A linear output unit can be used, as well, as long as possible negative outputs are forced to zero at test time.

Step 1 of the algorithm creates a partitioning of $\mathcal{T}$ into K subsets $C_1, \ldots, C_K$, where K is a hyper-parameter that shall undergo model selection (e.g., via cross-validation). Any clustering algorithm can be used to obtain the partitioning, although k-Means [4] is used hereafter due to its being a simplified version of the ML solution for the parameters of a GMM [4]. Steps 2–2.1 initialize $P^{(0)}(w_k), \boldsymbol{\mu}_k^{(0)}, \Sigma_k^{(0)}$ for $k = 1, \ldots, K$ based on the statistics of C_k as follows: $\boldsymbol{\mu}_k^{(0)}$ is defined as the average of the feature vectors in C_k, $\Sigma_k^{(0)}$ is defined as the empirical covariance matrix of C_k with respect to $\boldsymbol{\mu}_k^{(0)}$, and $P^{(0)}(w_k)$ is defined as the relative frequency of the feature vectors in C_k with respect to the overall cardinality N of $\mathcal{T}$, that is $N_k^{(0)}/N$ where $N_k^{(0)} = |C_k|$. The global parameter vectors $\boldsymbol{\mu}^{(0)}$ and $\Sigma^{(0)}$ are initialized in terms of $\boldsymbol{\mu}_k^{(0)}$ and $\Sigma_k^{(0)}$ (for $k = 1, \ldots, K$) in steps 3 and 4, respectively. Steps 5–5.3 iterate (for $t = 1, \ldots, T_{EM}$) the re-estimation of the overall mixing parameters, of $\boldsymbol{\mu}^{(t)}$, and of $\Sigma^{(t)}$ based on the component-specific re-estimates of $P^{(t)}(w_k)$, $\boldsymbol{\mu}_k^{(t)}$, and $\Sigma_k^{(t)}$ relying on Eqs. 3-5, thus realizing a hill-climbing procedure over the ML criterion. Step 6 is the core of the algorithm: it generates synthetic target outputs for actually training the GNN. For each training pattern $\mathbf{x}_i$ (for $i - 1, \ldots, N$) a one-dimensional target y_i is created exploiting the right-hand side of Eq. 1 applied over the parameters estimates yielded by step 5. It is now straightforward to define a supervised training set $\mathcal{S}$ (step 7 of the algorithm) by labeling each of the original unlabeled training patterns $\mathbf{x}_i$ with the corresponding, synthetically-generated target output

y_i, for $i = 1, \ldots, N$. Afterwards, the algorithm assumes two pseudo-procedures are available, namely `BP_Training`(net, $\mathcal{S}$), called in step 8, which realizes regular BP-based training of the network *net* over the supervised dataset $\mathcal{S}$, and `Function_Computed_By`(net), called in step 9, which is expected to return some representation of the function $\phi(\cdot)$ computed by *net*. The function $\phi(\cdot)$ forms the output of Algorithm 1 and it is eventually returned (step 10).

3 Experimental Demonstration

Throughout the experiments, a combination of grid-search and tree-structured Parzen estimator search [2] was used for the selection of the architectures and hyper-parameters of the different models. In particular, for the GNN the following parameters underwent search: number K of components, number of hidden layers, number of neurons per layer, form of the activation functions, number of training epochs, learning rate, mini-batch size, dropout, and optimizer (either Adam or RMSprop). Since the BP-based learning step within Algorithm 1 can be seen as accomplishing a regression task, the search was aimed at maximizing the R^2 score. The latter measures the proportion of the variance in the dependent variable that is predictable from the independent variables. It is computed as:

$$R^2 = 1 - \frac{\sum_{i=1}^{N}(y_i - \hat{y_i})^2}{\sum_{i=1}^{N}(y_i - \bar{y})^2} \tag{6}$$

where N is the number of samples, y_i is the target value, $\hat{y}_i$ is the predicted value, and $\bar{y}$ is the mean of the target values.

3.1 Evaluation Metrics

Beside the R^2 score, several popular evaluation metrics were used to assess the quality of the estimated PDF $\hat{p}(\cdot)$ yielded by any estimation algorithm with respect to the actual PDF $p(\cdot)$, namely:

- Mean Squared Error (MSE), that is $MSE = \frac{1}{N}\sum_{i=1}^{N}(p(\mathbf{x}_i) - \hat{p}(\mathbf{x}_i))^2$ where $\mathbf{x}_i$ represents the i-th random vector in a random sample $\{\mathbf{x}_1, \ldots, \mathbf{x}_N\}$ empirically drawn from $p(\cdot)$
- Kullback-Leibler (KL) divergence, i.e. $KL(p(\cdot)||\hat{p}(\cdot)) = \sum_i^N p(\mathbf{x}_i) \log\left(\frac{p(\mathbf{x}_i)}{\hat{p}(\mathbf{x}_i)}\right)$
- Maximum Error (Max Error), that is $\max_{i=1,\ldots,N} |p(\mathbf{x}_i) - \hat{p}(\mathbf{x}_i)|$
- Integrated Square Error (ISE), that is the integral (computed numerically) of the squared difference between $p(\cdot)$ and $\hat{p}(\cdot)$: $ISE = \int(p(\mathbf{x}) - \hat{p}(\mathbf{x}))^2 d\mathbf{x}$
- Explained variance score (EVS), defined as EVS $= 1 - \frac{\sum_{i=1}^{N}(p(\mathbf{x}_i)-\hat{p}(\mathbf{x}_i))^2}{\sum_{i=1}^{N}(p(\mathbf{x}_i)-\bar{p}(\cdot))^2}$ where $\bar{p}(\cdot)$ is the expectation of $p(\cdot)$ computed by averaging over $\mathbf{x}_1, \ldots, \mathbf{x}_N$.

3.2 Results, Round 1: Estimating an Exponential PDF

In the first experimental round we drew random samples $\mathcal{T}$ (for different values of N) from an exponential PDF, that is $p(x) = \lambda \exp(-\lambda x)$ with inverse scale $\lambda = 1.67$. This estimation task is only apparently innocuous, as pinpointed in [12], due to the peculiar form of $p(x)$ that hardly matches traditional parametric and non-parametric statistical models. The GNN was compared with respect to established density estimation techniques, including both statistical and ANN-based approaches. The statistical techniques used were a parametric model, that is the GMMs itself (allowing for a straight comparison between the performance yielded by the "teacher", i.e. the GMM, and the "student", i.e. the GNN) and two popular non-parametric approaches, namely the Parzen window (PW) [4] and the so-called k_n-nearest neighbor (k_n-NN) density estimator[1] [4]. The PNN was adopted as an ANN-based benchmark, due to its established overpowering other ANNs and machine learning algorithms for PDF estimation [12]. A quantitative comparison is reported in Table 1 in terms of the aforementioned metrics (R^2, MSE, Max error, ISE, KL, and EVS). The results in the Table are ordered for decreasing values of R^2. The best score obtained for each metric (i.e., column) and for each value of N is emphasized using boldface fonts. Some entries of the KL column are undefined due to the fact that the ANN-based estimator may yield a null value of the estimated PDF when fed with certain input patterns $\mathbf{x}_i$, and the KL divergence (Sect. 3.1) is just not defined when $\hat{p}(\mathbf{x}_i) = 0$.

The results allow us to draw the following conclusions: (1) both the ANN-based approaches outperformed all the established statistical techniques, by all metrics, proving that ANNs establish themselves as excellent candidates for PDF estimation; (2) the GNN yielded the highest scores by all metrics and all values of the training set sizes (for $N = 10, 50, 100, 200$), except for KL with $N = 50, 100, 200$ where the PNN yielded better performance; (3) the GNN (the "student") systematically yielded a dramatic improvement over the GMM (the "teacher"), proving the modeling capabilities of ANNs and, above all, proving that their generalization capabilities can actually ease the model convergence to a general probabilistic law that can reflects the nature of the training data sample and its actual probabilistic properties; (4) even when trained with as few as 10 training patterns, the GNN metrics overpowered the scores yielded by any statistical estimates based on any values of N (up to $N = 200$).

Figure 1 shows a graphical plot of the estimates of the exponential PDF with $N = 100$ yielded by the best GMM ($K = 26$) and the best GNN ($K = 13$), respectively. Both estimates are plotted against the actual PDF. It is seen that the GNN curve resulted in a close estimate to the underlying exponential, while the GMM presented issues when its argument x approached the origin of the axis. The noisy, piecewise-linear blue line shown in the right-hand portion (**b**) of Fig. 1 connects the blue dots that represent the individual training pairs (x_i, y_i) for $i = 1, \ldots, 100$. This highlights the fact that the learning and generalization capabilities of the GNN do extrapolate an overall smooth density function from

[1] Not to be confused with the usual k-nearest neighbor classifier.

Table 1. Estimation of the exponential PDF from samples of variable size N: the results are sorted in descending order of R^2 score.

Model	N	R^2	MSE	Max Error	ISE	KL	EVS
GNN	200	**0.9978**	**0.0004**	**0.0766**	**0.0116**	-	**0.997** 8
GNN	100	**0.9953**	**0.0007**	**0.0522**	**0.0233**	0.0095	**0.9954**
PNN	200	0.9949	0.0008	0.2080	0.0261	**0.0014**	0.9952
PNN	100	0.9936	0.0010	0.1900	0.0323	**0.0034**	0.9945
GNN	50	**0.9924**	**0.0012**	**0.0760**	0.0380	0.0212	**0.9926**
PNN	50	0.9911	0.0014	0.2115	0.0444	**0.0076**	0.9911
GNN	10	**0.9722**	**0.0037**	**0.1557**	**0.0668**	**0.0070**	**0.9727**
PW	50	0.9296	0.0113	0.7746	314.40	0.7020	0.9339
PNN	10	0.9291	0.0096	0.2059	0.1705	-	0.9291
PW	100	0.9247	0.0120	0.8547	300.90	0.7020	0.9263
k_n-NN	200	0.9206	0.0129	0.8886	326.50	0.7101	0.9211
PW	200	0.9180	0.0133	0.9090	308.30	0.7101	0.9193
PW	10	0.9142	0.0117	0.4570	74.60	0.3069	0.9146
k_n-NN	100	0.9138	0.0138	0.8010	314.80	0.7020	0.9219
k_n-NN	50	0.9078	0.0147	0.7351	316.26	0.7020	0.9271
GMM	200	0.9070	0.0151	1.1122	344.09	0.7101	0.9070
GMM	100	0.8478	0.0243	1.1380	343.15	0.7020	0.8479
GMM	50	0.7645	0.0376	1.0493	395.08	0.7020	0.7647
k_n-NN	10	0.6933	0.0420	1.2493	115.64	0.3069	0.7145
GMM	10	0.6079	0.0537	0.7735	91.1557	0.3069	0.6116

Table 2. Optimal number K of components in the GNN training algorithm as a function of the training set size N, along with the corresponding R^2 score.

N	K	R^2 Score
10	2	0.9722
50	8	0.9924
100	13	0.9953
200	25	0.9978

a set of otherwise individually noisy examples. Table 2 reports the optimal number K of components within the GNN training algorithm (as selected via the aforementioned search techniques) as a function of the number N of training patterns, and the corresponding R^2 score.

Figure 2 plots the R^2 yielded by the best GNN for the exponential PDF with $N = 100$ as a function of K (left-hand side of the figure) and of the number of hidden neurons (on the right). It is seen that the GNN reached high values

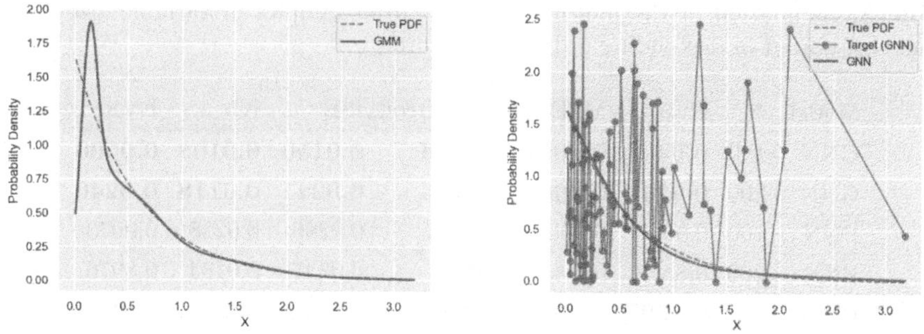

(a) Best GMM ($K = 26$) (b) Best GNN ($K = 13$)

Fig. 1. Best GMM-based (**a**) and best GNN-based (**b**) estimates of the exponential PDF with $N = 100$

of R^2 as soon as $K \geq 4$, and was not affected significantly by larger values for K, although there was a tendency to worsen the performance for $K \geq 15$. As for the impact of the architecture on R^2, it is seen that the number of hidden neurons affects the performance of the GNN to some extent, as expected of any ANN learning process. Trivial architectures (less then 5 hidden neurons) were unfit to the task. Some random fluctuation of the R^2 were observed throughout the overall increase in the number of neurons. More noise in the R^2 curve were observed for the larger architectures, in particular when more then 125 neurons were used. Nonetheless, the vast majority of the non-trivial GNN architectures reached for a high R^2 values regardless of the exact number of neurons.

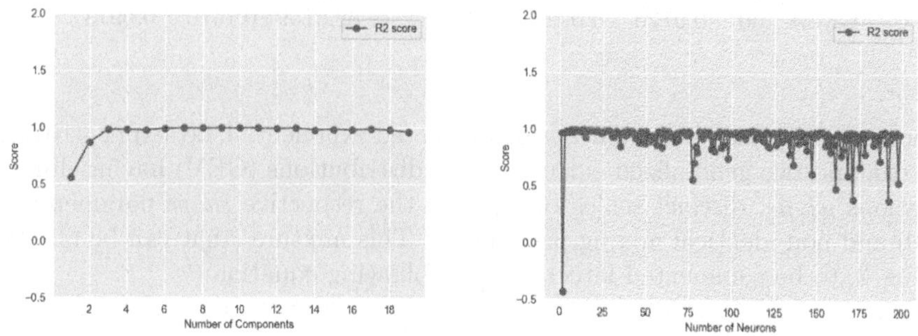

(a) (b)

Fig. 2. Estimates of the exponential PDF (with $N = 100$): R^2 curves as functions of the number K of components (**a**) and of the number of hidden neurons (**b**) in the GNN

3.3 Results, Round 2: Estimating a Multimodal PDF

Random samples $\mathcal{T} = \{x_1, \ldots, x_N\}$ of size N (for $N = 10, 50, 100, 200$) were drawn from a multimodal mixture density with 4 component PDFs (a standard

Table 3. Estimation of the multimodal PDF from samples of variable size N: the results are sorted in descending order of R^2 score.

Model	N	R^2	MSE	Max Error	ISE	KL	EVS
GNN	100	**0.9498**	**0.0002**	**0.0624**	**0.0150**	**0.0103**	**0.9500**
GNN	200	**0.9235**	**0.0002**	**0.0704**	**0.0241**	**0.0118**	**0.9240**
PNN	100	0.9047	0.0003	0.1163	0.0286	0.0228	0.9073
PNN	200	0.8899	0.0003	0.0943	0.0347	0.0134	0.8976
PNN	50	**0.8297**	**0.0005**	**0.0607**	**0.0454**	**0.0313**	**0.8341**
GNN	50	0.8035	0.0006	0.0642	0.0524	0.0427	0.8035
PW	100	0.7974	0.0006	0.1288	53.9312	0.1609	0.7998
PW	200	0.7872	0.0007	0.1427	58.5740	0.1627	0.7919
PNN	10	**0.7837**	**0.0006**	0.1010	**0.0492**	0.0555	**0.7838**
GMM	200	0.7536	0.0008	0.1737	68.8703	0.1627	0.7540
k_n-NN	200	0.7404	0.0008	0.1388	55.7257	0.1627	0.7438
k_n-NN	100	0.7298	0.0008	0.1069	59.8825	0.1609	0.7424
GNN	10	0.7250	0.0008	**0.0800**	0.0625	**0.0406**	0.7729
PW	50	0.6615	0.0010	0.1014	52.5236	0.1576	0.6620
k_n-NN	50	0.4752	0.0015	0.1541	64.3591	0.1576	0.5261
GMM	100	0.4636	0.0016	0.1521	69.2541	0.1609	0.4638
PW	10	0.4570	0.0016	0.0938	46.1844	0.1706	0.6277
k_n-NN	10	0.3616	0.0019	0.0941	20.1995	0.1706	0.4160
GMM	10	0.1837	0.0024	0.1113	43.9243	0.1706	0.4851
GMM	50	0.0720	0.0026	0.2248	82.3759	0.1576	0.0913

exponential, i.e. $\exp(-x)$ with $x > 0$, a reflected exponential, i.e. $\exp(x-10)$ with $x < 10$, and two generalized extreme-value distributions (GEV) having different locations μ_1, μ_2, distinct scales β_1, β_2, and the respective *shape* parameters set to 0) and non-identical mixing parameters. This mixture (that can be observed in Fig. 3, to be commented later) has the following equation:

$$p(x) = 0.20\exp(-x) + \sum_{i=1}^{2} \frac{P_i}{\beta_i} \exp\left(-\frac{x-\mu_i}{\beta_i}\right) \exp\left\{-\exp\left(-\frac{x-\mu_i}{\beta_i}\right)\right\} + 0.25\exp(x-10)$$

where $P_1 = 0.25$, $P_2 = 0.3$, $\mu_1 = 4.0, \mu_2 = 5.5$, $\beta_1 = 0.8$, and $\beta_2 = 0.7$. The skewness of the GEV makes it all the more difficult to estimate with Gaussian-based statistical approaches. Table 3 reports the results, ordered in decreasing values of R^2. Bold fonts are used to highlight the best results per each metric and per each value of N. Again, the ANN-based estimators outperformed the statistical approaches (that turned out to struggle with the present multimodal setup). The GNN yielded the best scores (including the KL, differently from the

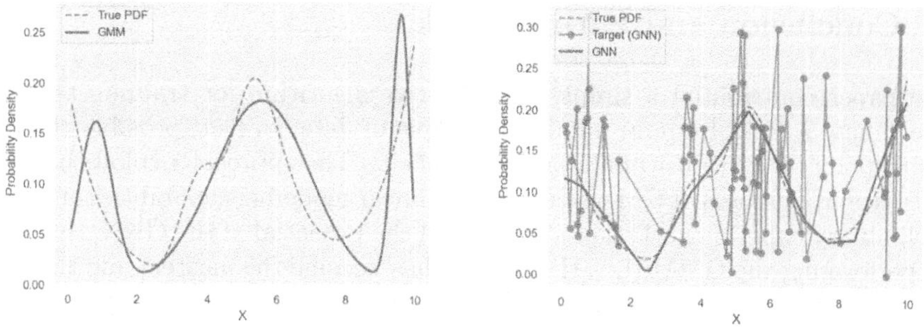

(a) Best GMM **(b)** Best GNN

Fig. 3. Best GMM-based (**a**) and GNN-based (**b**) estimates of the multimodal PDF with $N = 100$.

Table 4. Multimodal PDF: best GNN architectures for increasing values of N. *Legend:* "Layers" is the number of hidden layers, "Neurons" is the number of neurons in the 1st hidden layer, "AF" is short for "activation function", "Ada-Sigmoid" is a sigmoid with adaptive amplitude [8].

N	Layers	Neurons	Hidden AF	Output AF
10	2	54	ReLU	Ada-Sigmoid
50	3	24	Tanh	Ada-Sigmoid
100	4	26	Sigmoid	Linear
200	4	39	Tanh	Ada-Sigmoid

previous experimental round) for $N = 100$ and $N = 200$. For smaller datasets, the PNN mostly resulted in slightly better scores (the qualities of the PNN over small samples are established [12]). This is possibly related to the GMM (the "teacher") substantially under-achieving in the present PDF estimation task with $N = 10$ and $N = 50$. Nevertheless, the GNN still resulted in the best scores for some of the metrics when $N = 10$. All in all, based on the present results and those discussed in the previous Section, it is seen that all the non-parametric techniques (either statistical or ANN-based) overpowered the parametric model (the GMM). Figure 3 shows the estimates yielded by the best GMM (left) and by the best GNN (right) for $N = 100$, plotted against the actual multimodal PDF $p(x)$ and against the targets (blue line) generated by steps 6–6.1 of Algorithm 1. Table 4 reports the best GNN architectures (as selected by the search techniques mentioned above) for increasing values of N. Broadly speaking, the optimal depth of the GNN had a tendency to increase with the size of the training set. This was most likely caused by the proneness of ANNs to overfit small datasets whenever their architecture is too complex. Different forms for the activation functions turned up better for different numbers of layers. In general, the logistic sigmoid with adaptive amplitude (Ada-Sigmoid) [8] was selected as the best-performing output activation function.

4 Conclusion and Future Work

The paper contributed a simple and effective algorithm for training an ANN to estimate a PDF from an unlabeled sample of data. In fact, possibly the best selling point of the algorithm lies in its simplicity. The approach exploits both the statistical robustness of GMMs and the nonlinear modeling capabilities of ANNs. Preliminary experimental results showcased the potential of the GNN, the latter outperforming the GMM (i.e., the student has become the master) and the most popular non-parametric estimators. Likewise, the GNN compared quite favorably with the established ANN-based technique. These findings corroborate the idea that combining statistical approaches with ANNs can actually lead to better PDF estimates. On-going research is focusing on (1) extending the experimental investigation to multi-dimensional random vector distributions, and on (2) a study of the modeling and convergence capabilities of the GNN.

References

1. https://github.com/Duccioo/Gaussian-Mixture-Neural-Network
2. Akiba, T., et al.: Optuna: a next-generation hyperparameter optimization framework. In: Proceedings of the 25th ACM SIGKDD International Conference on Knowledge Discovery & Data Mining, KDD, pp. 2623–2631. ACM (2019)
3. Dempster, A.P., Laird, N.M., Rubin, D.B.: Maximum likelihood from incomplete data via the EM algorithm. J. Royal Stat. Soc. Ser. B **39**(1), 1–22 (1977)
4. Duda, R.O., Hart, P.E., Stork, D.G.: Pattern Classification, 2nd edn. Wiley, Hoboken (2001)
5. Magdon-Ismail, M., Atiya, A.: Density estimation and random variate generation using multilayer networks. IEEE Trans. Neural Netw. **13**(3), 497–520 (2002)
6. Modha, D.S., Fainman, Y.: A learning law for density estimation. IEEE Trans. Neural Netw. **5**(3), 519–23 (1994)
7. Pedregosa, F., et al.: Scikit-learn: machine learning in Python. J. Mach. Learn. Res. **12**, 2825–2830 (2011)
8. Trentin, E.: Networks with trainable amplitude of activation functions. Neural Netw. **14**(4–5), 471–493 (2001)
9. Trentin, E.: Soft-constrained neural networks for nonparametric density estimation. Neural Process. Lett. **48**(2), 915–932 (2018)
10. Trentin, E.: Asymptotic convergence of soft-constrained neural networks for density estimation. Mathematics **8**(4), 572 (2020)
11. Trentin, E.: Multivariate density estimation with deep neural mixture models. Neural Process. Lett. **55**(7), 9139–9154 (2023)
12. Trentin, E., Lusnig, L., Cavalli, F.: Parzen neural networks: fundamentals, properties, and an application to forensic anthropology. Neural Netw. **97**, 137–151 (2018)

Neural Decompiling of Tracr Transformers

Hannes Thurnherr(✉) and Kaspar Riesen

Institute of Computer Science, University of Bern, 3012 Bern, Switzerland
hannes.thurnherr@students.unibe.ch, kaspar.riesen@unibe.ch

Abstract. Recently, the transformer architecture has enabled substantial progress in many areas of pattern recognition and machine learning. However, as with other neural network models, there is currently no general method available to explain their inner workings. The present paper represents a first step towards this direction. We utilize *Transformer Compiler for RASP* (Tracr) to generate a large dataset of pairs of transformer weights and corresponding RASP programs. Based on this dataset, we then build and train a model, with the aim of recovering the RASP code from the compiled model. We demonstrate that the simple form of Tracr compiled transformer weights is interpretable for such a decompiler model. In an empirical evaluation, our model achieves exact reproductions on more than 30% of the test objects, while the remaining 70% can generally be reproduced with only few errors. Additionally, more than 70% of the programs, produced by our model, are functionally equivalent to the ground truth, and therefore a valid decompilation of the Tracr compiled transformer weights.

Keywords: Transformer · Interpretability · Decompiling · RASP

1 Introduction

The *transformer architecture* [1] is a type of neural network that was originally designed for machine translation but is now also successfully used for modelling many different pattern recognition tasks. This notably includes advanced language understanding [2] but also the modelling of non-sequential data such as images [3]. The transformer architecture differentiates itself from other architectures by using a mixture of self-attention and MLP-layers. It can be used as both a decoder-only variant or in an encoder-decoder configuration where the encoder augments the decoder forward pass using cross-attention.

Interpretability is a subfield of machine learning that is concerned with the problem of understanding the internals of neural networks [4]. Especially the interpretability of transformer-based models has seen growing attention in the past few years. While some progress has been made, e.g. [5–7], transformer neural networks are still largely regarded as black boxes. That is, interpretation relies on manual work which makes it hard to scale (for instance, manually translating billions of model weights and activations into human-readable descriptions is

C. Y. Suen et al. (Eds.): ANNPR 2024, LNAI 15154, pp. 25–36, 2024.
https://doi.org/10.1007/978-3-031-71602-7_3

hardly feasible – even if one knew how to do so). The fact that current state-of-the-art systems are not interpretable, means that they could have unacceptable failure modes, which may only be discovered after deployment.

Attempts to automate at least parts of the complete interpretation pipeline have been made [8,9]. However, most of these approaches only focus on the analysis of a subcomponent of the network. In the present paper, we propose an end-to-end framework, where a decompiler model is responsible for the entire interpretation process of a complete neural network (taking all of its parameters into account). In particular, we investigate whether it is possible to train a deep-learning model to automate the translation of a set of compiled transformer weights, into *RASP*[1] code [10]. RASP is a programming language that consists of functions that correspond to various components of transformers and is designed to be computationally equivalent to transformers. RASP code could thus open up a way of translating transformation processes into a form that is more readable for humans.

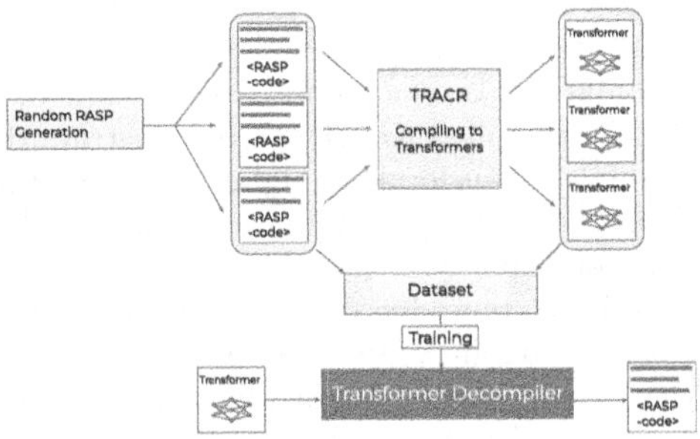

Fig. 1. The core concept of the present method is based on the idea that Tracr can be used to generate training data for an automated end-to-end interpretability system

One of the major challenges of the present research is the generation of a large amount of training data. To date, only a few examples of pairs of network weights and corresponding descriptions are available (from which a neural network could learn to translate between the two). To address this challenge, we use the *Transformer Compiler for Rasp* (Tracr) [11] to generate a novel set of training data. In particular, we use Tracr to produce a large set of pairs of transformer weights and corresponding RASP programs. We hypothesize that using these pairs to train a *Transformer Decompiler Model* could represent a first step towards a holistic and fully automated solution for transformer interpretability. In Fig. 1, the proposed approach of the present paper is visually summarized.

The remainder of the paper is organized as follows. Next, in Sect. 2, we describe in detail how the dataset, used to train our model, is engineered. Then, in Sect. 3, we outline the novel method for processing the RASP transformer weight pairs and training the decompiler model. In Sect. 4, we present and discuss the results of an experimental evaluation. Finally, in Sect. 5, we draw conclusions and discuss possible future research activities.

[1] *Restricted Access Sequence Processing Language.*

2 Pairs of RASP Code and Transformer Weights

This section describes how we generate random but functional RASP code and how these RASP programs are then filtered and translated into transformer weights. This development of a novel dataset consisting of pairs of RASP code and the corresponding transformer weights is crucial for training the proposed model.

We use Algorithm 1 to create random, compileable RASP programs. This algorithm processes the primitives *rasp.tokens* and *rasp.indices* using the RASP functions *Select()*, *Aggregate()*, *SelectorWidth()*, *Map()* and *SequenceMap()* in a way that accounts for the compatibility of functions and the different datatypes. We achieve this by basing the algorithm around a pool of available inputs, which is, in turn, initialized containing only the two primitives *rasp.tokens* and *rasp.indices* (Line 2 of Algorithm 1)

Algorithm 1 starts by randomly selecting one of the five RASP functions to use next (Line 6). This random selection is weighted in accordance with how frequently the functions appear in a collection of manually written RASP programs. We then iterate through all of the parameters of the function (Line 7), randomly choosing a variable from the pool of available inputs and test whether it is of the datatype needed for the current parameter (Line 8 and 9). If we find that one or more of the parameters of the currently selected function cannot be satisfied by any of the currently available inputs, we choose a different function (Line 11). If this is the case for all of the functions, we restart the generation of the program (Line 12 and 13).

Algorithm 1 Random RASP Program Generation

```
1: available_inputs ← Initialize with primitives rasp.tokens and rasp.indices
2: available_functions ← Select(), Aggregate(), SelectorWidth(), Map(), and SequenceMap()
3: available_lambdas ← 2-input lambdas, 1-input lambdas and 2-input, boolean output lambdas
4: phase ← 0
5: while not converged to one output do
6:     Select a function based on its probability from available_functions
7:     for each parameter of the function do
8:         if appropriate input available for the parameter then
9:             Use inputs from available_inputs and available_lambdas
10:        else
11:            Select a different function from available_functons
12:            if no function is compatible then
13:                Restart the program generation
14:            end if
15:        end if
16:    end for
17:    Add function's output to available_inputs
18:    if phase = 0 and termination criterion is met then
19:        phase ← 1
20:    end if
21:    if phase = 1 then
22:        Remove used inputs from available_inputs
23:    end if
24: end while
25: return The final set of operations forming the RASP program
```

Once a suitable input for all of the function parameters has been found, the output of this function is represented by a new entry in the pool of available inputs (Line 17). Note that some functions, like *Map()* or *Select()*, take lambda functions as some of their inputs. These are not stored in the pool of available inputs but in external lists (Line 3).

In Algorithm 1, there are two phases (*phase 0* and *phase 1*). In *phase 0*, we expand the number of available inputs by leaving used variables in the pool of available inputs while adding new variables to this pool. The duration of *phase 0* is controlled via the termination criterion (Line 18). In our implementation, we check whether the length of $(available\_inputs - 2)$ is greater than the duration of *phase 0*. Note that this criterion could be varied to guide the length of the generated programs. In *phase 1*, variables are removed from the available inputs once they are used. Since each function has only one output but typically multiple inputs, this process eventually converges to a single variable, representing the output of the program as a whole.

The output of Algorithm 1 is a computational graph represented as a list of nodes, each of which has a function and three inputs. Inputs, in turn, are represented by integers, denoting either the node they originate from, a lambda function, or an empty token for functions that take fewer than three inputs. Figure 2 visually summarizes the process of Algorithm 1.

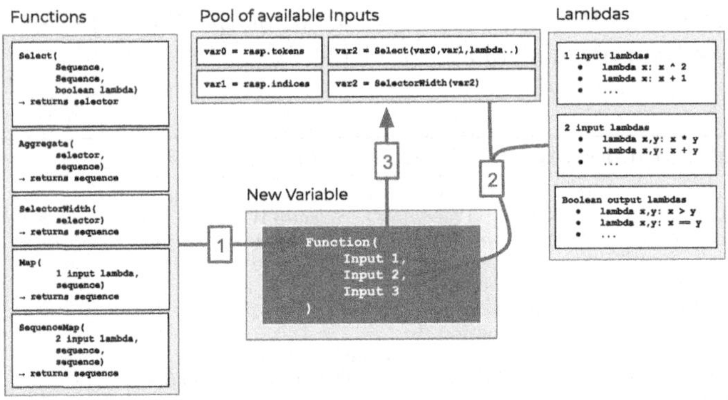

Fig. 2. Visualization of the three major steps of Algorithm 1 that are repeated until the pool of available inputs converges to one entry. 1: Select a function; 2: Fill the function with variables; 3: Add the newly created variable to the pool of available inputs

As not all programs produced by Algorithm 1 are suitable for our purpose, we use a rejection sampling strategy. This means we test each program for undesired properties and only keep the ones that pass all the tests. To identify as many flaws or invalidities as possible, we run each program on a set of test inputs and employ the built-in Tracr validator. Properties that are filtered out relate to some limitations of RASP, like the inability to process float sequences in the *Aggregate()* function and limitations of the Tracr compiler, which does not

a: Error during compiling

```
var0 = rasp.tokens
var1 = rasp.indices
var2 = SequenceMap(lambda x, y: x - y, var1, var1)
var3 = SequenceMap(lambda x,y:x/y if y!=0 else 0,var1, var0)
var4 = Select(var2, var2, lambda x, y: x > y)
var5 = Aggregate(var3, var4)
```

b: More than two SequenceMap() uses:

```
var0 = rasp.tokens
var1 = rasp.indices
var2 = Select(var0, var1, lambda x, y: x == y)
var3 = Select(var1, var0, lambda x, y: x < y)
var4 = SelectWidth(var2)
var5 = SequenceMap(lambda x, y: x + y, var4, var4)
var6 = SequenceMap(lambda x, y: (x + y) / 2, var5, var5)
var7 = SequenceMap(lambda x, y: x ** (1/y), var6, var6)
var8 = Aggregate(var7, var3)
```

c: More than 15 lines

```
var0 = rasp.tokens
var1 = rasp.indices
var2 = Select(var0, var0, lambda x, y: x > y)
var3 = Select(var0, var0, lambda x, y: x > y)
var4 = Select(var1, var1, lambda x, y: x > y)
var5 = SelectWidth(var2)
var6 = Select(var5, var5, lambda x, y: x == y)
var7 = SelectWidth(var4)
var8 = Select(var7, var7, lambda x, y: x == y)
var9 = SelectWidth(var6)
var10 = Select(var9, var9, lambda x, y: x != y)
var11 = SelectWidth(var3)
var12 = Select(var11, var11, lambda x, y: x > y)
var13 = SelectWidth(var10)
var14 = Map(lambda x: 1 if x > 0 else 0, var13)
var15 = Aggregate(var14, var8)
var16 = Aggregate(var15, var12)
```

d: Validator found Issue: Categorical Aggregate with Selector with width > 1

```
var0 = rasp.tokens
var1 = rasp.indices
var2 = Select(var0, var1, lambda x, y: x == y)
var3 = Select(var1, var0, lambda x, y: x < y)
var4 = SelectWidth(var2)
var5 = SequenceMap(lambda x, y: x + y, var4, var4)
var6 = SequenceMap(lambda x, y: (x + y)/2, var5, var5)
var7 = SequenceMap(lambda x, y: x**(1/y), var6, var6)
var8 = Aggregate(var7, var3)
```

Fig. 3. Four examples of generated programs that are rejected

compile RASP code correctly if it involves the aggregation of non-categorical sequences using a selector with a width that is greater than one.

In Fig. 3, we show four examples of generated programs that are rejected due to different reasons:

a: The Tracr compiler can throw errors on certain valid RASP programs. Such programs are rejected.

b: Multiple uses of the function *SequenceMap()* can cause the compilation to take an unacceptably long time (several minutes rather than less than a second). We deal with this by filtering out programs with more than two occurrences of *SequenceMap()*.

c: The training set of programs is limited to 15 lines of code which amounts to a maximum of 60 program tokens.

d: The inbuilt Tracr validator filters programs where aggregation of non-categorical values using a selector wider than one occurs.

In total, we produce a dataset that consists of about 533,000 programs and corresponding transformers[2]. The produced programs contain between 5 and 12 lines of RASP code which equates to between 20 and 48 tokens. The corresponding transformers consist of between 8 and 42 weight matrices and therefore between 8 and 42 tokens.

To more closely examine the distribution of the programs, we run tests on a subset of 10,000 programs taken from the dataset. For instance, to determine the function[3] of each program, we generate a set of 1,000 input sequences. Programs which generate the same output to all of these inputs are then deemed to implement the same function.

During this evaluation, we observe that approximately 60% of the programs generate the same output for the 1,000 inputs as at least one other program in the dataset. However, the fact that certain functions are equal with respect to

[2] The generated data is publicly available upon request.

[3] By function we refer to the algorithm that is implemented by the RASP code.

their input-output relationship does not necessarily mean that they are internally equal to each other. Actually, we observe that only about 5% of all programs refer to duplicates (i.e., pairs of programs with identical RASP code strings). However, it is worth mentioning that the small input space of the transformers that are compiled from the RASP programs could mean that two different RASP programs compile to the same transformer weights, leading to some ambiguity in the dataset, where multiple outputs (i.e. programs) would be correct for one input (i.e. weights).

3 The Transformer Decompiler Method (TraDe)

3.1 Format of the Model Input

The formal representation of the RASP code (of the five component functions of the RASP code), into which the Tracr transformer weights are translated, is a crucial step in our procedure. It might be possible to use standard text tokenization [12] for this vectorization task. However, this would require the model to learn to distinguish valid RASP code from a very large space of possible outputs. Furthermore, since it is necessary to reverse the vectorization process, the application of commonly used methods for the vectorization of graphs, such as message passing [13], is not directly applicable to the computational graphs of the RASP program.

Based on these considerations, we employ a series of one-hot-encoded vectors (four per line of RASP code) to vectorize RASP code. Each series represents a specific part of the line, viz. the function and the three possible inputs to this function. Depending on the function, these numbers are interpreted differently. For instance, if the first vector of a line represents the function *SequenceMap()*, the next vector will be interpreted as the one-hot-encoded position of the lambda in the list of lambdas that produces one output from two inputs. In Fig. 4, the process of RASP vectorization is visualized

It is also necessary to bring the weights of the Tracr transformer into a form in which they can be fed into the decompiler. This task is not trivial as transformer weights are not organized sequentially. However, as demonstrated by vision transformers [3], the transformer architecture can also process sequences of tokens arranged sequentially, despite their original non-sequential configuration. To retain some of the structure of the transformer, we use a matrix-based tokenization[4]. That is, each token corresponds to a weight matrix in the compiled transformer. The matrix is flattened and concatenated with a small vector representing some information about the matrix such as shape, position in the transformer (head and layer) and function of the matrix (embedding, MLP, query, key or value). As the sizes of the matrices vary strongly, the resulting

[4] We also experiment with other tokenization schemes, such as layer-based tokenization or the naive flattening and partitioning into tokens, of all weights. Matrix-based tokenization emerged as superior in early experiments, though only by a small margin.

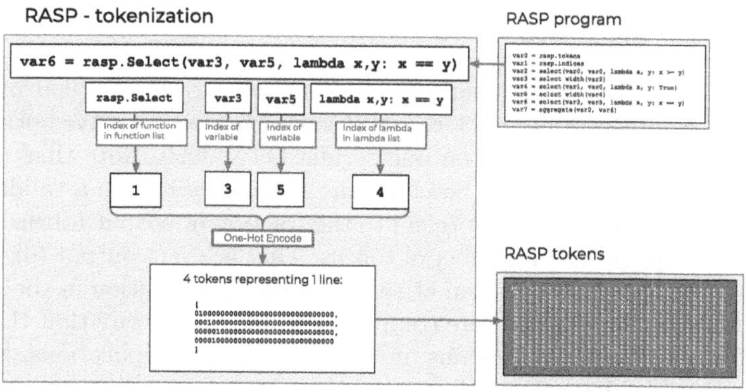

Fig. 4. Illustration of the vector representation of the RASP code. It works by dividing each line into four components, which are in turn represented by a one-hot-encoded vector, denoting one of the options for this component

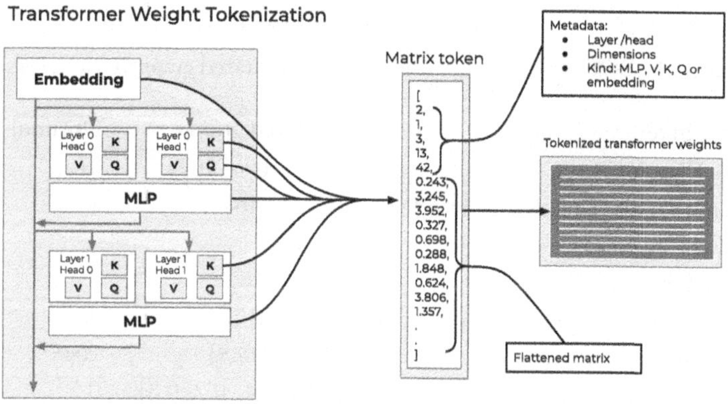

Fig. 5. Illustration of the process of translating the weights of a compiled transformer into a set of tokens

vectors are padded with zeros to achieve equal length. In Fig. 5, the process of transformer weight tokenization is visualized.

3.2 Decompiler Model

Our aim is to solve the problem of translating from the modality of transformer weights to the modality of RASP code. To this end, we adapt an existing system that also translates between modalities; namely the Whisper speech-to-text models by OpenAI [14]. Similar to Whisper, we employ an encoder-decoder transformer architecture. The encoder looks at the model input, in our case the tokenized transformer weights, and guides the decoder which produces the next RASP token based on all of the previously produced ones.

4 Experimental Evaluation

To evaluate our model, we generate an additional dataset of 1,000 programs, independent from the original 533K samples used for training. We optimize the hyperparameters of our model on one Nvidia GTX 3090. Note that the optimization of hyperparameters is based on the *token-accuracy* on a validation set of 28K samples. Token-accuracy refers to the fraction of output tokens that are correct relative to the total number of tokens. That is, every output token of our model is compared with the token at the corresponding position in the program that the transformer weights were compiled from. We are aware that the usefulness of this token-accuracy remains unclear for practical applications. However, it seems plausible that when a majority of the tokens are correct, this allows for the extraction of useful features from the RASP code, like for instance, the causal flow through the network.

The evaluated hyperparameters as well as the best-performing values for all hyperparameters are summarized in Table 1.

Table 1. Best performing values for all evaluated hyperparameters.

Hyperparameter	Explored Range	Optimal Value
Feature Dimension	{512, 1024, 2024}	512
Heads per Layer (encoder & decoder)	{4,6,. . .,16}	16
Number of layers:		
Encoder Layers	{4,6,. . .16}	4
Decoder Layers	{4,6,. . .,16}	4
Feedforward Dimension	{1024, 2048}	2048
Dropout	{0.01, 0.1, 0.2, 0.3}	0.2
Input Dimension ($\mathbf{x}$)		2000
Output Dimension ($\mathbf{y}$)		32

First, we evaluate the trained decompiler model on our independently generated test set in the non-autoregressive mode, in which it predicts the next token based on the ground-truth prefix.

In Fig. 6 we show the relative proportion of test programs that can be reproduced with a certain token-accuracy. We find that about 30% of all programs are reproduced identically to the ground truth (i.e., with a token accuracy of 100%)[5]. Approximately 85% of all test programs achieve a token-accuracy of 90% and overall we can report that all programs achieve at least 68% token-accuracy in this mode.

Moreover, in this mode we observe that 60% of the generated code is actually valid and runs without compilation error and 41% of the output programs are

[5] We name the proportion of output programs that are identical to the ground truth *sequence equality*.

functionally equivalent to the ground truth. This means that they represent the same input-output relations as the RASP code from which the transformer weights are originally compiled (an example of an output that does not match the ground truth but is functionally equivalent to the ground truth is shown in Fig. 7).

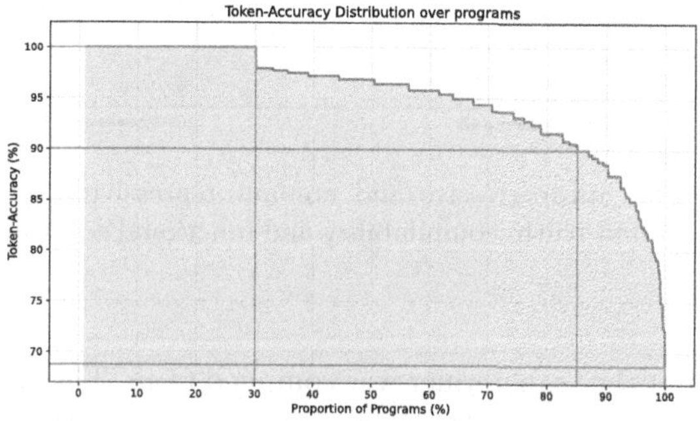

Fig. 6. Our model decompiles more than 30% of programs with 100% token-accuracy, with the rest of the programs being reproduced with at least 68% token-accuracy

Next, we test the model in autoregressive mode, in which it produces each token based on the tokens it previously produced. In this mode, the relative number of programs that achieve 100% token accuracy (i.e., sequence equality) drops from 30% to 26%. However, 91% of the outputs are now compilable (rather than 60% as achieved in the non-autoregressive mode). Remarkably, 73% (rather than 41%) of the output programs are functionally equivalent to the ground truth

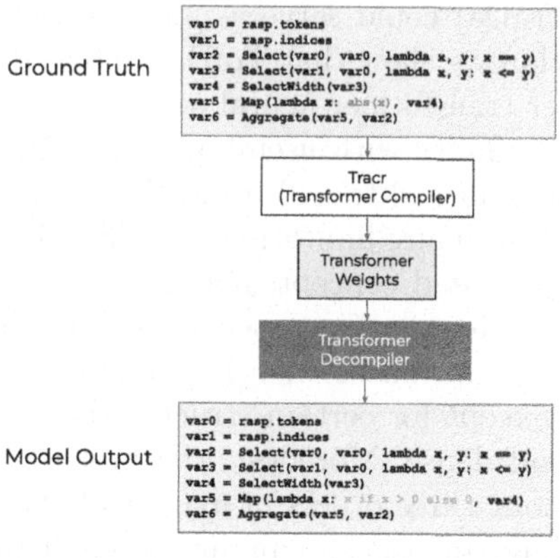

Fig. 7. These programs are identified as erroneous reproduction. However, they are functionally equivalent. The `var4` variable is produced by a `SelectWidth()` function, which only outputs values equal to or greater than zero. When applied to such values, the functions `abs(x)` and `x if x > 0 else 0` are equivalent

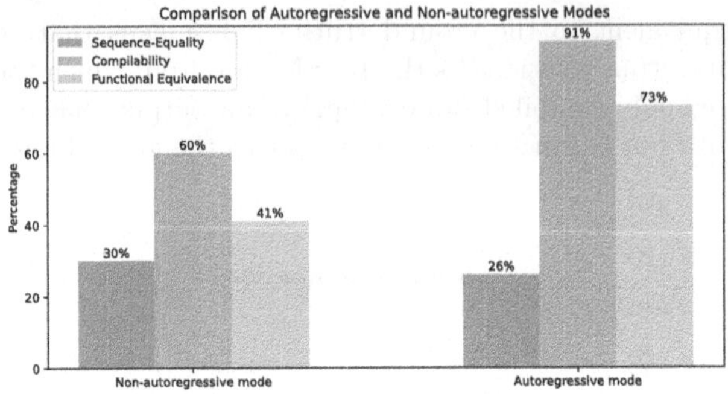

Fig. 8. Comparison of autoregressive and non-autoregressive mode across sequence equivalence to the ground truth, compilability and functional equivalence to the ground truth

RASP program that the transformer was compiled from. This could be explained by the model not seeing its previous outputs, but only the beginning of the ground-truth-program in the non-autoregressive mode. Seeing its own outputs for previous lines might allow the model to act according to the decisions it made earlier in the generation process.

The relative amount of outputs that achieve sequence equality, compilability, as well as functional equivalence are summarized in Fig. 8 for both modes (non-autoregressive and autoregressive).

5 Conclusion

To date, there are no general solutions available for the automatic interpretation of neural network models. In the present paper, we suggest that the Transformer Decompiler Method (TraDe) could serve as an approach to address this issue. In an empirical evaluation we show that TraDe enables a model to interpret the weights of other, smaller transformer neural networks and translate them into a more human-readable modality with useful accuracy. Our work thus represents a significant step towards an end-to-end framework for better interpretability.

However, there are still many limitations standing in the way of any practical application of the proposed concept. For instance, the transformer weights resulting from the Tracr compilation process are very different to transformer weights resulting from an optimization using stochastic gradient descent. The former is very sparse (except for certain structured elements), while the latter is very unstructured and dense. Moreover, the best-performing variant of the decompiler model is about three orders of magnitude larger than the models it is capable of decompiling (in terms of parameters). If this ratio is not significantly reduced, the application to modern, large transformer models will remain impossible. Lastly, though the step from weight matrices to RASP code is a large improvement in terms of interpretability, it would be wrong to call the produced RASP programs human-readable. Even with the simple 5 to 12 lines of RASP code, it can take some minutes for a human to determine what sequence operation the algorithm implements.

Possible future work is concerned with reducing (or eliminating) the above-mentioned limitations. For instance, by taking the current decompiler model on a set of compressed Tracr models, it might be possible to adapt the decompiling Tracr transformer weights to the task of decompiling learned transformer weights (compressed Tracr models contain matrices that are trained with gradient descent, and thus might more closely reflect realistic weights). Another option might be to reduce the task of the decompiler from the reproduction of all RASP code to the detection of certain features, like backdoors [15] or deceptive tendencies [16,17], or to the analysis of a sub-component of the transformer [18]. Last but not least, it could also be interesting to see which weight matrix is attended to the most when producing a certain piece of RASP code.

Acknowledgement. This study was funded by the *Hasler Foundation Switzerland* (grant number 23085).

Disclosure of Interests. The authors have no competing interests to declare that are relevant to the content of this article.

References

1. Vaswani, A., et al.: Attention is all you need. In: Advances in Neural Information Processing Systems, vol. 30 (2017)
2. Radford, A., et al.: Improving language understanding by generative pre-training, Tim Salimans (2018)
3. Dosovitskiy, A., et al.: An image is worth 16×16 words: transformers for image recognition at scale. arXiv preprint arXiv:2010.11929 (2020)
4. Zhang, Yu., Tiňo, P., Leonardis, A., Tang, K.: A survey on neural network interpretability. IEEE Trans. Emerg. Top. Comput. Intell. **5**(5), 726–742 (2021)
5. Bricken, T., et al.: Towards monosemanticity: decomposing language models with dictionary learning. Transf. Circ. Thread, 2 (2023)
6. Olsson, C., et al.: In-context learning and induction heads. arXiv preprint arXiv:2209.11895 (2022)
7. Bills, S., et al.: Language models can explain neurons in language models (2023). https://openaipublic.blob.core.windows.net/neuron-explainer/paper/index. Accessed 14 May 2023
8. Ahmad, O., Béreux, N., Baret, L., Hashemi, V., Lecue, F.: Causal analysis for robust interpretability of neural networks. In: Proceedings of the IEEE/CVF Winter Conference on Applications of Computer Vision, pp. 4685–4694 (2024)
9. Conmy, A., Mavor-Parker, A., Lynch, A., Heimersheim, S., Garriga-Alonso, A.: Towards automated circuit discovery for mechanistic interpretability. In: Advances in Neural Information Processing Systems, vol. 36 (2024)
10. Weiss, G., Goldberg, Y., Yahav, E.: Thinking like transformers. In: International Conference on Machine Learning, pp. 11080–11090. PMLR (2021)
11. Lindner, D., Kramár, J., Farquhar, S., Rahtz, M., McGrath, T., Mikulik, V.: Tracr: Compiled transformers as a laboratory for interpretability. In: Advances in Neural Information Processing Systems, vol. 36 (2024)

12. Mullen, L.A., Benoit, K., Keyes, O., Selivanov, D., Arnold, J.: Fast, consistent tokenization of natural language text. J. Open Source Softw. **3**(23), 655 (2018)
13. Vignac, C., Loukas, A., Frossard, P.: Building powerful and equivariant graph neural networks with structural message-passing. Advances in Neural Information Processing Systems, vol. 33, pp. 14143–14155 (2020)
14. Radford, A., Kim, J.W., Xu, T., Brockman, G., McLeavey, C., Sutskever, I.: Robust speech recognition via large-scale weak supervision. In: International Conference on Machine Learning, pp. 28492–28518. PMLR (2023)
15. Chen, X., Liu, C., Li, B., Lu, K., Song, D.: Targeted backdoor attacks on deep learning systems using data poisoning. arXiv preprint arXiv:1712.05526 (2017)
16. Hagendorff, T.: Deception abilities emerged in large language models. arXiv preprint arXiv:2307.16513 (2023)
17. Scheurer, J., Balesni, M., Hobbhahn, M.: Large language models can strategically deceive their users when put under pressure. In: ICLR 2024 Workshop on Large Language Model (LLM) Agents
18. Langosco, L., Alex, N., Baker, W., Quarel, D.J., Bradley, H., Krueger, D.: Towards meta-models for automated interpretability (2024). Manuscript in preparation

Pitfalls in Processing Infinite-Length Sequences with Popular Approaches for Sequential Data

Michele Casoni[1(✉)], Tommaso Guidi[1], Matteo Tiezzi[1], Alessandro Betti[2], Marco Gori[1], and Stefano Melacci[1]

[1] DIISM, University of Siena, 52100 Siena, Italy
{m.casoni,t.guidi}@student.unisi.it,
{matteo.tiezzi,marco.gori,stefano.melacci}@unisi.it
[2] Scuola Alti Studi, IMT, 55100 Lucca, Italy
alessandro.betti@imtlucca.it

Abstract. One of the enduring challenges for the Machine Learning community is developing models that can process and learn from very long data sequences. Transformer-based models and Recurrent Neural Networks (RNNs) have excelled in processing long sequences, yet face challenges in transitioning to processing infinite-length sequences online, a crucial step in mimicking human learning over continuous data streams. While Transformer models handle large context windows efficiently, they suffer from quadratic computational costs, motivating research into alternative attention mechanisms. Conversely, RNNs, particularly Deep State-Space Models (SSMs), have shown promise in long sequence tasks, outperforming Transformers in certain benchmarks. However, current approaches are limited to finite-length sequences, which are pre-buffered and randomly shuffled to cope with stochastic gradient descent. This paper addresses the fundamental gap in transitioning from offline-processing of a dataset of sequences to online-processing of possibly infinite-length sequences, a scenario often neglected in existing research. Empirical evidence is presented, demonstrating the performance and limits of existing models. We highlight the challenges and opportunities in learning from a continuous data stream, paving the way for future research in this area.

Keywords: Recurrent Neural Networks · Transformers · Long Sequences · Online Learning · Learning over Time

1 Introduction

In recent years, Transformer-based models have emerged as the forefront of various tasks across natural language processing (NLP), computer vision, and beyond, particularly excelling in scenarios with ample training data [5,17,32]. These models possess the inherent capability to process large context windows

C. Y. Suen et al. (Eds.): ANNPR 2024, LNAI 15154, pp. 37–48, 2024.
https://doi.org/10.1007/978-3-031-71602-7_4

containing thousands of tokens efficiently, thanks to the attention mechanism they are based on and parallel computation of the output [27]. Additionally, in Long Sequence Time-Series Forecasting (LSTF), Transformer-based networks have showcased remarkable performances. However, the quadratic computational cost of attention with respect to the input length represents one of the main drawbacks when processing long sequences, inspiring many recent works to introduce alternative definitions of attention [9,17]. Alternatively to the formidable performance of Transformer-based architectures, Recurrent Neural Networks (RNNs) have demonstrated competitive results in NLP tasks, especially with the recent rise of Deep State-Space Models (SSMs). As the focus shifts towards evaluating a broader range of long sequences tasks, benchmarks like the Long Range Arena (LRA) [24] present challenges encompassing sequences ranging from 1,000 to 16,000 elements across various domains, like natural language and mathematics. Notably, models such as SSMs and Linear Recurrent Units (LRUs) have outperformed Transformers in these tasks [15,33].

Despite these advancements, a crucial gap necessitates further exploration: the transition from *offline* processing of *long* sequences to *online* processing of possibly *infinite-length* sequences. Indeed, while the ultimate goal of mimicking the human capacity to learn over sequential-data inspired several researchers, all the aforementioned approaches are designed to *offline* process datasets of *finite-length* sequences, mostly learning by Back-Propagation Through Time (BPTT) with stochastic sampling of the examples. This is in contrast with the way humans learn from perceptual stimuli, which are intrinsically continuous over time and not pre-buffered finite-length sequences randomly shuffled to cope with stochastic gradient descent [3,6,26]. Moreover, most of the popular research activities in the field of Transformers and RNNs are not considering the case of learning *online* from a continuous, possibly infinite, stream of data (i.e., infinite-length sequence), without resets or segmentations [1,2,13,25]. This is also in contrast with the motivations behind seminal works in the field of recurrent models, such as the online learning algorithm of Williams and Peng [29], and the original formulation of LSTMs [8], amongst others.[1]

This paper addresses this gap by exploring the fundamental problem of transitioning from processing long to infinite-length sequences online, showing how the direct application of existing models to the considered setting is not straightforward, yielding unsatisfactory performance. A more detailed description of the problem is given in Sect. 2. Popular approaches for dealing with sequential data are described in Sect. 3, together with the notation that is used for presenting them, and concrete experience on these models processing infinite-length sequences online is provided in Sect. 4. An overall discussion on the results is included in Sect. 5, while Sect. 6 is left for conclusions.

---

[1] From [29]: *"an on-line algorithm, designed to be used to train a network while it runs; no manual state resets or segmentations of the training stream is required"*. From [8]: LSTMs were introduced with a learning algorithm that unlike BPTT is *"local in space and time"*, where *"there is no need to store activation values observed during sequence processing in a stack with potentially unlimited size"*.

2 Processing Infinite-Length Sequences

Nowadays, Transformer-based models represent the state-of-the art in several tasks where abundant training data is available, ranging from the domain of natural language (question answering, summarization, classification, ...) to the one of vision and beyond [5]. These models are intrinsically able to process context windows with thousands of tokens. In the case of LSTF, Transformer-based networks have been recently shown to effectively reach outstanding results (Informer, [32]), also challenged by more ad-hoc architectures (FSNet, [18]). When it comes to the evaluation of a larger spectrum of long sequence tasks, the Long Range Arena (LRA) benchmark [24] includes problems with sequences composed of 1,000 to 16,000 elements, in the context of natural language, images, mathematics and others. SSMs and LRUs yield state-of-the art performance [14,15,22], overcoming Transformers [33]. Two of the most evident directions for future research consist in further improving the quality of SSMs/LRUs, and enabling recurrence-based Transformers to reach and overcome the performance of quadratic-complexity attention. However, there is a major gap that requires more specific studies to go beyond existing technologies, summarized in the following question: what is the transition from processing *long* sequences to *infinite-length* sequences?

Let us consider a sequence of patterns $(u_t)_{t=1}^{L} := (u_1, \ldots, u_L)$, where L is an integer that represents the sequence length and $u_t \in \mathbb{R}^{d_{\text{in}}}$ is the t-th element of the sequence. Dealing with learning online in a lifelong manner, from a single, continuous stream of sequential data $(u_1, u_2, \ldots, u_\infty)$ (i.e., data that is not i.i.d., the order of data does matter, having a role in the decision process) is characterized by two main requirements: (*i.*) to compute meaningful gradients at time t that depend on properties of the data processed so far, without unrolling the network over the whole (potentially infinite) input sequence (in the most extreme case only in function of information at time t and $t-1$), updating the model parameters at each t; (*ii.*) to have the capability of adapting to the properties of the streamed data, not forgetting the different skills that are gained over time, as commonly stressed in the context of continual learning. When considering property (*ii.*), we fall in a setting which can be considered an instance of Online Continual Learning with a strong intersection to Streaming Learning, as recently summarized in [7]. Notice that this is different from the case of works that focus on continual learning from streams of finite-length sequences.

To our best knowledge, the setting described in this Section is largely unexplored. In fact, despite many progresses in the related fields, requirements (*i.*) and (*ii.*) are still already hard to fulfil when faced in an independent manner, thus yielding an even more difficult scenario when joined together. However, it represents the case of learning from the visual data that hit our eyes, the sounds we continuously perceive, the video stream from an always-on camera, language, and so on and so forth. Common benchmarks on sequential data, such as LRA, are based on collections of long sequences, thus not perfectly suited for the aforementioned setting, while they are good opportunities to evaluate the fulfilment of requirement (*i.*). It is also practical to work in the typical setting of online learning problems, where all the data is used for training purposes and errors are accumulated while learning (i.e., u_t is processed and an output at time t is compared with the available ground truth to update the accuracy estimate; then, the output is used to evaluate a loss function, compute gradients at t and update the model). LSTF is more flexible from this point of view than classification problems on long sequences, since it offers ground truth at each t. Moreover, while the model is not expected to learn new tasks, the properties of the series might change over time thus partially allowing to explore requirement (*ii.*). However, deeply evaluating property (*ii.*) is still beyond what can be concretely done with these datasets. Of course, one might concatenate the sequences taken from offline databases and evaluate a learning problem on such long, single stream, that would be however artificial, due to the intrinsic discontinuity when switching from a sequence to the following one. In Sect. 3, we describe some models and learning algorithms that are commonly employed in literature for dealing with sequential data.

3 Popular Approaches for Sequential Data

In this Section we describe some popular models and algorithms in the literature for dealing with long sequence processing. The problem of capturing long range dependencies has been treated for years by researchers, starting from using RNNs. Recurrent models process one pattern at-a-time, following the sequence order, progressively (sequentially) encoding the information into d_s context units whose output values model a *state* $x_t \in \mathbb{R}^{d_s}$. While in standard feed-forward networks hidden neurons develop internal representations of the current input patterns, in recurrent models the state encodes the temporal properties of the input data that was sequentially provided so far. The state is further processed for generating an *output* signal $y_t \in \mathbb{R}^{d_{out}}$, $d_{out} \geq 1$. The Elman's architecture [4] was one of the pillars to the foundations of classic RNNs, and it is based on the following equations,

$$x_t = \sigma(Ax_{t-1} + Bu_t), \quad y_t = \sigma_{out}(Cx_t + Du_t), \tag{1}$$

where we considered a *state matrix* $A \in \mathbb{R}^{d_s \times d_s}$, an *input matrix* $B \in \mathbb{R}^{d_s \times d_{in}}$, an *output matrix* $C \in \mathbb{R}^{d_{out} \times d_s}$ and a *feedthrough matrix* $D \in \mathbb{R}^{d_{out} \times d_{in}}$.[2] We denote with σ and σ_{out} non-linearities on the state and output computations, respectively, often selected to be a tanh or sigmoid activations. If σ is the identity function, then we say the RNN model is *linear*. When σ_{out} is the identity function, linear RNNs are said to be in *state-space model form* [21]. There are several books, surveys and reviews in which the reader can find further details and properties of RNN-based modes [10,20,31]. Let us collect the model parameters in $\theta := \{A, B, C, D\}$. RNNs are usually trained by Backpropagation Through Time (BPTT) [19], the de-facto standard algorithm to learn

[2] $D \neq 0$ introduces a skip connection that was not included in the original Elman network [4].

θ by gradient descent, where the chain-rule is applied over the unfolded network in order to compute gradients [28]. Let us introduce the instantaneous loss function $\ell(y_t, \hat{y}_t)$, which quantifies to what degree the predicted output y_t matches some supervised/target output $\hat{y}_t$. Given a sequence and the empirical risk $\mathcal{L}(\theta) = \frac{1}{L}\sum_{t=1}^{L} \ell(y_t, \hat{y}_t)$, it can be shown that the gradient w.r.t. θ is a sum of products of partial gradients,

$$\frac{\partial \mathcal{L}}{\partial \theta} = \frac{1}{L}\sum_{t=1}^{L} \frac{\partial \ell(y_t, \hat{y}_t)}{\partial y_t} \frac{\partial y_t}{\partial x_t} \sum_{j=1}^{t} \Big(\prod_{s=j}^{t} \frac{\partial x_s}{\partial x_{s-1}} \Big) \frac{\partial x_{j-1}}{\partial \theta}, \tag{2}$$

where a central role is played by the term $\prod_{s=j}^{t} \partial x_s / \partial x_{s-1}$, which *transports* the error back in time from step t to step j. We can rewrite this term in the form of a product of Jacobi matrices [16],

$$\prod_{s=j}^{t} \frac{\partial x_s}{\partial x_{s-1}} = \prod_{s=j}^{t} A^T \mathrm{diag}(\sigma'(x_{s-1})), \tag{3}$$

where diag$(\cdot)$ converts a vector into a diagonal matrix, and σ' computes the derivative of the activation function σ in Eq. (1) in an element-wise fashion. Unless the partial terms $\partial x_s / \partial x_{s-1}$ are close to 1, the product in Eq. (3) could explode or vanish. In details, in the simplified case of a linear model (i.e., setting σ to the identity function) a sufficient condition for long-term components to vanish is that of having the largest eigenvalue of the state matrix A smaller than one, i.e., $\lambda < 1$. Conversely, $\lambda > 1$ is a necessary condition for gradients to explode, causing older states to have little contribution. See [16] for further details. When gradients undergo vanishing during their backward propagation through the network, the critical *credit assignment* of backpropagation is compromised. In fact, information regarding minor state changes in the distant past loses its ability to influence future states. Conversely, when gradients explode, optimization algorithms encounter substantial difficulties in traversing the cost surface. In order to make RNNs better suited to handle longer sequences, popular variants were proposed and widely used in the scientific literature and several real-world applications, such as Long Short-Term Memories (LSTMs), Gated Recurrent Units (GRUs), and others [31]. In order to prevent these issues, the Truncated BPTT (TBPTT) [29] limits the gradient flow after a fixed number of time steps, unfortunately resulting in a structural inability to capture dependencies beyond the designated time window.

We now leverage a more general form of Eq. (1), comprising a transition function $F(\cdot)$ and readout function $G(\cdot)$, defined as follows for notational convenience:

$$x_t = F(x_{t-1}, u_t, \theta^F) \qquad y_t = G(x_t, u_t, \theta^G) \tag{4}$$

where, referring to Eq. (1), $\theta^F := [A, B]$ and $\theta^G := [C, D]$. BPTT requires to cache the neuron activations for every element within the sequence processed by the model, in order to be able to perform gradient computation. Such cache grows linearly with the sequence length. In contrast, Real-time Recurrent Learning

(RTRL) [30] does not require the storage of past activations, and was proposed as an *online* alternative to BPTT, which enables weight updates promptly after each new input is processed, provided that external error feedback to the model output is accessible for each input. For every timestep t, one can define the *influence matrix* M_t as

$$M_t = \frac{\partial x_t}{\partial \theta^F}, \tag{5}$$

the derivatives of the current state x_t with respect to the parameters θ^F of the transition function F in Eq. (4). Unpacking Eq. (5) by one time step, it is possible to prove the following recursive formula for M_t

$$M_t = J_t M_{t-1} + \bar{M}_t, \tag{6}$$

where $J_t = \frac{\partial x_t}{\partial x_{t-1}}$ is the Jacobian of the actual state w.r.t. the previous state and $\bar{M}_t = \frac{\partial x_t}{\partial \theta^F_t}$ is called the *immediate influence*. It is possible to obtain the derivatives of the loss function with respect to θ^F by

$$\frac{\partial \ell_t}{\partial \theta^F} = \frac{\partial \ell_t}{\partial x_t} \frac{\partial x_t}{\partial \theta^F} = \bar{c}_t M_t, \tag{7}$$

where $\bar{c}_t = \frac{\partial \ell_t}{\partial x_t}$ is called the *immediate credit assignment vector*. It is natural to learn θ^G online, because only information at present time t is required to calculate the gradient $\frac{\partial \ell_t}{\partial \theta^G}$. RTRL suggests to propagate the partials $\frac{\partial x_t}{\partial x_{t-1}}$ and $\frac{\partial x_t}{\partial \theta}$ from timestep t to $t+1$. This algorithm is online and past-facing, which is defined by the fact that only previously computed quantities are used in the computations. RTRL is also deterministic and provides the solution in closed form. Since RTRL requires, for each time step, the storage of M_t, which involves the gradients of each component of the state (d_s elements) with respect to all the parameters involved in the state computation (i.e., $|\theta^F| = d_s \cdot d_{\text{in}} + {d_s}^2$), its memory complexity is $O(d_s(d_s \cdot d_{\text{in}} + {d_s}^2)) \sim O({d_s}^3)$ and, for the computation of $J_t M_{t-1}$, its time complexity is $O({d_s}^4)$. Because of RTRL high memory and time complexities, researchers have recently focused on finding approximations for reducing its computational cost. For example, Unbiased Online Recurrent Optimization (UORO) [23] is a stochastic approximation of RTRL. Decomposing M_t as the outer product of two tensors of lower rank A_t and B_t, such that

$$M_t^{kij} = A_t^k B_t^{ij}, \tag{8}$$

UORO provides stochastic approximations for A_t and B_t defined through a random vector $\nu \in \mathbb{R}^{d_s}$, which satisfies $\mathbb{E}[\nu^i \nu^j] \propto \delta_{ij}$ and $\mathbb{E}[\nu^i] = 0$. More precisely,

$$\begin{aligned} A_t^k &= \rho_0 \sum_{k'} J_t^{kk'} A_{t-1}^{k'} + \rho_1 \nu^k \\ B_t^{ij} &= \rho_0^{-1} B_{t-1}^{ij} + \rho_1 \sum_{k'} \nu^{k'} \bar{M}_t^{k'ij}, \end{aligned} \tag{9}$$

where ρ_0 and ρ_1 are positive constants. It is possible to prove (see [12]) that the resulting outer product is an unbiased estimator for M_t. This algorithm is still online and past-facing, but the memory and time complexities are reduced to $O({d_s}^2)$.

Long sequences can also be processed using Transformer-based architectures, *stateless* architectures fully based on attention. The essence[3] of the attention-based encoding process (which in RNNs is Eq. (1)) becomes

$$y_t = \sum_{i=1}^{L} \frac{\text{sim}\,(q_t, k_i)}{\sum_{j=1}^{L} \text{sim}\,(q_t, k_j)} v_i^T, \quad \begin{aligned} q_z &= q(u_z, W_q) \\ k_z &= k(u_z, W_k) \\ v_z &= v(u_z, W_v) \end{aligned} \tag{10}$$

being sim () a similarity function, such as $\text{sim}\,(a, b) = \exp\,(a^T b/\sqrt{|a|})$, and q, k, v are three projections depending on learnable matrices W_q, W_k, W_v, respectively. The cost of computing y_t for L steps is quadratic, $O(L^2)$. In Transformers, each y_t can be computed in parallel to the other ones, and this became an attractive property for GPU-oriented implementations. Because of the quadratic time complexity of attention, we focus on a specific Transformer-based model called Informer [32], which leverages on a *ProbSparse* attention mechanism given by

$$y_t = \sum_{i=1}^{L} \frac{\text{sim}\,(\bar{q}_t, k_i)}{\sum_{j=1}^{L} \text{sim}\,(\bar{q}_t, k_j)} v_i^T, \tag{11}$$

where, given an integer u, $\bar{q}_t$ is of the same size of q_t and it only contains the Top-u queries under the sparsity measurement $M(q_t; \mathbf{t})$ defined as

$$M(q_t; \mathbf{t}) = \ln \sum_{j=1}^{L} \text{sim}\,(q_t, k_j) - \frac{1}{L} \sum_{j=1}^{L} \frac{q_t^T k_j}{\sqrt{|q_t|}}. \tag{12}$$

This attention mechanism achieves $O(L \ln L)$ in time complexity and memory usage. Transformers performances are also challenged by more ad-hoc architectures, like FSNet [18]. This model is based on a Temporal Convolutional Network (TCN) backbone with M layers, characterized by $\{\theta_m\}_{m=1}^{M}$ respectively, and two auxiliary features: a per-layer adapter for fast adaptation to abrupt changes in the input (*Fast* part) and a sparse associative memory interaction, which facilitates learning of recurring patterns (*Slow* part). The per-layer adapter uses the Exponential Moving Average (EMA) of the backbone's gradient $\mathbf{g}_t^m$ with a smoothing coefficient γ to smooth out online training's noises and to capture the temporal information in time series, i.e.,

$$\hat{\mathbf{g}}_t^m = \gamma \hat{\mathbf{g}}_t^m + (1 - \gamma)\mathbf{g}_t^m. \tag{13}$$

Here, $\mathbf{g}_t^m$ denotes the gradient of the m-th layer at time t, while $\hat{\mathbf{g}}_t^m$ is the EMA gradient. Then, it maps the smoothed gradients to adaptation coefficients,

[3] Transformers usually include multiple multi-head attention-based layers interleaved by FFNs. We considered a simple head, with no further projections, to better connect it to the RNN case.

subsequently flattening the gradient EMA into a vector and splitting the gradient vector into a given number of chunks. Moreover, the interaction with an external memory is triggered only if the cosine distance between $\hat{\mathbf{g}}_t^m$ and $\hat{\mathbf{g}}_t'^m$ (another gradient EMA with a smaller coefficient $\gamma' < \gamma$) is smaller than a given threshold.

4 Experiments

In order to provide the reader with a concrete experience, we report the results we achieved in LSTF (selected for the reasons mentioned in Sect. 2), using real-world (ETT, ECL, WTH) and artificial (Synth, S-A, S-G) datasets,[4] inheriting the same setting of [18], where time-series are split into three disjoint parts, preserving the data order.[5] The initial segment of the sequence ($\approx 20\%$) is a training set, followed by a small ($\approx 5\%$) validation set and the remainder of the sequence acts as a test set. These last two parts are still used for online learning, accumulating errors as previously described. We used Mean Squared Error (MSE) and Mean Absolute Error (MAE) for evaluating the models, which are the standard metrics used in LSTF. Some models require a finite-length sequence as input, and in such cases we segmented the time-series using a sliding Window (different sizes, chosen accordingly to previous experiences on this problem [18]). We exploited the AdamW [11] optimizer (we also evaluated vanilla GD), and the same architectures and optimal parameter values suggested by the respective authors in their experiences with these data (when available) as starting configurations. We cross-validated new values around them (learning rates, γ in FSNet, *label_len* param up to window size in Informer). We compared results with the ones of a Dummy model which, at time t, simply outputs the data received at time $t-1$. In what follows, we describe the results obtained for each model.[6] We start from Vanilla RNNs with online computation of the gradients (RTRL, UORO [12]). In this case (like the one of LSTMs described later), we used simpler models, with 8, 16, 32, 128 or 256 hidden units, and the following learning rates: 10^{-4}, 10^{-3}, 10^{-2}. Results are reported in Table 1 and show that Vanilla RNNs trained online overcome the Dummy model only on S-A and S-G datasets. On ECL, the performances of Vanilla RNNs differ from the ones of the Dummy by two orders of magnitude, which become three and six in the case of Synth dataset, for RTRL and UORO respectively. Since UORO is an approximation of RTRL, it is natural that it gives worse or comparable results w.r.t. RTRL, not better ones.

---

[4] For a detailed description of each dataset, see [32], except for Synth, which is a sinusoidal signal at frequency of 2.8 mHz, with 36000 samples.

[5] We used/adapted the code of the authors of [18,32], and the library of [12].

[6] In FSNet, Synth data, we considered the second-best sets of parameter values, since the best ones were yielding numerical errors.

Table 1. Mean Squared Error (MSE) and Mean Absolute Error (MAE) obtained with a Vanilla RNN trained online (RTRL, UORO) and the Dummy model on the (long) test portion of the time series. Results that overcome the Dummy model are marked with †.

Data →		**Synth**		**ETTh1**		**ECL**		**WTH**		**S-A**		**S-G**	
Model ↓	Window	*MSE*	*MAE*	*MSE*	*MAE*	*MSE*	*MAE*	*MSE*	*MAE*	*MSE*	*MAE*	*MSE*	*MAE*
RNN (RTRL)	-	$9.2e^{-3}$	0.051	0.381	0.412	125.481	1.022	0.337	0.397	*1.373*†	*0.936*†	*1.297*†	*0.900*†
RNN (UORO)	-	2.723	0.506	0.381	0.412	277.188	1.352	3.329	0.910	*1.373*†	*0.936*†	*1.297*†	*0.900*†
Dummy	-	$3.0e^{-6}$	$1.6e^{-3}$	0.363	0.355	6.747	0.376	0.207	0.188	1.430	0.951	1.377	0.924

Considering Vanilla LSTMs with TBPTT (with a sliding window of 1, 10, 60, and 100 samples), we report in Table 2 results obtained with and without resetting the state when moving from a TBPTT-window to the following one. Also in this case, we considered simple models with 8, 16, 32, 128 or 256 hidden units, and the following learning rates: 10^{-4}, 10^{-3}, 10^{-2}. We remark that results obtained with or without reset are similar in values and orders of magnitude. It is possible to see that, unlike RNNs trained online, LSTMs are capable to overcome the Dummy model also on ETTh1 and WTH datasets, but metrics reveal the limits of LSTMs in LSTF on Synth and ECL, which differ from the Dummy ones with the same orders of magnitude recorded in RNNs.

Table 2. Mean Squared Error (MSE) and Mean Absolute Error (MAE) obtained with a Vanilla LSTM trained with TBPTT (with and without resetting the state when moving from a TBPTT-window to the following one) and the Dummy model on the (long) test portion of the time series. Results that overcome the Dummy model are marked with †.

Data →		**Synth**		**ETTh1**		**ECL**		**WTH**		**S-A**		**S-G**	
Model ↓	Window	*MSE*	*MAE*	*MSE*	*MAE*	*MSE*	*MAE*	*MSE*	*MAE*	*MSE*	*MAE*	*MSE*	*MAE*
LSTM	1	$4.0e^{-3}$	0.011	*0.301*†	0.362	179.994	1.189	*0.187*†	0.257	*1.222*†	*0.884*†	*1.262*†	*0.888*†
	10	$2.7e^{-3}$	0.011	0.435	0.448	140.911	1.051	0.210	0.287	*1.230*†	*0.886*†	*1.201*†	*0.864*†
	60	$1.4e^{-3}$	0.009	*0.339*†	0.392	141.825	1.053	0.212	0.288	*1.216*†	*0.882*†	*1.213*†	*0.868*†
	100	$1.7e^{-3}$	0.010	*0.290*†	0.357	265.110	1.515	0.212	0.286	*1.215*†	*0.883*†	*1.217*†	*0.870*†
LSTM (No Reset)	1	$2.0e^{-3}$	0.009	*0.347*†	0.392	261.114	1.491	*0.206*†	0.280	*1.205*†	*0.879*†	*1.209*†	*0.866*†
	10 .	$1.1e^{-3}$	0.011	0.377	0.410	256.850	1.493	*0.200*†	0.273	*1.226*†	*0.884*†	*1.202*†	*0.864*†
	60	$1.2e^{-3}$	0.008	*0.316*†	0.374	145.238	1.089	*0.204*†	0.276	*1.269*†	*0.901*†	*1.195*†	*0.862*†
	100	$1.1e^{-3}$	0.008	0.368	0.409	271.016	1.505	*0.204*†	0.278	*1.229*†	*0.886*†	*1.265*†	*0.888*†
Dummy	-	$3.0e^{-6}$	$1.6e^{-3}$	0.363	0.355	6.747	0.376	0.207	0.188	1.430	0.951	1.377	0.924

We also report the performances on LSTF of more recent models, more specifically of Informer [32] and FSNet [18]. Instead of offline-pretraining the model on the training set as in [18] and then move to the online processing of the rest of the series, we process the whole time-series in an online manner from the very beginning, simulating lifelong online learning. In Table 3, it is clear that FSNet overcomes the Dummy on each dataset, except for Synth, with a difference of four orders of magnitude, while Informer has the same behavior of LSTMs (missing ECL too).

Table 3. Mean Squared Error (MSE) and Mean Absolute Error (MAE) obtained with FSNet and Informer trained online and the Dummy model on the (long) test portion of the time series. Results that overcome the Dummy model are marked with †. FSNet and Informer were also trained Offline, shuffling data, for comparison.

Data →		**Synth**		**ETTh1**		**ECL**		**WTH**		**S-A**		**S-G**	
Model ↓	Window	*MSE*	*MAE*	*MSE*	*MAE*	*MSE*	*MAE*	*MSE*	*MAE*	*MSE*	*MAE*	*MSE*	*MAE*
FSNet	1	0.028	0.084	*0.326*†	0.358	*3.470*†	0.546	*0.182*†	0.241	*1.370*†	*0.930*†	1.419	0.938
	10	0.041	0.085	*0.265*†	*0.331*†	*3.338*†	0.552	*0.181*†	0.241	*1.275*†	*0.901*†	*1.311*†	*0.907*†
	60	0.054	0.094	*0.260*†	*0.328*†	*3.276*†	0.549	*0.180*†	0.241	*1.349*†	*0.924*†	*1.310*†	*0.902*†
	100	0.046	0.087	*0.257*†	*0.326*†	*3.218*†	0.542	*0.181*†	0.243	*1.337*†	*0.923*†	*1.340*†	*0.919*†
Informer	1	$9.7e^{-5}$	0.006	*0.357*†	0.380	309.722	1.511	*0.166*†	0.224	*1.204*†	*0.877*†	*1.076*†	*0.821*†
	10	$3.7e^{-4}$	0.010	*0.287*†	*0.354*†	311.685	1.512	*0.165*†	0.225	*1.116*†	*0.843*†	*1.299*†	*0.899*†
	60	$1.4e^{-3}$	0.021	*0.286*†	0.355	311.709	1.507	*0.165*†	0.224	*1.121*†	*0.843*†	*1.327*†	*0.911*†
	100	$4.2e^{-4}$	0.011	*0.289*†	0.357	310.627	1.515	*0.165*†	0.224	*1.097*†	*0.835*†	*1.327*†	*0.910*†
FSNet (Offline)	60	0.029	0.088	*0.275*†	*0.338*†	*3.143*†	0.472	*0.176*†	0.236	*1.349*†	*0.923*†	*1.278*†	*0.896*†
Informer (Offline)	60	$1.4e^{-3}$†	0.019	*0.321*†	0.372	114.962	0.930	*0.148*†	0.198	*1.093*†	*0.832*†	*1.046*†	*0.808*†
Dummy	-	$3.0e^{-6}$	$1.6e^{-3}$	0.363	0.355	6.747	0.376	0.207	0.188	1.430	0.951	1.377	0.924

To provide another useful reference, we considered the original setup in [18], where offline-training (multiple epochs, shuffled data, window size 60) is exploited, before processing the rest of the series as in the already described experiences. In the last two rows of Table 3 we clearly see that the quality of the results increases when compared to the same window-sized models in the upper portion of the table.

5 Discussion

Of course, this must not be intended as an exhaustive comparison of the state-of-the art, but just a tangible experience to show the challenging nature of the discussed learning setting. Results of Tables 1, 2 and 3 show that a Dummy model, which simply outputs as prediction the data received at the previous time instant, is better than other tested methods in several cases (in 48.14% of the results), while in the remaining cases (marked with †) the gap in terms of performance of the compared models is not so evident, despite their large structural differences. As a matter of fact, RNNs and LSTMs are way simpler models than FSNet or Informer, and we also compared sliding-window-based models against models that never reset their internal state. In general, most of the results are relatively far from the ideal optimum (i.e., zero). All these considerations suggest that there is room for developing new models which are better at capturing regularities on the input stream in an online manner. The small size of the validation part of the series, w.r.t. the test one (typical of lifelong learning problems), is another potential issue to address with models that self-adapt their hyper-parameters over time, since the cross-validated values might turn out to not perform well on the long run.

6 Conclusions

Despite the advancements on Transformer-based architectures and RNNs, a critical gap exists in transitioning from processing long sequences to infinite-length one, particularly in an online learning setting. While current approaches excel in finite-length sequence learning, they fall short when dealing with continuous, possibly infinite streams of data without resets or segmentations. This work addressed this gap by exploring the fundamental challenge of online learning from infinite-length sequences. Results indicate that while more complex models like FSNet and Informer may have structural advantages, simpler models like RNNs and LSTMs perform comparably in many cases. However, none of the models achieve the ideal optimum, indicating the need for novel approaches that better capture regularities in input streams online. In summary, this study sheds light on the challenging nature of transitioning from processing long to infinite-length sequences online and suggests avenues for developing more effective models capable of handling continuous data streams.

References

1. Betti, A., Casoni, M., Gori, M., Marullo, S., Melacci, S., Tiezzi, M.: Neural time-reversed generalized riccati equation. In: Proceedings of the AAAI Conference on Artificial Intelligence, vol. 38, pp. 7935–7942 (2024)
2. Betti, A., et al.: Continual learning through Hamilton equations. In: Conference on Lifelong Learning Agents, vol. 199, pp. 201–212. PMLR (2022)
3. Betti, A., Gori, M., Melacci, S.: Cognitive action laws: the case of visual features. IEEE Trans. Neural Netw. Learn. Syst. **31**(3), 938–949 (2020). https://doi.org/10.1109/TNNLS.2019.2911174
4. Elman, J.L.: Finding structure in time. Cogn. Sci. **14**(2), 179–211 (1990)
5. Anil, R.G., et al.: Gemini: a family of highly capable multimodal models (2023)
6. Gori, M., Melacci, S.: Collectionless artificial intelligence. arXiv 2309.06938 (2023)
7. Gunasekara, N., Pfahringer, B., Gomes, H.M., Bifet, A.: Survey on online streaming continual learning. In: Proceedings of the IJCAI, pp. 6628–6637 (2023)
8. Hochreiter, S., Schmidhuber, J.: Long short-term memory. Neural Comput. **9**(8), 1735–1780 (1997)
9. Katharopoulos, A., Vyas, A., Pappas, N., Fleuret, F.: Transformers are RNNs: fast autoregressive transformers with linear attention. In: International Conference on Machine Learning, pp. 5156–5165 (2020)
10. Lipton, Z.C., Berkowitz, J., Elkan, C.: A critical review of recurrent neural networks for sequence learning. arXiv:1506.00019 (2015)
11. Loshchilov, I., Hutter, F.: Decoupled weight decay regularization. arXiv preprint arXiv:1711.05101 (2017)
12. Marschall, O., Cho, K., Savin, C.: A unified framework of online learning algorithms for training recurrent neural networks. J. Mach. Learn. Res. **21**(1), 5320–5353 (2020)
13. Marullo, S., Tiezzi, M., Betti, A., Faggi, L., Meloni, E., Melacci, S.: Continual unsupervised learning for optical flow estimation with deep networks. In: Conference on Lifelong Learning Agents, pp. 183–200. PMLR (2022)

14. Massaroli, S., et al.: Laughing hyena distillery: extracting compact recurrences from convolutions. In: Advances in NeurIPS (2023). https://openreview.net/forum?id=OWELckerm6
15. Orvieto, A., et al.: Resurrecting recurrent neural networks for long sequences. arXiv preprint arXiv:2303.06349 (2023)
16. Pascanu, R., Mikolov, T., Bengio, Y.: On the difficulty of training recurrent neural networks. In: International Conference on ML, pp. 1310–1318 (2013)
17. Peng, H., Pappas, N., Yogatama, D., Schwartz, R., Smith, N., Kong, L.: Random feature attention. In: ICLR (2021). https://openreview.net/forum?id=QtTKTdVrFBB
18. Pham, Q., Liu, C., Sahoo, D., Hoi, S.: Learning fast and slow for online time series forecasting. In: ICLR (2022)
19. Rumelhart, D.E., et al.: Learning internal representations by error propagation (1985)
20. Salehinejad, H., Sankar, S., Barfett, J., Colak, E., Valaee, S.: Recent advances in recurrent neural networks. arXiv preprint arXiv:1801.01078 (2017)
21. Schäfer, A.M., Zimmermann, H.G.: Recurrent neural networks are universal approximators. In: Kollias, S.D., Stafylopatis, A., Duch, W., Oja, E. (eds.) ICANN 2006. LNCS, vol. 4131, pp. 632–640. Springer, Heidelberg (2006). https://doi.org/10.1007/11840817_66
22. Smith, J.T., Warrington, A., Linderman, S.W.: Simplified state space layers for sequence modeling. preprint arXiv:2208.04933 (2022)
23. Tallec, C., Ollivier, Y.: Unbiased online recurrent optimization. arXiv preprint arXiv:1702.05043 (2017)
24. Tay, Y., et al.: Long range arena: a benchmark for efficient transformers. arXiv preprint arXiv:2011.04006 (2020)
25. Tiezzi, M., Marullo, S., Faggi, L., Meloni, E., Betti, A., Melacci, S.: Stochastic coherence over attention trajectory for continuous learning in video streams. In: Proceedings of IJCAI-22, pp. 3480–3486 (2022)
26. Tiezzi, M., Melacci, S., Betti, A., Maggini, M., Gori, M.: Focus of attention improves information transfer in visual features. In: Advances in NeurIPS, vol. 33, pp. 22194–22204 (2020)
27. Vaswani, A., et al.: Attention is all you need. In: Advances in NeurIPS, vol. 30 (2017)
28. Werbos, P.J.: Backpropagation through time: what it does and how to do it. Proc. IEEE **78**(10), 1550–1560 (1990)
29. Williams, R.J., Peng, J.: An efficient gradient-based algorithm for on-line training of recurrent network trajectories. Neural Comput. **2**(4), 490–501 (1990)
30. Williams, R.J., Zipser, D.: A learning algorithm for continually running fully recurrent neural networks. Neural Comput. **1**(2), 270–280 (1989)
31. Yu, Y., Si, X., Hu, C., Zhang, J.: A review of recurrent neural networks: LSTM cells and network architectures. Neural Comput. **31**(7), 1235–1270 (2019)
32. Zhou, H., et al.: Informer: beyond efficient transformer for long sequence time-series forecasting. In: Proceedings of the AAAI Conference on AI, vol. 35(12), pp. 11106–11115 (2021). https://doi.org/10.1609/aaai.v35i12.17325, https://ojs.aaai.org/index.php/AAAI/article/view/17325
33. Zucchet, N., Meier, R., Schug, S., Mujika, A., Sacramento, J.: Online learning of long range dependencies. In: NeurIPS (2023)

Robust Clustering with McDonald's Beta-Liouville Mixture Models for Proportional Data

Oussama Sghaier, Manar Amayri, and Nizar Bouguila(✉)

Concordia Institute for Information Systems Engineering (CIISE), Concordia University, Montreal, QC, Canada
{oussama.sghaier,manar.amayri,nizar.bouguila}@concordia.ca

Abstract. In this work, we study the problem of determining a proportional data structure in the absence of prior information about the number of clusters. We use finite mixture models to represent the data samples. These models are based on the McDonald's Beta-Liouville distribution, which is a novel distribution that combines the salient characteristics of the Liouville and McDonald's Beta distributions. When compared to common distributions like the Dirichlet, it provides higher flexibility and a more robust covariance structure. But, one major problem in mixture modeling is figuring out how many clusters to have. In this instance, the minimum message length (MML) technique is used to determine the number of clusters. In addition, as part of the Expectation Maximization procedure, the complexity of the mixture model - that is, the total number of components - can be automatically and concurrently determined while estimating the parameters. Our suggested approach is utilized in medical settings, specifically to focus medication for individuals with heart disease based on clinical data and analyze breast tissue taking into account histological scans. When it comes to data with strictly bounded values and an asymmetric distribution, the McDonald's Beta-Liouville mixture model outperforms the Gaussian mixture model (GMM) and the Dirichlet mixture model (DMM). The novel McDonald's Beta-Liouville mixing model (McDonald's BLMM) is used to model two real-world datasets. The results demonstrate that this model performs well and is more accurate than the GMM and DMM.

Keywords: Clustering · minimum message length · mixture models · McDonald's Beta-Liouville

1 Introduction

The creation of digital data, including text, images, and videos, has skyrocketed in recent years. Therefore, effective tools for clustering and analyzing complex data are desperately needed. Among the many clustering approaches now in

C. Y. Suen et al. (Eds.): ANNPR 2024, LNAI 15154, pp. 49–60, 2024.
https://doi.org/10.1007/978-3-031-71602-7_5

use, finite mixture models are gaining popularity because they provide a formal framework for unsupervised learning [5]. They have shown to be successful in a number of fields, such as image processing and machine learning. The Dirichlet family of mixture models, which includes the generalized Dirichlet, Beta-Liouville, and Dirichlet mixtures [8,12], has been studied as a more accurate and flexible choice for data modeling than the standard Gaussian mixture models [6,11]. This adjustment is the result of the Gaussian distribution's shortcomings when handling non-Gaussian data. Because of its rigidity, it is thought to be inappropriate for proportional data. However, because of the significant interdependencies between the random variables, the Dirichlet is less reliable in practical applications. Interestingly, the Dirichlet distribution is included as a specific instance in the Beta-Liouville distribution, which has fewer parameters than the generalized Dirichlet distribution. Furthermore, it has been shown [2] that Beta-Liouville mixture models outperform both Dirichlet and generalized Dirichlet mixtures. We decided to broaden the Beta-Liouville distribution in our investigation and develop a brand-new distribution known as the McDonald's Beta-Liouville. Using McDonald's Beta to model data on the [0,1] support, we then created its mixture models for the clustering task [13]. In comparison to Dirichlet and Beta-Liouville, it contains more parameters. Data modeling benefits from increased degrees of freedom as a result. Furthermore, the additional shape factors can change the tail weights, simultaneously change the skewness and kurtosis, and increase the entropy of the distribution. In mixture models, determining the number of consistent components that best explain the data is another difficult problem. This objective has been approached in a number of ways. These methods can be divided into three groups from a computational perspective: resampling methods, stochastic methods, and deterministic methods [7]. Among the stochastic procedures are Markov Chain Monte Carlo (MCMC) techniques, which have several applications to mixed models. To determine the number of clusters, cross-validation methods [14] and resampling approaches have also been used. The deterministic strategy is the final type of approaches and can be further classified into two groups. The first one includes the Bayesian approximation criteria, such as the Laplace empirical criteria (LEC) and Schwarz's Bayesian information criterion (BIC). Techniques from the second class, such as Akaike's information criterion (AIC) and the minimum message length (MML) and mixture minimum description length (MMDL), are based on information/coding theory principles. A more comprehensive review of selection techniques is provided in [1,5]. In our study, we are interested in deterministic methods, particularly MML.

In this paper, we introduce an MML criterion for determining the number of components in McDonald's Beta-Liouville mixture models. The structure of the paper is as follows: We offer an extensive overview of McDonald's Beta-Liouville distribution in Sect. 2. Section 3 presents the derivation of the MML expression for a McDonald's Beta-Liouville mixture. The complete estimation and selection algorithm are outlined in Sect. 4. The results of the experiments are discussed in Sect. 5. Finally, we conclude our work in Sect. 6.

2 McDonald's Beta-Liouville Distribution

A K-dimensional vector $\boldsymbol{X} = (X_1, \ldots, X_K)$ follows a Liouville distribution with parameters $(\alpha_1, \cdots, \alpha_K)$ and density generator $g(.)$ if its pdf (probability density function) is defined by [9]:

$$p(\boldsymbol{X}|\alpha_1, \cdots, \alpha_K) = g(u) \prod_{i=1}^{K} \frac{X_i^{\alpha_i - 1}}{\Gamma(\alpha_i)} \tag{1}$$

where $u = \sum_{i=1}^{K} X_i < 1$, and $0 < X_i < 1, i = 1, \cdots, K$. One common choice of the generator function is:

$$g(u) = \frac{\Gamma(\sum_{i=1}^{K} \alpha_i)}{u^{\sum_{i=1}^{K} \alpha_i - 1}} f(u) \tag{2}$$

where $f(.)$ is the pdf of the variable u, as a result, we can obtain a new expression of the pdf of Liouville distribution:

$$p(\boldsymbol{X}) = \frac{\Gamma(\sum_{i=1}^{K} \alpha_i)}{u^{\sum_{i=1}^{K} \alpha_i - 1}} f(u) \prod_{i=1}^{K} \frac{X_i^{\alpha_i - 1}}{\Gamma(\alpha_i)} \tag{3}$$

The Beta distribution is a useful choice for u [4] since it has two shape parameters that let it approach any arbitrary distribution. However, in this specific context, the extended form of the Beta distribution, dubbed McDonald's Beta (McDonald's Beta), has three parameters, as opposed to the two in the standard version [13]. As a result, it can fit data with more flexibility. Because of its additional feature, McDonald's Beta can accurately represent skewness and kurtosis in data, which is highly helpful for modeling real-world observations [13]. Moreover, the additional shape factors modify tail weights and increase the produced distribution's entropy. We use the McDonald's Beta distribution to model the study's random variable u, whose pdf is given by:

$$f(u|\alpha, \beta, \lambda) = \frac{\lambda u^{\alpha\lambda - 1}(1 - u^{\lambda})^{\beta - 1}}{B(\alpha, \beta)} \tag{4}$$

with:

$$B(a, b) = \frac{\Gamma(a)\Gamma(b)}{\Gamma(a + b)} \tag{5}$$

represents the Beta function and $\Gamma(.)$ denotes the Gamma function, u is a scalar where $0 \leq u \leq 1$. The three shape parameters are: α, β, and λ. The expression of the pdf for our suggested distribution for work, the McDonald's Beta-Liouville distribution, is obtained by considering $u = \sum_{k=1}^{K} X_k$ and using the McDonald's Beta as the density function for u in Eq. (2). We then inject Eq. (4) in Eq. (3). The following is the entire expression:

$$p(\boldsymbol{X}|\alpha_1,\cdots,\alpha_K,\alpha,\beta,\lambda) = \frac{\Gamma(\sum_{k=1}^{K}\alpha_k)\Gamma(\alpha+\beta)}{\Gamma(\alpha)\Gamma(\beta)}\lambda \times$$

$$\left(\sum_{k=1}^{K} X_k\right)^{(\alpha\lambda-\sum_{k=1}^{K}\alpha_k)} \left[1-\left(\sum_{k=1}^{K} X_k\right)^{\lambda}\right]^{\beta-1} \prod_{k=1}^{K} \frac{X_k^{\alpha_k-1}}{\Gamma(\alpha_k)} \tag{6}$$

Let $\mathcal{X} = (\boldsymbol{X}_1,\cdots,\boldsymbol{X}_N)$ be a set of training examples where N is the number of samples. Each sample $\boldsymbol{X}_i(i=1,\cdots,N)$ is represented by a finite mixture. A McDonald's Beta-Liouville mixture with M components for each X_i is defined as:

$$p(\boldsymbol{X}_i|\Theta) = \sum_{j=1}^{M} p_j p(\boldsymbol{X}_i|\boldsymbol{\theta}_j) \tag{7}$$

where $\Theta = (p_1,\cdots,p_M,\boldsymbol{\theta}_1,\cdots,\boldsymbol{\theta}_M)$ is the set of parameters, $\boldsymbol{\theta}_j = (\alpha_{j1},\cdots,\alpha_{jK},\alpha_j,\beta_j,\lambda_j)$, $p_j > 0, j = 1,\cdots,M$ are the mixing parameters $(\sum_{j=1}^{M} p_j = 1)$.

3 The MML Criterion for a McDonald's Beta-Liouville Mixture

From an information-theoretic standpoint, the MML method is based on grading statistical models according to their ability to compress a message that contains the data. The number of clusters that make up the optimal mixture is thus found by minimizing the amount of information needed to move a collection of data $\mathcal{X} = (\boldsymbol{X}_1,\cdots,\boldsymbol{X}_N)$ from a transmitter to a recipient. An explicit prior distribution over parameter values is used in the MML principle due to its strong relationship to Bayesian inference [15]. The minimization of the subsequent objective function serves as the foundation for the MML technique [15]:

$$MessLen \approx -\log(h(\Theta)) - \log(P(\mathcal{X}|\Theta)) + \frac{1}{2}\log(|F(\Theta)|) + \frac{N_p}{2}\left(1+\log(\frac{1}{12})\right) \tag{8}$$

where $P(\mathcal{X}|\Theta)$ is the likelihood, $h(\Theta)$ is the prior probability $F(\Theta)$ is the expected Fisher information matrix and $|F(\Theta)|$ denotes its determinant, and N_p is the number of parameters to be estimated, which equals in our case to $M(4 \times K) - 1$.

Determining the Fisher information matrix analytically is typically difficult [5]. As recommended in [10], we consider an approximated Fisher information matrix, which is generated after the vectors in the data set are allocated to the different clusters. The determinant of the Fisher information matrix is equal to the product of the determinants of the information matrix for each component and the information matrix of the mixing parameter vector P, assuming that the components are independent.

$$|F(\Theta)| = |F(P)| \prod_{j=1}^{M} |F(\boldsymbol{\theta}_j)| \tag{9}$$

where $F(\boldsymbol{\theta}_j)$ is the Fisher information about the component j of a single McDonald's Beta-Liouville distribution, and $F(P)$ is the Fisher information about the vector of mixing parameters. It should be noted that $\sum_{j=1}^{M} p_j = 1$ is satisfied by the mixing parameters for $F(P)$. Consequently, the extended Bernoulli process can be conceptualized as a series of trials where each trial has M alternative outcomes, each of which is labeled as the first, second,..., M^{th} cluster. The parameters p_1, p_2,..., p_M determine a multinomial distribution that represents the number of trials for the jth cluster. In this case, the determinant of the Fisher information matrix is as follows [3]:

$$F(P) = \frac{N^{M-1}}{\prod_{j=1}^{M} p_j} \tag{10}$$

where N is the number of data elements. Let us consider the jth cluster $\mathcal{X}_j = (\boldsymbol{X}_l, \cdots, \boldsymbol{X}_{l+n_j-1})$ for $F(\boldsymbol{\theta}_j)$ of the mixture, where the number of components in cluster j is n_j, and $l < N$. The jth cluster's selection allows us to shorten the notation without sacrificing generality. Let $\log(p(\mathcal{X}_j|\boldsymbol{\theta}_j)) = L(\mathcal{X}_j|\theta_j)$.

$$L(\mathcal{X}_j|\theta_j) = n_j \left(\log \Gamma(\sum_{k=1}^{K} \alpha_{jk}) + \log \lambda_j + \log \Gamma(\alpha_j + \beta_j) - \log \Gamma(\alpha_j) - \log \Gamma(\beta_j) - \sum_{k=1}^{K} \log \Gamma(\alpha_{jk}) \right) + (\alpha_j \lambda_j - \sum_{k=1}^{K} \alpha_{jk}) \times \sum_{i=l}^{l+n_j-1} \left(\log \sum_{k=1}^{K} X_{ik} \right) + \sum_{i=l}^{l+n_j-1} \sum_{k=1}^{K} \log X_{ik} + (\beta_j - 1) \sum_{i=l}^{l+n_j-1} \log \left(1 - \left(\sum_{k=1}^{K} X_{ik} \right)^{\lambda_j} \right)$$

Given the assumption of independence among the parameters of θ_j, the Fisher information for a solitary component can be expressed as follows:

$$F(\boldsymbol{\theta}_j) = F(\alpha_{j1}, \cdots, \alpha_{jK}) F(\alpha_j, \beta_j, \lambda_j) \tag{11}$$

where $F(\alpha_{j1}, \cdots, \alpha_{jK}) = -H(\alpha_{j1}, \cdots, \alpha_{jK})$ with $H(\alpha_{j1}, \cdots, \alpha_{jK})$ is the Hessian matrix for the parameters $(\alpha_{j1}, \cdots, \alpha_{jK})$ and $F(\alpha_j, \beta_j, \lambda_j) = -H(\alpha_j, \beta_j, \lambda_j)$ with $H(\alpha_j, \beta_j, \lambda_j)$ is Hessain matrix of $(\alpha_j, \beta_j, \lambda_j)$. The whole Hessian matrix can be written in the following way: $H(\theta_j)$ = block-diag$(H(\alpha_{j1}, \cdots, \alpha_{jK}), H(\alpha_j, \beta_j, \lambda_j))$ with:

$$H(\alpha_{j1}, \cdots, \alpha_{jK}) = \frac{\partial^2 L(\mathcal{X}_j|\theta_j)}{\partial \alpha_{jk_1} \partial \alpha_{jk_2}} \quad k_1, k_2 = 1 \cdots, K \tag{12}$$

$$H(\alpha_j, \beta_j, \lambda_j) = \begin{pmatrix} \frac{\partial^2 L(\mathcal{X}_j|\theta_j)}{\partial^2 \alpha_j} & \frac{\partial^2 L(\mathcal{X}_j|\theta_j)}{\partial \alpha_j \partial \beta_j} & \frac{\partial^2 L(\mathcal{X}_j|\theta_j)}{\partial \alpha_j \partial \lambda_j} \\ \frac{\partial^2 L(\mathcal{X}_j|\theta_j)}{\partial \beta_j \partial \alpha_j} & \frac{\partial^2 L(\mathcal{X}_j|\theta_j)}{\partial^2 \beta_j} & \frac{\partial^2 L(\mathcal{X}_j|\theta_j)}{\partial \beta_j \partial \lambda_j} \\ \frac{\partial^2 L(\mathcal{X}_j|\theta_j)}{\partial \lambda_j \partial \alpha_j} & \frac{\partial^2 L(\mathcal{X}_j|\theta)}{\partial \lambda_j \partial \beta_j} & \frac{\partial^2 L(\mathcal{X}_j|\theta_j)}{\partial^2 \lambda_j} \end{pmatrix} \tag{13}$$

The first derivatives of $L(\mathcal{X}_j|\theta_j)$ are given by:

$$\frac{\partial L(\mathcal{X}_j|\theta_j)}{\partial \alpha_{jk}} = n_j \left(\Psi(\sum_{l=1}^{K}\alpha_{jl}) - \Psi(\alpha_{jk})\right) + \sum_{i=l}^{l+n_j-1}\left(\log(X_{ik}) - \log(\sum_{k=1}^{K}X_{ik})\right) \tag{14}$$

$$\frac{\partial L(\mathcal{X}_j|\theta_j)}{\partial \alpha_j} = (\Psi(\alpha_j+\beta_j) - \Psi(\alpha_j)) + \sum_{i=l}^{l+n_j-1}\lambda_j \log(\sum_{k=1}^{K}X_{ik}) \tag{15}$$

$$\frac{\partial L(\mathcal{X}_j|\theta_j)}{\partial \beta_{jk}} = (\Psi(\alpha_j+\beta_j) - \Psi(\beta_j)) + \sum_{i=l}^{l+n_j-1}\log\left(1-\left(\sum_{k=1}^{K}X_{ik}\right)^{\lambda_j}\right) \tag{16}$$

$$\frac{\partial L(\mathcal{X}_j|\theta_j)}{\partial \lambda_j} = \frac{1}{\lambda_j} + \alpha_j \sum_{i=l}^{l+n_j-1}\log(\sum_{k=1}^{K}X_{ik}) - \sum_{i=l}^{l+n_j-1}(\beta_j-1)\frac{\left(\log\sum_{k=1}^{K}X_{ik}\right)\left(\sum_{k=1}^{K}X_{ik}\right)^{\lambda_j}}{1-\left(\sum_{k=1}^{K}X_{ik}\right)^{\lambda_j}} \tag{17}$$

The second and mixed derivatives of $L(\mathcal{X}_j|\theta_j)$ are given by:

$$\frac{\partial^2 L(\mathcal{X}_j|\theta_j)}{\partial \alpha_{jk_1}\partial \alpha_{jk_2}} = \begin{cases} n_j\left(\Psi'(\sum_{k=1}^{K}\alpha_{jk}) - \Psi'(\alpha_{jk})\right) \\ \qquad\qquad if \quad k_1 = k_2 = k \\ n_j\Psi'(\sum_{k=1}^{K}\alpha_{jk}) \quad otherwise \end{cases} \tag{18}$$

$$\frac{\partial^2 L(\mathcal{X}_j|\theta_j)}{\partial \alpha_{jk}\partial \alpha_j} = \frac{\partial^2 L(\mathcal{X}|\Theta)}{\partial \alpha_{jk}\partial \beta_j} = \frac{\partial^2 L(\mathcal{X}|\Theta)}{\partial \alpha_{jk}\partial \lambda_j} = 0 \tag{19}$$

$$\frac{\partial^2 L(\mathcal{X}_j|\theta_j)}{\partial^2 \alpha_j} = n_j\left(\Psi'(\alpha_j+\beta_j) - \Psi'(\alpha_j)\right) \tag{20}$$

$$\frac{\partial^2 L(\mathcal{X}_j|\theta_j)}{\partial^2 \beta_j} = n_j\left(\Psi'(\alpha_j+\beta_j) - \Psi'(\beta_j)\right) \tag{21}$$

$$\frac{\partial^2 L(\mathcal{X}_j|\theta_j)}{\partial^2 \lambda_j} = -\sum_{i=l}^{l+n_j-1}(\beta_j-1)\frac{\left(\log\sum_{k=1}^{K}X_{ik}\right)^2\left(\sum_{k=1}^{K}X_{ik}\right)^{\lambda_j}}{\left[1-\left(\sum_{k=1}^{K}X_{ik}\right)^{\lambda_j}\right]^2} - \frac{n_j}{\lambda_j^2} \tag{22}$$

$$\frac{\partial^2 L(\mathcal{X}_j|\theta_j)}{\partial \alpha_j\partial \beta_j} = \frac{\partial^2 L(\mathcal{X}|\Theta)}{\partial \beta_j\partial \alpha_j} = n_j\Psi'(\alpha_j+\beta_j) \tag{23}$$

$$\frac{\partial^2 L(\mathcal{X}_j|\theta_j)}{\partial \alpha_j\partial \lambda_j} = \frac{\partial^2 L(\mathcal{X}|\Theta)}{\partial \lambda_j\partial \alpha_j} = \sum_{i=l}^{l+n_j-1}\log\left(\sum_{k=1}^{K}X_{ik}\right) \tag{24}$$

$$\frac{\partial^2 L(\mathcal{X}_j|\theta_j)}{\partial \beta_j\partial \lambda_j} = \frac{\partial^2 L(\mathcal{X}_j|\theta_j)}{\partial \lambda j\partial \beta j} = -\sum_{i=l}^{l+n_j-1}\frac{\left(\log\sum_{k=1}^{K}X_{ik}\right)\left(\sum_{k=1}^{K}X_{ik}\right)^{\lambda_j}}{1-\left(\sum_{k=1}^{K}X_{ik}\right)^{\lambda_j}} \tag{25}$$

where Ψ and Ψ' are the digamma and trigamma functions, respectively. $h(\Theta)$ has an impact on the performance of the MML criterion. In the lack of further knowledge regarding the characteristics of the mixture, we represent the parameters of the different components as a priori independent from the mixing probabilities, that is:

$$h(\Theta) = h(P) \prod_{j=1}^{M} h(\boldsymbol{\theta}_j) \tag{26}$$

We define:

$$h(\Theta) = (M-1)!(6K!)^M \prod_{j=1}^{M} f(\alpha_j, \beta_j, \lambda_j)^{-3} \left(\sum_{k=1}^{K} f(\alpha_{jk}) \right)^{-K} \tag{27}$$

where $f(\alpha_{jk}) = e^{6 \frac{\sum_{l=1}^{K} \hat{\alpha}_{jl}}{\hat{\alpha}_{jk}}}$, $f(\alpha_j, \beta_j, \lambda_j) = e^{6 \frac{(\hat{\alpha}_j + \hat{\beta}_j + \hat{\lambda}_j)^2}{\hat{\alpha}_j \hat{\beta}_j \hat{\lambda}_j}}$, and the hat on parameters indicates estimated quantities.

4 Estimation and Selection Algorithm

When dealing with mixture models, one of the most important tasks that must be performed is estimating the relevant parameters. In this study, we will apply the Maximum Likelihood approach (ML), which involves estimating the parameter values that maximize the likelihood function denoted by $P(\mathcal{X}|\Theta)$:

$$P(\mathcal{X}|\Theta) = \prod_{i=1}^{N} \sum_{j=1}^{M} p_j \frac{\Gamma(\sum_{k=1}^{K} \alpha_{jk}) \Gamma(\alpha_j + \beta_j)}{\Gamma(\alpha_j)\Gamma(\beta_j)} \lambda_j$$
$$\times \left(\sum_{k=1}^{K} X_{ik} \right)^{(\alpha_j \lambda_j - \sum_{k=1}^{K} \alpha_{jk})} \left[1 - \left(\sum_{k=1}^{K} X_{ik} \right)^{\lambda_j} \right]^{\beta_j - 1} \prod_{k=1}^{K} \frac{X_{ik}^{\alpha_{jk}-1}}{\Gamma(\alpha_{jk})} \tag{28}$$

To anticipate the mixture parameters, ML can be used with EM method through an iterative process [5]. During the Expectation stage, we define $Z_i = (Z_{i1}, \cdots, Z_{iM})$ to find the cluster of the vector $\boldsymbol{X}_i$. We define, in particular, $\boldsymbol{Z}_{ij} = 1$ if and only if $\boldsymbol{X}_i$ is a member of class j, and 0 otherwise. At the conclusion of the Expectation step, we replace the missing data $Z = (Z_1, \cdots, Z_N)$ with the expected values of the class assignments and the posterior probabilities ($\hat{Z}_{ij} = \frac{p_j p(\boldsymbol{X}_i|\boldsymbol{\theta}_j)}{\sum_{j=1}^{M} p_j p(\boldsymbol{X}_i|\boldsymbol{\theta}_j)}$). The parameter estimations are updated using the Maximization step. The concept entails optimizing the subsequent expression:

$$L(\mathcal{X}|\Theta) = \sum_{i=1}^{N} \sum_{j=1}^{M} \hat{Z}_{ij} \log(p_j p(\boldsymbol{X}_i|\boldsymbol{\theta}_j)) \tag{29}$$

We start with a predetermined set of initial estimations and estimate the parameters using the Newton-Raphson approach:

$$\boldsymbol{\theta}_j^{(t+1)} = \boldsymbol{\theta}_j^{(t)} - H(\boldsymbol{\theta}_j^{(t)})^{-1} \frac{\partial L(\mathcal{X}|\Theta^{(t)})}{\partial \boldsymbol{\theta}_j^{(t)}} \tag{30}$$

where $H(\boldsymbol{\theta_j}^{(t)})^{-1}$ is the inverse of the Hessian matrix [2]. For the p_j, a closed form expression can be given by:

$$p_j = \frac{1}{N} \sum_{i=1}^{N} \hat{Z}_{ij} \tag{31}$$

We decided to initialize the parameters in our work with random values drawn from a uniform distribution. Furthermore, starting at random with values that lie inside a specified range reduces the likelihood that we will become stuck in local minima or maxima. This uncertainty allows for early study of a broader search field, which may provide in better results. After obtaining the estimate equations, Algorithm 1 contains the entire learning algorithm:

Algorithm 1 Expectation Maximization Algorithm

1: For each candidate M:
2: Initialization
3: Expectation step: Compute the posterior probabilities: $\hat{Z}_{ij} = \frac{p_j p(\boldsymbol{X_i}|\boldsymbol{\theta_j})}{\sum_{j=1}^{M} p_j p(\boldsymbol{X_i}|\boldsymbol{\theta_j})}$
4: Maximization step:
  a) Update $\boldsymbol{\theta_j}$ using Eq(30), $j = 1, \cdots, M$
  b) Update p_j using Eq(31), $j = 1, \cdots, M$
5: Calculate the associate message length using Eq(8)
6: Select the optimal model M^* such that: $M^* = \arg\min_M MessLen(M)$

5 Experimental Results

This section examines our model's performance and contrasts it with Gaussian finite mixture models and McDonald's Beta finite mixture models. We apply our suggested approach to two applications namely breast tissue analysis and patient targeting for heart disease in order to verify its robustness. We use the min-max method to normalize our datasets since we assume that all input values fall within the range of (0,1). In order to assess the robustness of the model, we also take into account four metrics: accuracy, precision, recall, and F1-score.

5.1 Breast Cancer Targeting Treatment

The World Health Organization (WHO) reported in December 2020 that there has been a notable shift in the worldwide cancer landscape, with 2.26 million instances of breast cancer being identified annually. Breast cancer is now the most common cancer. Because to daily habits like smoking and drinking, as well as insufficient physical activity, the number of identified cases has increased over the years. For this portion of the experiment, we used a publicly accessible dataset that had labels for benign and malignant conditions to test our technique. We divided the tissues into two clusters.

The results depicted in Table 1 illustrate the performance metrics of different models evaluated on the Breast Cancer dataset (UCI repository of machine learning databases). Among these models, McDonald's Beta-Liouville Mixture Model (McDonald's BLMM) emerges as the top-performing method across all evaluated metrics. Notably, it achieves an accuracy of 91.36%, an F1-score of 88.8%, a precision of 81.25%, and a recall of 97.9%. In comparison, other models such as Gaussian Mixture Model (GMM), Dirichlet Mixture Model (DMM), and McDonald's Beta Mixture Model (McDonald's Beta MM) demonstrate lower performance scores across the board. These results underscore the superior performance of McDonald's BLMM in accurately classifying breast cancer instances, highlighting its potential for improving diagnostic accuracy and patient care in clinical settings. Figure 1 shows that our algorithm was able to find the optimal number of clusters. The confusion matrix for breast cancer is shown in Fig. 2 when we use the number of clusters that MML found using McDonald's BLMM. In this confusion matrix, only 59 samples were misidentified, yielding an accuracy rate of 91.36%.

Table 1. Results on Breast Cancer dataset

	GMM	DMM	McDonald's Beta MM	McDonald's BLMM
Accuracy	85.35	84.91	89.31	**91.36**
F1-score	82.57	76.53	86.55	**88.80**
Precision	70.74	**84.0**	77.3	81.25
Recall	90.16	70.29	**98.32**	97.9

5.2 Heart Disease Targeting Treatment

We will cluster anonymized patient data who have been diagnosed with heart disease in this section of the work. We examine information from Long Beach, California's V. A. Medical Center (UCI repository of machine learning databases). In addition to the goal, there are nine attributes and 303 instances in this 5-component dataset that is openly accessible. The patient's age, gender, type of chest pain, resting blood pressure (measured in millimeters of mercury on hospital admission), serum cholesterol (measured in milligrams per deciliter), fasting blood sugar level > 120 mg per deciliter ($1 =$ true; $0 =$ false), resting electrocardiogram results, maximum heart rate attained, and exercise-induced angina ($1 =$ yes; $0 =$ no) are the attributes that describe the dataset. The optimal number of clusters found by our approach is demonstrated by Fig. 3. Although our algorithm was able to detect the optimal number of clusters which is 5, we choose to reduce the labels to only two labels by considering the label 0 the normal label and the rest labels are the anomaly labels. This strategy was done to make the work more simple. The table presented in Table 2 showcases the performance metrics of various models evaluated on the Heart Disease dataset. Among these models, McDonald's BLMM stands out as the top-performing method across several key metrics. Notably, it achieves an accuracy of 74.29%, an F1-score of 52.17%, a precision of 49.18%, and a recall of 55.55%. Comparatively, other

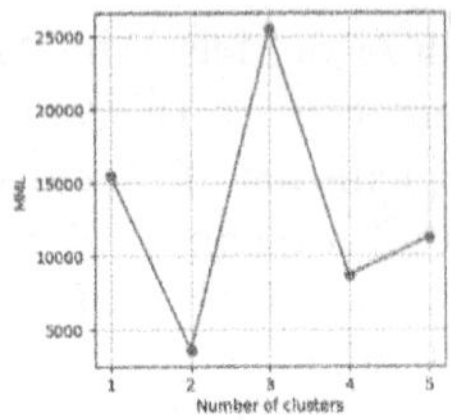

Fig. 1. McDonald's BLMM with MML for Breast Cancer data

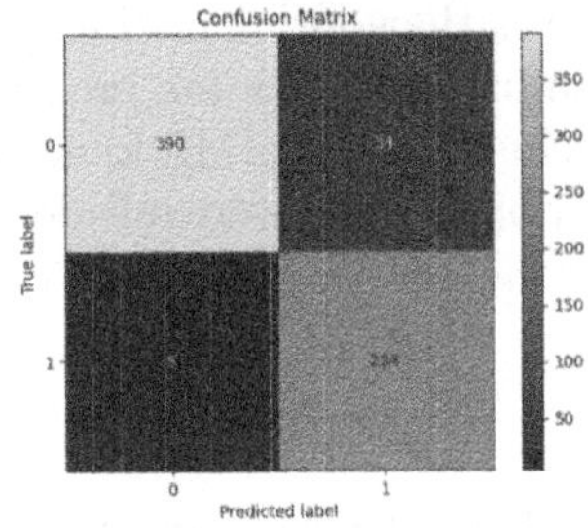

Fig. 2. Confusion Matrix for Breast Cancer data in case of McDonald's BLMM

Table 2. Results on Heart Disease dataset

	GMM	DMM	McDonald's Beta MM	McDonald's BLMM
Accuracy	57.47	65.42	71.96	**74.29**
F1-score	40.52	45.58	**56.52**	52.17
Precision	31.31	37.8	46.42	**49.18**
Recall	57.4	57.4	**72.22**	55.55

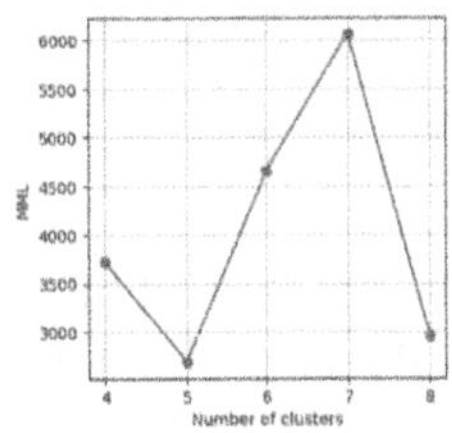

Fig. 3. McDonald's BLMM with MML for Heart Disease data

models such as GMM, DMM, and BLMM demonstrate inferior performance scores across the board. For instance, GMM achieves an accuracy of 57.47%, an F1-score of 40.52%, a precision of 31.31%, and a recall of 57.4%, while DMM achieves an accuracy of 65.42%, an F1-score of 45.58%, a precision of 37.8%, and a recall of 57.4%.

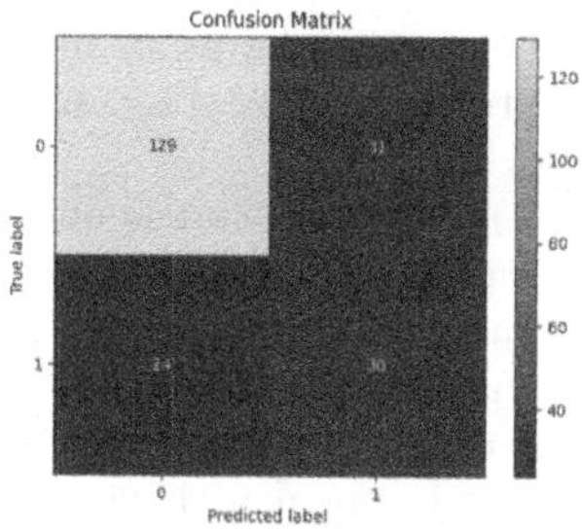

Fig. 4. Confusion Matrix for Heart Disease data in case of McDonald's BLMM

Figure 4 displays the confusion matrix for Heart Disease based on the number of clusters that MML identified using McDonald's BLMM. With only 55 samples incorrectly detected in this confusion matrix, the accuracy rate is 74.29%.

6 Conclusion

In this study, we provide a new approach for data clustering, based on the Minimum Message Length (MML) criterion and McDonald's Beta-Liouville mixing models to find the optimal number of components. We examined actual medical datasets, providing a comprehensive assessment of the recommended methodology. Experiments with data showed that the system could recognize complex structures in a variety of situations. McDonald's Beta Liouville mixing models showed robustness, scalability, and adaptability to a range of real-world scenarios when paired with the MML criterion. The ability of McDonald's Beta-Liouville mixture models to uncover latent patterns in a range of data distributions and cluster data accurately demonstrate their considerable potential for managing the complexities of real database systems. Future research could concentrate on developing an online learning approach and broadening the proposed model in order to solve the crucial problem of dynamic data clustering.

Acknowledgments. The completion of this research was made possible thanks to the Natural Sciences and Engineering Research Council of Canada (NSERC) and a Startup Grant from Concordia University.

References

1. Baxter, R.A., Oliver, J.J.: Finding overlapping components with MML. Stat. Comput. **10**(1), 5–16 (2000)
2. Bouguila, N.: Infinite liouville mixture models with application to text and texture categorization. Pattern Recognit. Lett. **33**(2), 103–110 (2012)
3. Bouguila, N., Almakadmeh, K., Boutemedjet, S.: A finite mixture model for simultaneous high-dimensional clustering, localized feature selection and outlier rejection. Expert Syst. Appl. **39**(7), 6641–6656 (2012)

4. Bouguila, N., Elguebaly, T.: A fully bayesian model based on reversible jump MCMC and finite beta mixtures for clustering. Expert Syst. Appl. **39**(5), 5946–5959 (2012)
5. Bouguila, N., Fan, W. (eds.): Mixture Models and Applications. USL, Springer, Cham (2020). https://doi.org/10.1007/978-3-030-23876-6
6. Bouguila, N., Ziou, D.: Online clustering via finite mixtures of dirichlet and minimum message length. Eng. Appl. Artif. Intell. **19**(4), 371–379 (2006)
7. Bouguila, N., Ziou, D.: High-dimensional unsupervised selection and estimation of a finite generalized dirichlet mixture model based on minimum message length. IEEE Trans. Pattern Anal. Mach. Intell. **29**(10), 1716–1731 (2007)
8. Fan, W., Bouguila, N.: Learning finite beta-liouville mixture models via variational bayes for proportional data clustering. In: Rossi, F. (ed.) IJCAI 2013, Proceedings of the 23rd International Joint Conference on Artificial Intelligence, Beijing, China, 3-9 August 2013, pp. 1323–1329. IJCAI/AAAI (2013)
9. Fang, K.W.: Symmetric Multivariate and Related Distributions. CRC Press, Boca Raton (2018)
10. Figueiredo, M., Jain, A.: Unsupervised learning of finite mixture models. IEEE Trans. Pattern Anal. Mach. Intell. **24**(3), 381–396 (2002)
11. Guo, J., Amayri, M., Fan, W., Bouguila, N.: Beta-liouville and inverted beta-liouville based predictive models for occupancy detection using small training data. In: 2022 IEEE Symposium Series on Computational Intelligence (SSCI), pp. 223–230 (2022)
12. Guo, J., Amayri, M., Fan, W., Bouguila, N.: Liouville-based predictive models for occupancy estimation using small training data. IEEE Internet Things J. **10**(21), 18876–18889 (2023)
13. Manoj, C., Wijekoon, P., Yapa, R.D.: The mcdonald generalized beta-binomial distribution: a new binomial mixture distribution and simulation based comparison with its nested distributions in handling overdispersion. Int. J. Stat. Probab. **2**(2), 24–30 (2013)
14. Smyth, P.: Model selection for probabilistic clustering using cross-validated likelihood. Stat. Comput. **10**(1), 63–72 (2000)
15. Wallace, C.S.: Statistical and Inductive Inference by Minimum Message Length. Springer, Cham (2005). https://doi.org/10.1007/0-387-27656-4

Evaluating Support Vector Machines with Multiple Kernels by Random Search

Shigeo Abe(✉)

Kobe University, Rokkodai, Nada, Kobe, Japan
abe@kobe-u.ac.jp
www2.kobe-u.ac.jp/~abe

Abstract. The minimal complexity support vector machine with a three-kernel structure (ML1 SVM (MK)) has a reduced number of hyperparameters for multiple kernels and is optimized in three-stage cross-validation: In the first stage, the hyperparameter values for the L1 SVM (MK) are determined fixing the hyperparameter values for multiple kernels. In the second stage the hyperparameter values for the multiple kernels are determined, and in the third stage, the value of the hyperparameter of the ML1 SVM (MK) to control the maximum margin is determined. In this paper, we examine whether this training strategy and the reduction of hyperparameters do not deteriorate generalization performance due to falling in the local minimum. To do this we compare the results with cross-validation by random search of ML1 SVM (MK) with the full three-kernel structure. By computer experiments, we find that the generalization abilities by the three-stage cross-validation for the reduced kernel structure is statistically comparable to or better than random search for the two-class problems and most of the multiclass problems tested.

1 Introduction

Model selection is very important in realizing the support vector machine (SVM) [1,2] with high generalization abilities. The most widely used method is grid search in which cross-validation (CV) at the grid points of hyperparameter values is performed and the hyperparameter values that realize the highest accuracy by cross-validation are selected. However, because grid search is time consuming, several methods such as genetic based algorithms [3] and random search [4,5] have been proposed.

Random search is a simple model selection strategy and is shown to be statistically comparable to grid search while with less computational burden [6]. Random search is terminated when the best cross-validation accuracy is not improved for a consecutive period. To speed up random search further, the improvement of stopping condition is considered [7].

To improve generalization abilities of the SVM, in [8–13] we developed the minimum complexity SVM (MSVM), which is a fusion of the SVM and the minimal complexity machine [14], and which results in maximizing the minimum

C. Y. Suen et al. (Eds.): ANNPR 2024, LNAI 15154, pp. 61–72, 2024.
https://doi.org/10.1007/978-3-031-71602-7_6

margin and minimizing the maximum margin. To improve the generalization ability of the MSVM further, we developed ML1 SVM (MK) [15], which is the minimum complexity L1 SVM with multiple kernels. Using the RBF (Radial Basis Function) kernels, the ML1 SVM (MK) has originally eleven hyperparameters. Therefore, grid search is very time consuming in determining the optimum hyperparameter values. To speed up model selection, we reduced the number of hyperparameters to five. But still time consuming compared to two hyperparameters for the L1 SVM. To alleviate this, we developed three-stage cross-validation: In the first stage, we determine the hyperparameter values for the L1 SVM (MK), i.e., the γ value for the RBF kernels and the margin parameter value C, fixing the hyperparameter values for multiple kernels. In the second stage we determine the hyperparameter values for the multiple kernels relating to the kernel ratio and the kernel distance. And in the third stage, we determine the value of the hyperparameter of the ML1 SVM (MK) to minimize the maximum margin. The three-stage cross-validation works to reduce the number of cross-validation trials and according to the performance evaluation for benchmark data sets, the ML1 SVM (MK) performed better than the ML1 SVM. However, because the number of hyperparameters is reduced and grid search is done in three stages, the optimum parameter values may fail to be obtained.

In this paper, we evaluate generalization abilities of the ML1 SVM (MK) by random search and compare the results with those by the three-stage grid search. The purpose of comparison is two-fold: to examine that 1) the three-stage grid search is comparable in generalization abilities to random search; and that 2) the reduced three-kernel structure is comparable in generalization abilities to the standard three-kernel structure. To realize comparison, we use the standard three-kernel structure in random search. If the generalization abilities of the both search methods are comparable, the above two goals are satisfied.

To make comparisons fair, we set the minimum number of random search trials to the number of trials in the three-stage grid search. We calculate the average accuracies and the standard deviations for the test data sets and compare the statistical difference of the two search methods by the t test. We also compare the average accuracies by cross-validation and examine whether the difference of accuracies is caused by over-fitting or under-fitting. The speed of random search is measured by the iteration count when the cross-validation accuracies are last updated.

In Sect. 2, we briefly summarize the architecture of the ML1 SVM (MK) and its model selection by three-stage cross-validation. Then we discuss multiple kernels, and model selection by three-stage grid search and random search. In Sect. 3, we compare the ML1 SVM (MK) and L1 SVM optimized by grid search and random search using two-class and multiclass problems.

2 Minimal Complexity L1 Support Vector Machines with Multiple Kernels

In this section, we summarize the architecture of the ML1 SVM (MK) and the three-stage grid search according to [15]. We also discuss random search for the ML1 SVM (MK) with a standard three-kernel structure.

2.1 Architecture

To classify two-class data, we consider determining the following separating hyperplane in the l-dimensional feature space that is mapped from the input vector $\mathbf{x}$ by the mapping function $\phi(\mathbf{x})$:

$$D(\mathbf{x}) = \mathbf{w}^\top \phi(\mathbf{x}) + b = 0, \tag{1}$$

where $\mathbf{w}$ is the l-dimensional coefficient vector and b is the bias term.

The ML1 SVM, which is a fusion of the MCM and the L1 SVM, simultaneously maximizes the minimum margin between the separating hyperplane and the training samples and minimizes the maximum margin. Minimization of the maximum margin leads to minimizing the radius of the hypersphere that encloses all the training data and thus it will improve the generalization ability.

The architecture of the ML1 SVM is given by

$$\min Q(\mathbf{w}, b, h, \xi) = C_h\, h + \frac{1}{2}\, \|\mathbf{w}\|^2 + C \sum_{i=1}^{M} \xi_i \tag{2}$$

$$\text{s.t. } y_i\, (\mathbf{w}^\top\, \phi(\mathbf{x}_i) + b) + \xi_i \geq 1,\ \xi_i \geq 0,\ i = 1, \ldots, M, \tag{3}$$

$$h \geq y_i\, (\mathbf{w}^\top\, \phi(\mathbf{x}_i) + b),\ i = 1, \ldots, M, \quad h \geq 1, \tag{4}$$

where $\{\mathbf{x}_i, y_i\}$ $(i = 1, \ldots, M)$ are M training data and their labels, $\mathbf{x}_i$ are the n-dimensional input vectors and $y_i = 1$ for Class 1 and -1 for Class 2, ξ_i are the slack variables for the ith inequality constraints, $\xi = (\xi_1, \ldots, \xi_M)^\top$, C is the (minimum) margin parameter to control the minimum margin, C_h is the maximum margin parameter to control the maximum margin, and h is the upper bound of the constraints. We call (4) the upper bound constraints.

Excluding (4) and $C_h\, h$ in (2), we obtain the soft margin L1 SVM, which maximizes the minimum margin, while allowing some training samples to be misclassified. The excluded parts work to minimize the maximum margin by minimizing the upper bound h. The tradeoff between minimization and maximization is controlled by parameter C_h.

The dual form of the ML1 SVM is given by

$$\max Q(\alpha) = \sum_{i=1}^{M} (\alpha_i - \alpha_{M+i}) - \frac{1}{2} \sum_{i,j=1}^{M} (\alpha_i - \alpha_{M+i}) \times (\alpha_j - \alpha_{M+j})\, y_i\, y_j\, K(\mathbf{x}_i, \mathbf{x}_j) \tag{5}$$

$$\text{s.t.} \sum_{i=1}^{M} y_i\, (\alpha_i - \alpha_{M+i}) = 0, \tag{6}$$

$$C_h \geq \sum_{i=1}^{M} \alpha_{M+i}, \tag{7}$$

$$C \geq \alpha_i \geq 0,\ C_h \geq \alpha_{M+i} \geq 0,\ i = 1, \ldots, M, \tag{8}$$

where α_i $(i = 1, \ldots, 2M)$ are the Lagrange multipliers, $\alpha = (\alpha_1, \ldots, \alpha_{2M})^{\top}$, $K(\mathbf{x}_i, \mathbf{x}_j)$ is the kernel, and $K(\mathbf{x}_i, \mathbf{x}_j) = \phi^{\top}(\mathbf{x}_i)\, \phi(\mathbf{x}_j)$.

The Lagrange multipliers α_i $(i = 1, \ldots, M)$ are associated with the inequality constraints given by (3) and work to maximize the minimum margin, and α_i $(i = M{+}1, \ldots, 2M)$ are associated with (4) for minimizing the maximum margin. Therefore, if we delete variables α_{M+i} $(i = 1, \ldots, M)$ and C_h from the above optimization problem, we obtain the original L1 SVM.

We separate (6) into

$$\sum_{i=1}^{M} y_i\, \alpha_i = 0, \quad \sum_{i=1}^{M} y_i\, \alpha_{M+i} = 0, \tag{9}$$

so that we can optimize $\alpha_i\,(i = 1, \ldots, M)$ and $\alpha_{M+i}\,(i = 1, \ldots, M)$, alternatingly.

To make the two subprograms that optimize $\alpha_i\,(i = 1, \ldots, M)$ and $\alpha_{M+i}\,(i = 1, \ldots, M)$ similar, we change the inequality constraint (7) to the equality constraint. This architecture is called ML1$_{\text{v}}$ SVM and the original architecture with (9), ML1 SVM.

As for the details of the training methods including the KKT (Karush-Kuhn-Tucker) conditions, refer to [10,11].

The decision function given by (1) is rewritten as follows:

$$D(\mathbf{x}) = \sum_{i \in S} \alpha_i\, y_i\, K(\mathbf{x}_i, \mathbf{x}) + b, \tag{10}$$

where S is the set of support vector indexes.

According to the computer experiments [10,11], the ML1$_{\text{v}}$ SVM generalized better than the ML1 SVM. Therefore, in the following study, we use the ML1$_{\text{v}}$ SVM.

2.2 Multiple Kernels

In the ML1$_\text{v}$ SVM and L1 SVM, we replace a single kernel to the following multiple kernel:

$$K(\mathbf{x},\mathbf{x}') = \sum_{i=1}^{K} \beta_i\, K_i(\mathbf{x},\mathbf{x}'), \quad \sum_{i=1}^{K} \beta_i = 1,\ \beta_i \geq 0,\ i = 1,\ldots,K, \tag{11}$$

where K is the number of kernels, $K_i(\mathbf{x},\mathbf{x}')$ are the ith kernels, and β_i are their coefficients.

Because the RBF kernel works relatively well for a wide range of applications, in the following study we use the RBF kernel:

$$K_i(\mathbf{x},\mathbf{x}') = \exp(-\gamma_i\,\|\mathbf{x}-\mathbf{x}'\|^2), \tag{12}$$

where γ_i are the ith width parameters to control the width of the spread.

To make model selection easy, we reduce the number of hyperparameters as small as possible. Namely, we assume that $K = 3$ and $\beta_1 = \beta_3$. Then

$$\beta_1 = \frac{1}{2}(1-\beta_2). \tag{13}$$

Also, we assume that

$$\gamma_1 = \frac{\gamma_2}{2^L}, \quad \gamma_3 = 2^L\,\gamma_2, \tag{14}$$

where L is a positive integer.

We call the ML1$_\text{v}$ SVM and L1 SVM with the above multiple kernel, ML1$_\text{v}$ SVM (MK) and L1 SVM (MK), respectively. We also call the kernel given by (11) standard multiple kernel and the three kernels with restrictions given by (13) and (14) reduced three kernels.

2.3 Model Selection

In the following, we discuss the three-stage grid search for the ML1$_\text{v}$ SVM with the reduced three-kernel structure and random search for the ML1$_\text{v}$ SVM with the standard three-kernel structure.

Three-Stage Grid Search. In model selection of the ML1$_\text{v}$ SVM (MK), we need to determine the values of γ_2, C, β_2, L, and C_h. Let γ_2^i ($i = 1,\ldots,N_\gamma$), C_i ($i = 1,\ldots,N_C$), β_2^i ($i = 1,\ldots,N_{\beta_2}$), L_i ($i = 1,\ldots,N_L$), and C_h^i ($i = 1,\ldots,N_{C_h}$), be grid points of grid search.

To speed up model selection by grid search, we carry out three-stage model selection: in the first stage, we select the values of γ_2^i ($i \in 1,\ldots,N_\gamma$) and C_j ($j = 1,\ldots,N_C$) by grid search with $C_h = 0$, $\beta_2 = \beta_2^1$, and $L = L_1$. Namely, for a fixed γ_2^i, we evaluate the cross-validation accuracy changing C_j from $j = 1$ to N_C. If the current cross-validation accuracy is larger than the previous one, we update γ_2^{opt} and C_{opt} by γ_2^i and C_j respectively. We repeat this procedure for $i = 1$ to N_γ. If the cross-validation accuracy obtains 100% or it does not improve twice, we move to cross-validation for γ_2^{j+1}. This speed-up strategy is also used in the following second and third stages.

In the second stage, we determine the values of C_j ($j = 1, \ldots, N_C$), β_2^k ($k = 1, \ldots, N_{\beta_2}$) and L_l ($l = 1, \ldots, L$) by grid search with $C_h = 0$. The value of C is already determined in the first stage, but here we re-optimize it for improvement in the generalization ability. (We may fix the value of C determined in the first stage. As for the difference of the generalization abilities when the C value is re-optimized in the second stage, please see [15].) The parameter values with the highest cross-validation accuracy are C_{opt}, β_2^{opt}, and L_{opt}.

In the third step we determine the value of C_h^m ($m = 1, \ldots, N_{C_h}$) by cross validation. The parameter value C_h^m with the highest cross-validation accuracy is C_h^{opt}.

For the ML1$_\mathrm{v}$ SVM (MK), the C_h value associated with the highest accuracy among the candidate C_h values is selected as the parameter value. Therefore, there may be a case where the accuracy by cross-validation for the ML1$_\mathrm{v}$ SVM (MK) is lower than that for the L1 SVM (MK). Usually this is avoided by assigning a small value to the smallest value of candidate C_h values.

After model selection, we train the ML1$_\mathrm{v}$ SVM (MK) with $\gamma = \gamma_2^{\mathrm{opt}}$, $C = C_{\mathrm{opt}}$, $\beta_2 = \beta_2^{\mathrm{opt}}$, $L = L_{\mathrm{opt}}$, and $C_h = C_h^{\mathrm{opt}}$ for calculating the accuracies for the training and test data sets.

Random Search. Model selection by random search is terminated when a stopping condition is satisfied after the prescribed number of random search iterations. Let i_{min} and i_{max} be the minimum and the maximum iteration counts of random search, respectively. Then, for the ith ($i_{\mathrm{max}} \geq i \geq i_{\mathrm{min}}$) iteration count of random search, if

$$\frac{i}{2} > i_{\mathrm{cvmax}} \tag{15}$$

is satisfied model selection is terminated, where i_{cvmax} is the iteration count when the maximum cross-validation accuracy up to the ith search is obtained [4,5].

To clarify whether the reduced three-kernel structure degrades the generalization ability compared to the standard three-kernel structure, we use the standard three-kernel structure instead of the reduced three-kernel structure with (13) and (14).

Let a hyperparameter value for grid search be P_i ($i = 1, \ldots, N_p$). Then for random search, we randomly select a P value in $[P_1, P_{N_p}]$, where P_1 is the minimum value and the P_{N_p} is the maximum value of P in cross-validation. For γ, C, and C_h, we set small and large values to the minimum and maximum values, respectively. To make the selection frequency of small values comparable to that of large values, we generate the random value by

$$\exp(R\,(\ln P_{N_p} - \ln P_1) + \ln P_1), \tag{16}$$

where R is a random value uniformly distributed in [0,1].

For β_i ($i = 1, 2, 3$), we randomly generate β_i values in [0,1] and normalize the sum to 1:

$$\beta_i \leftarrow \frac{\beta_i}{\sum_{j=1}^{3} \beta_j} \quad i = 1, 2, 3. \tag{17}$$

To make comparison of grid search and random search fair, we set $i_{\rm min}$ to the number of grid points in grid search and set $i_{\rm max} = 10\, i_{\rm min}$. This means that random search is iterated at least for the number of grid points.

3 Performance Evaluation

The main purpose of performance evaluation is to clarify whether the reduced three-kernels is better than or at least comparable to standard three kernels and three-stage grid search is good enough for model selection. The comparison is made between grid search and random search for the ML1$_{\rm v}$ SVM (MK) and between those for the L1 SVM.

In grid search, we selected γ and C (C_h) from $\{0.01, 0.1, 0.5, 1, 5, 10, 15, 20, 50, 100, 200\}$, and $\{0.1, 1, 10, 50, 100, 500, 1000, 2000\}$, respectively. And for β_2, $\{0, 0.1, 0.2, 0.3, 0.4, 0.5, 0.6, 0.7, 0.8, 0.9\}$ and for L, $\{1, 2, 3\}$. The numbers of grid points are $11 \times 8 = 88$ for the first stage, and $8 \times 10 \times 3 = 240$ for the second stage and 8 for the third stage. Therefore, $i_{\rm min} = 88$ for the L1 SVM and 88+240+8=336 for the ML1$_{\rm v}$ SVM (MK).

For each training data set, we performed fivefold cross-validation by both random search and grid search, trained the classifiers with the selected hyperparameter values, and tested the accuracies for the test data set associated with the training data set. We repeated this procedure for all the 100 or 20 training and test data sets, and analyzed the statistical difference of the test data accuracies for the ML1$_{\rm v}$ SVM (MK) and the L1 SVM by the t test with the confidence level of 95%. The two-class and multiclass problems used in the experiment are the same as those used in [15].

Table 1 shows the average accuracies and the standard deviations of the test data for the 13 two-class problems. In the first row, in I/Tr/Te, I, Tr, and Te show the number of inputs, the number of training data, and the number of test data, respectively. "ML1$_{\rm v}$ (MK)-r" and "ML1$_{\rm v}$ (MK)-g" denote that ML1$_{\rm v}$ SVM (MK) was optimized by random search and grid search, respectively.

For each problem, the better average accuracy is in bold. The "Average (B)" row denotes the average accuracy of the 13 problems and the number of better accuracies between random search and grid search of the same classifier.

We performed statistical analysis by the t test between ML1$_{\rm v}$ SVM (MK)-r and ML1$_{\rm v}$ SVM (MK)-g and between L1 SVM-r and L1 SVM-g, but there was no statistical difference with the confidence level of 95% as shown in the last row of the table, where in "W/T/L", W, T, and L denote that the ML1$_{\rm v}$ SVM (MK)-r (or L1 SVM-r) is statistically better than the ML1$_{\rm v}$ SVM (MK)-g (or L1 SVM-g), W times, comparable to, T times, and worse than, L times.

Now compare the results for ML1$_v$ SVM (MK)-g and ML1$_v$ SVM (MK)-r. There is no statistical difference between the two for the 13 problems but from the Average (B) measure, the average accuracy of the 13 problems for the ML1$_v$ SVM (MK)-g is better than that for the ML1$_v$ SVM (MK)-r and the average accuracy for each of the 13 problems for the ML1$_v$ SVM (MK)-g is better 11 times. Therefore, although there is no statistical difference, the ML1$_v$ SVM (MK)-g performed better than ML1$_v$ SVM (MK)-r.

For the L1 SVM-r and L1 SVM-g also, there is no statistical difference between the two for the 13 problems but from the Average (B) measure, the average accuracy of the 13 problems for the L1 SVM-g is slightly better than that for the L1 SVM-r although the number of times that one classifier performed better than the other is the same for L1 SVM-r and L1 SVM-g. Therefore, there is not much difference of performance between L1 SVM-r and L1 SVM-g.

Comparing the single kernel and multiple kernels, the ML1$_v$ SVM (MK)-g performed better than L1 SVM-g from the Average measure. But the ML1$_v$ SVM (MK)-r performed worse than L1 SVM-r. It seems that random search is not good at handling standard multiple kernels compared to grid search with reduced multiple kernels.

Table 1. Average accuracies and the standard deviations of the test data for the two-class problems

Problem I/Tr/Te	ML1$_v$ (MK)-r	ML15$_v$ (MK)-g	L1-r	L1-g
Banana 2/400/4900	89.151±0.641	**89.217**±0.576	**89.202**±0.632	89.151±0.790
Cancer 9/200/77	73.325±4.615	**73.455**±4.658	**73.299**±4.669	72.935±4.649
Diabetes 8/468/300	76.080±2.029	**76.170**±2.074	76.243±1.933	**76.347**±1.823
Flare-solar 9/666/400	66.817±2.201	**67.060**±2.063	67.023±2.194	**67.063**±2.102
German 20/700/300	75.887±2.131	**76.107**±2.122	75.843±2.159	**76.060**±2.140
Heart 13/170/100	82.010±3.442	**82.750**±3.582	82.350±3.681	**82.900**±3.610
Image 18/1300/1010	97.564±0.521	**97.772**±0.359	**97.396**±0.504	97.158±0.410
Ringnorm 20/400/7000	98.164±0.349	**98.212**±0.463	**98.167**±0.292	98.136±0.342
Splice 60/1000/2175	**88.982**±0.702	88.924±0.756	**88.931**±0.824	88.892±0.842
Thyroid 5/140/75	94.973±2.578	**95.307**±2.426	95.427±2.372	95.427±2.468
Titanic 3/150/2051	77.406±0.862	**77.419**±0.751	**77.423**±0.736	77.361±0.773
Twonorm 20/400/7000	97.342±0.294	**97.392**±0.253	97.327±0.325	**97.389**±0.265
Waveform 21/400/4600	**89.825**±0.599	89.812±0.792	89.581±0.799	**89.722**±0.689
Average (B)	85.194 (2)	**85.354** (11)	85.247 (6)	**85.272** (6)
W/T/L	—	0/13/0	—	0/13/0

Now examine the result from the cross-validation accuracies. Table 2 shows the cross-validation accuracies for the ML1$_v$ SVM (MK) and L1 SVM optimized by random search and grid search. The table also shows the average i_{cvmax} when (15) is satisfied.

For the L1 SVM, random search performed better than grid search both from the Average and B measures. This is more evident for the $\mathrm{ML1_v}$ SVM (MK). Both for the $\mathrm{ML1_v}$ SVM (MK) and L1 SVM, the better performance by random search did not contribute in improving the performance for the test data. Namely, overfitting occurred.

For the two-class problems, the above results show that the three-stage cross-validation by grid search performed fairly well.

The average i_{cvmax} was increased by using multiple kernels because of the increase in the complexity of the SVM structure. But the average i_{cvmax} for the 13 problems is 117.4 and is still smaller than the number of grid points of 336. Therefore, if i_{cvmax} is detected before (15) is satisfied, model selection by random search could be faster than grid search.

Table 2. Average cross-validation accuracies and average iterations by random search for the two-class problems

Problem	$\mathrm{ML1_v}$ (MK)-r	i_{cvmax}	$\mathrm{ML1_v}$ (MK)-g	L1-r	i_{cvmax}	L1-g
Banana	**90.853**	125.3	90.750	90.522	40.6	**90.528**
Cancer	**76.255**	148.3	76.225	**75.930**	38.6	75.715
Diabetes	**78.274**	149.7	78.254	**78.038**	34.8	78.015
Flare-solar	**67.680**	81.2	67.641	67.625	18.0	67.584
German	**76.909**	161.2	76.769	**76.511**	47.2	76.481
Heart	**84.906**	118.7	84.876	**84.782**	31.6	84.441
Image	97.608	145.4	**97.635**	97.200	74.8	**97.204**
Ringnorm	**98.830**	115.5	98.785	98.585	32.3	**98.588**
Splice	**89.195**	118.6	88.925	88.770	30.8	88.650
Thyroid	**97.850**	81.9	97.686	97.414	26.8	**97.507**
Titanic	**79.507**	49.6	79.260	**79.393**	11.0	79.387
Twonorm	**98.263**	89.8	98.180	98.138	28.3	98.067
Waveform	**91.425**	140.6	91.260	**90.885**	42.3	90.797
Average (B)	**86.735** (12)	117.4	86.634(1)	**86.446** (9)	35.2	86.382 (4)

To examine how the accuracies change as the iterations proceed, we optimized the $\mathrm{ML1_v}$ SVM (MK) with $i_{\mathrm{min}} = i_{\mathrm{max}}$ and changed i_{max}. Figures 1 and 2 show the results for the image and thyroid problems, respectively. In Fig. 1, for the image problem, the average cross-validation accuracy by random search (CV-r) reaches that by grid search (CV-g) after 500 iterations. However, the average test accuracy by random search (Test-r) does not reach that by grid search (Test-g). While in Fig. 2, for the thyroid problem, CV-r exceeds CV-g after 150 iterations. And the Test-r reaches Test-g at 40 iterations, and afterwords it decreases and at 200 iterations, it begins to increase. These results indicate that generalization ability fluctuates according to the stopping condition.

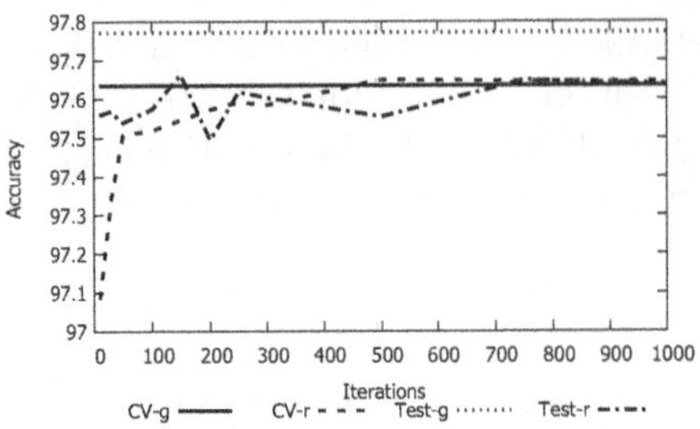

Fig. 1. Convergence of accuracies for the image problem

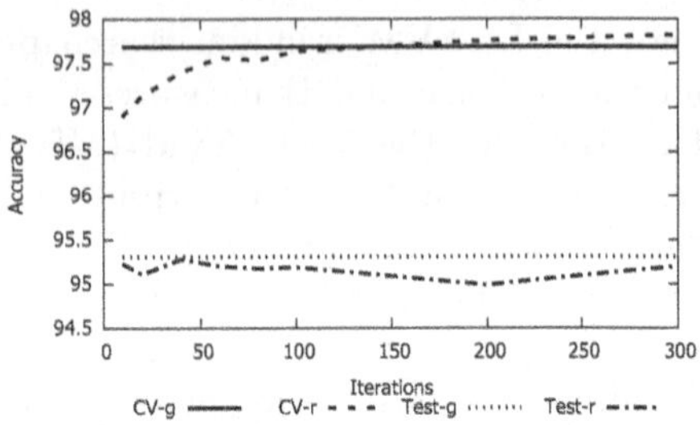

Fig. 2. Convergence of accuracies for the thyroid problem

Table 3 shows the accuracies of the test data for the multiclass problems. In the first row "C" in I/C/Tr/Te denotes the number of classes. In the MNIST (r) problem, the original roles of training and test sets are switched to reduce cross-validation time. Because cross-validation of the MNIST (r) and letter problems were very slow, we set $i_{\min} = i_{\max} = 50$. The '+' and '−' signs attached to the accuracies by random search indicate that the accuracies by random search are statistically better than and worse than that by grid search, respectively.

For the L1 SVM, the results by random search and grid search are comparable with the same numbers of larger average accuracies, although the average accuracy by grid search is slightly better and grid search shows the statistically better accuracy for the thyroid problem.

For the ML1$_v$ SVM (MK), the average accuracies by random search are statistically worse than by grid search for the thyroid and satimage problems, but for the hiragana-13 problem, the average accuracy is statistically better. From Average (B) measure, although the average accuracy by random search is worse, the number of better accuracies is larger.

Table 3. Accuracies of the test data for the multiclass problems

Problem I/C/Tr/Te	ML1$_v$ (MK)-r	ML1$_v$?(MK)-g	L1-r	L1-g
Numeral 12/10/810/820 [2]	99.567±0.300	**99.591**±0.276	**99.646**±0.286	99.598±0.291
Thyroid 21/3/3772/3428 [16]	$^-$97.277±0.274	**97.487**±0.233	$^-$97.356±0.328	**97.535**±0.215
Blood cell 13/12/3097/3100 [2]	**94.771**±0.315	94.579±0.346	94.369±0.307	**94.390**±0.269
Hiragana-50 50/39/4610/4610 [2]	**99.777**±0.081	99.741±0.078	99.728±0.089	**99.734**±0.084
Hiragana-13 13/38/8375/8356 [2]	$^+$**99.837**±0.054	99.774±0.070	99.713±0.070	**99.720**±0.062
Hiragana-105 105/38/8375/8356 [2]	**99.977**±0.027	99.965±0.031	**99.969**±0.029	99.964±0.030
Satimage 36/6/4435/2000 [16]	$^-$91.822±0.448	**92.330**±0.566	**91.645**±0.566	91.608±0.616
USPS 256/10/7291/2007 [17]	**97.877**±0.285	97.845±0.321	97.743±0.318	**97.783**±0.324
MNIST(r) 784/10/10000/60000 [18]	**97.399**±0.072	97.383±0.079	**97.093**±0.070	97.091±0.072
Letter 16/26/16000/4000 [16]	**98.041**±0.166	98.035±0.171	**97.880**±0.173	97.815±0.153
Average (B)	97.635 (7)	**97.673** (3)	97.514 (5)	**97.524** (5)
W/T/L	—	1/7/2	—	0/9/1

Now examine the result from the cross-validation accuracies shown in Table 4. The accuracies for the L1 SVM are comparable by random search and grid search except for the thyroid problem. This causes the lower average accuracy for the test data set by random search.

The accuracies for the $ML1_v$ SVM (MK) are comparable by random search and grid search except for the thyroid, hiragana-13, and satimage problems. Accuracies of the test data sets for these problems also show the statistical differences. For the thyroid and satimage problems, the iterations by random search may not be enough to reach to the cross-validation accuracy obtained by grid search.

For the MNIST (r) and letter problems, the average accuracies by random search are better than those by grid search. Therefore, the setting of $i_{\min} = i_{\max} = 50$ was sufficient for these problems.

Compared to the i_{cvmax} for the L1 SVM, that for the $ML1_v$ SVM increased by four times on average.

Table 4. Average cross-validation accuracies and average iterations by random search for the multiclass problems

Problem	$ML1_v$ (MK)-r	i_{cvmax}	$ML1_v$ (MK)-g	L1-r	i_{cvmax}	L1-g
Numeral	**99.815**	81.8	99.747	**99.747**	16.9	99.741
Thyroid	97.215	157.6	**97.443**	97.226	78.0	**97.418**
Blood cell	**95.003**	192.3	94.806	**94.608**	39.1	94.583
Hiragana-50	**99.729**	148.7	99.671	**99.610**	32.9	99.608
Hiragana-13	**99.838**	183.1	99.745	99.639	65.3	**99.640**
Hiragana-105	**99.973**	105.6	99.954	99.949	26.9	**99.950**
Satimage	92.052	206.3	**92.468**	**91.775**	64.1	91.723
USPS	**97.913**	126.8	97.829	**97.671**	22.6	97.657
MNIST(r)	**97.114**	22.1	97.100	**96.801**	14.5	96.780
Letter	**97.692**	20.1	97.663	**97.473**	13.8	97.426
Average (B)	97.634 (8)	124.4	**97.643** (2)	97.450 (7)	37.4	**97.453** (3)

4 Conclusions

In this paper we compared performance of the $ML1_v$ SVM with the reduced three-kernel structure optimized by three-stage cross-validation to that of the $ML1_v$ SVM with the standard three-kernel structure optimized by random search. According to the computer experiments for two-class and multiclass problems, both architectures showed comparable performance. Therefore, it is concluded that random search worked for the standard three-kernel structures and three-stage cross-validation did not deteriorate generalization performance compared to random search.

Acknowledgment. This work was supported by JSPS KAKENHI Grant Number 22K04154.

References

1. Vapnik, V.N.: Statistical Learning Theory. Wiley, New York, NY (1998)
2. Abe, S.: Support Vector Machines for Pattern Classification, 2nd edn. Springer, UK (2010)
3. Lorena, A.C., De Carvalho, A.C.: Evolutionary tuning of SVM parameter values in multiclass problems. Neurocomputing **71**(16), 3326–3334 (2008)
4. Bergstra, J., Bengio, Y.: Random search for hyper-parameter optimization. J. Mach. Learn. Res. **10**, 281–305 (2012)
5. Goodfellow, I., Bengio, Y., Courville, A.: Deep Learning. The MIT Press, Cambridge (2016)
6. Mantovani, R.G., Rossi, A.L.D., Vanschoren, J., Bischl, B., de Carvalho, A.C.P.L.F.: Effectiveness of random search in SVM hyper-parameter tuning. In: Proceedings of International Joint Conference on Neural Networks, pp. 1–8 (2015)
7. Florea, A.C., Andonie, R.: A dynamic early stopping criterion for random search in SVM hyperparameter optimization. In: Iliadis, L., Maglogiannis, I., Plagianakos, V. (eds.) AIAI 2018. IAICT, vol. 519, pp. 168–180. Springer, Cham (2018). https://doi.org/10.1007/978-3-319-92007-8_15
8. Abe, S.: Analyzing minimal complexity machines. In: Proceedings of International Joint Conference on Neural Networks, pp. 1–8. Budapest, Hungary (2019)
9. Abe, S.: Soft upper-bound minimal complexity LP SVMs. In: Proceedings of International Joint Conference on Neural Networks, pp. 1–7 (2021)
10. Abe, S.: Minimal complexity support vector machines. In: Schilling, F.-P., Stadelmann, T. (eds.) ANNPR 2020. LNCS (LNAI), vol. 12294, pp. 89–101. Springer, Cham (2020). https://doi.org/10.1007/978-3-030-58309-5_7
11. Abe, S.: Minimal complexity support vector machines for pattern classification. Computers **9**, 88 (2020)
12. Abe, S.: Soft upper-bound support vector machines. In: Proceedings of International Joint Conference on Neural Networks, pp. 1–8 (2022)
13. Abe, S.: Do minimal complexity least squares support vector machines work?. In: El Gayar, N., Trentin, E., Ravanelli, M., Abbas, H. (eds.) Artificial Neural Networks in Pattern Recognition, ANNPR 2022, LNCS, vol. 13739, pp. 53–64. Springer, Cham (2023). https://doi.org/10.1007/978-3-031-20650-4_5
14. Jayadeva. Learning a hyperplane classifier by minimizing an exact bound on the VC dimension. Neurocomputing **149**, 683–689 (2015)
15. Abe, S.: Training minimal complexity support vector machines with multiple kernels. In: Proceedings of 2023 IEEE International Conference on Systems, Man and Cybernetics (SMC 2023), pp. 913–919, Honolulu, Hawaii, October 2023
16. Asuncion, A., Newman, D.J.: UCI machine learning repository (2007). http://www.ics.uci.edu/~mlearn/MLRepository.html
17. USPS Dataset. https://www.kaggle.com/bistaumanga/usps-dataset
18. LeCun, Y., Cortes, C.: The MNIST database of handwritten digits. http://yann.lecun.com/exdb/mnist/

Applications in Medical and Health Sciences

Automatic Interpretation of ^{18}F-Fluorocholine PET/CT Findings in Patients with Primary Hyperparathyroidism: A Novel Dataset with Benchmarks

Natasha Sharma[1], Sona Balogova[2,3], Lucia Noskovicova[2], Françoise Montravers[3], Jean-Noel Talbot[3,4], and Edmondo Trentin[1](✉)

[1] Dipartimento di Ingegneria dell'Informazione e Scienze Matematiche, Università di Siena, Siena, Italy
trentin@dii.unisi.it

[2] Department of Nuclear Medicine, Comenius University and St. Elisabeth Oncology Institute and Bory Hospital a.s., Bratislava, Slovakia

[3] Service de Médecine Nucléaire, Hôpital Tenon, Paris, France

[4] Institut National des Sciences et Techniques Nucléaires (INSTN), Saclay, France

Abstract. Primary hyperparathyroidism (pHPT) is a prevalent endocrine disorder characterized by excessive production of parathyroid hormone due to hyperactive parathyroid glands. This paper aims to enhance the diagnostic accuracy of PET/CT using ^{18}F-fluorocholine (FCH) by employing radiomic image analysis to differentiate hyperfunctioning parathyroid glands (HPTG) from thyroid gland (TG) normal or adenomatous tissue. To this aim, the paper contributes a novel, real-life dataset (made publicly available online for download) along with several benchmarks. First, we collected and labeled FCH PET/CT images from 92 patients with pHPT, extracting 56 s-order and higher-order radiomic features using the software LIFEx. These features were analyzed in 98 HPTG and 91 TG findings, comparing clinical characteristics via non-parametric Wilcoxon rank sum tests. Then, several variants of pattern recognition models (k-nearest neighbor and random forest) and deep neural networks were applied to the task of discriminating between HPTG and TG over the dataset in order to fix baseline results for the Community to challenge. Moreover, two ensemble methods are proposed that combine the aforementioned classifiers, achieving an area under the curve of up to 92.30%. In conclusion, the present integration of radiomic features and machine learning provides a promising approach to the task, setting a benchmark for future research in the field.

Keywords: Primary hyperparathyroidism · ^{18}F-fluorocholine PET/CT · deep neural network · ensemble method · stacking

1 Introduction

Primary hyperparathyroidism (pHPT) is an endocrine disorder caused by hyperfunction of one or more parathyroid glands that overproduce parathyroid hor-

C. Y. Suen et al. (Eds.): ANNPR 2024, LNAI 15154, pp. 75–86, 2024.
https://doi.org/10.1007/978-3-031-71602-7_7

mone (PTH) and is the third most common endocrine disease affecting 2–3 times more often women than men [14]. The curative treatment of pHPT consists of the surgical extirpation of hyperfunctioning parathyroid glands (HPTG). The pre-operative localization of HPTG using functional imaging methods of nuclear medicine and ultrasonography provides valuable diagnostic information enabling the optimization of surgical targeting. ^{99m}Tc-sestamibi scintigraphy and then SPECT/CT have been for several decades the reference functional imaging modality for localizing [14]. A decade ago, some of us proposed ^{18}F-fluorocholine (FCH) PET/CT in that aim [10]. Compared to ^{99m}Tc-sestamibi scintigraphy or SPECT/CT, this alternative functional imaging modality of nuclear medicine offers higher spatial resolution, shorter scanning time and lower radiation dose [3,13]. FCH PET/CT has now gained a wide recognition in this setting [13,16]. As well as ^{99m}Tc-sestamibi, FCH is frequently taken-up by the normal thyroid tissue usually with a low intensity, but more intensely in case of thyroid anomaly [11]. Given the high prevalence of nodular goitre in the population, the accurate interpretation of the FCH PET/CT findings requires facing the following major tasks: (1) differentiation of a HPTG from a nodular thyroid lesion; (2) differentiation of a HPTG from normal or nodular ectopica thyroid tissue; (3) precise localization of FCH foci within or outside the thyroid bed; (4) differentiation of an ectopic HPTG from inflammatory lymph nodes, thymus, neoplastic lesions.

One way to optimize the diagnostic accuracy of the interpretation of FCH PET/CT findings in patients with confirmed pHPT may be radiomic image analysis [20]. Radiomics is a quantitative approach to medical imaging that aims to enhance existing data available to clinicians through advanced mathematical analysis, e.g., using the mathematical extraction of the spatial distribution of signal intensity and the interrelationships of pixels (or voxels). The underlying assumption is that the biomedical images contain specific information that is imperceptible to the human eye. Radiomic analysis can be performed on medical images from magnetic resonance imaging (MRI), computed tomography (CT) as well as positron emission tomography (PET) [20].

The paper contributes a preliminary investigation of radiomics via pattern recognition (PR) techniques, deep neural networks (DNNs), and a mixture of PR and DNNs, fixing baseline results for the Community to use as references for future research in the field. The experimental analysis relies on a novel dataset that we collected, pre-processed, and subjected to proper feature extraction and selection. The dataset is contributed to the Community, as well, and made publicly available to download (see Sect. 2). The paper is structured as follows: Sect. 2 presents the data collection process, along with the feature extraction and selection procedures applied in order to obtain the novel dataset. Section 3 reports the experimental design (Sect. 3.1) and the benchmarks (Sect. 3.2), surveying the baseline results yielded by the different algorithms applied to the present classification task and the outcome of the corresponding evaluation metrics. Finally, conclusions are drawn in Sect. 4.

2 Data Collection and Dataset Preparation

The dataset (called *ParaThyroid-Classifier*) and the corresponding documentation and license are available on GitHub at the URL https://github.com/nssharmaofficial/ParaThyroid-Classifier. The data were collected either at the Oncological Institute of St. Elizabeth, Bratislava (Slovakia), to the extent of 27 FCH PET/CTs, or at hôpital Tenon, Paris (France), to the extent of 65 FCH PET/CTs. The FCH preparation injected prior to PET/CT has been manufactured by an industrial radiopharmaceutical firm according to its marketing authorization. The patients gave their informed consent for a further use of their anonymized FCH PET/CT images in non-profit clinical research performed by our medical and academic institutions. The dataset embraces a total of 92 patients with primary hyperparathyroidism, namely: 63 female patients having average age and body weight of 57 years and 70 kg, respectively; 29 male patients with average age and body weight of 49 years and 76 kg. The raw data consisted in patient-specific 3D PET/CT scans that underwent segmentation, feature extraction/selection, and pre-processing as follows.

A segmentation of the regions of interest was first performed manually with the help of automatic 3D segmentation in the freely available software LIFEx, which allows the calculation of conventional, textural, and shape elements of diagnostic images [12]. A total of 189 findings (i.e., areas corresponding to abnormal glands) were segmented, encompassing 91 lesions of the thyroid gland and 98 lesions of the parathyroid gland. First-order textural features were extracted from the intensity frequency histogram, which represents the intensity frequency distribution of a single voxel in the circumscribed lesion [1]. The features include the standard deviation of the voxel-specific distribution, the skewness, the kurtosis, the histogramentropy and the histogramenergy [1]. This also includes conventional PET parameters such as SUVmax, SUVmin, SUVmean, MTV, and total lesion choline uptake (TLCU) corresponding to TLG in ^{18}F-fluorodeoxyglucose PET/CT [12]. Then, a total of 56 s- and higher-order radiomic elements (based on GLCM, GLRLM, NGTDM, and GLSZM matrices) were extracted (the unfamiliar reader is referred to [1].).

Next, data preprocessing was accomplished in order to explore potential associations between specific radiomic features and the class of each gland over the dataset. To achieve this, we compared the distributions of the radiomic features between two groups: thyroid glands (denoted as "0") and HPTG (denoted as "1"). The first step involved assessing the Normality of the data on a feature-by-feature basis. The Shapiro-Wilk test [7] was chosen for this purpose, due to its sensitivity to deviations from Normality, particularly with limited sample sizes. The Shapiro-Wilk test revealed that a substantial portion of the radiomic features under consideration did not follow a Normal distribution [7]. Among the radiomic features examined, only 5 exhibited a Normal distribution, while the remaining 51 deviated from Normality. Given the non-Normal distribution of the data, the non-parametric Wilcoxon rank-sum test (Mann-Whitney U test) [7] was utilized for comparing distributions. After applying the Bonferroni correction for multiple testing, it was found that 31 variables exhibited significant differences in their distributions, meeting the adjusted significance level ($\alpha = 0.000892$), thereby strongly rejecting the null hypothesis of no difference. Conversely, 25 variables did not show significant differences, indicating no discernible impact of disease status on the distribution of these particular variables.

Feature selection was accomplished next, relying on a wrapping method (generally expected to result in better predictive accuracy than filtering methods) [4], namely the robust Boruta algorithm (which detects element importance by creating shadow functions) [9]. Eventually, 16 most-significant features were selected by the Boruta algorithm, mostly drawn from the statistics of the randomic higher-order features of the aforementioned GLCM and NGTDM matrices. A complete list is available in the documentation of the dataset at the aforementioned URL. Finally, the individual features retained were normalized such that they ended up being zero-centered and having unit variance.

3 Methods and Benchmarks

This section reports on the methods used for fixing the benchmarks on the novel dataset and the corresponding baselines. Section 3.1 presents the experimental design and the performance evaluation strategies adopted. Section 3.2 reports the experimental results.

Table 1. Average accuracies yielded by the k-NN model over th e 10-fold cross-validation procedure for increasing values of k.

k	average accuracy (%)
1	71.83
3	74.02
5	74.10
7	75.82
9	73.84
11	73.66
13	71.49

3.1 Experimental Design

First of all, the average accuracy yielded by the different models (reported in the next Section) was evaluated by means of a 10-fold cross-validation procedure. Architectures and hyperparameters were selected using grid search. Other model-specific performance measures (specificity, sensitivity, area under the curve, and p-value) were then computed using a fixed partitioning of the data into a training + validation set (80% of the data) and a test set (the remaining 20%). This

Table 2. Results obtained with k-NN over the fixed training - validation - test partitioning of the dataset.

k	specificity	sensitivity	AUC	p-value
1	66.7	73.7	70.3	0.066
3	66.7	78.9	81.6	0.916
5	66.7	73.7	79.5	0.337
7	77.8	73.7	82.2	reference
9	77.8	73.7	82.6	0.827
11	72.2	79.0	83.8	0.550
13	72.2	79.0	83.9	0.561

resulted in a total of 152 observations (73 lesions of the thyroid gland, 79 lesions of the parathyroid) for training and validation, and a total of 37 observations for test (18 lesions of the thyroid gland, 19 lesions of the parathyroid), respectively.

3.2 Benchmark Results

A first baseline was set relying on the k-nearest neighbor (k-NN) algorithm [6,18]. Table 1 reports the average accuracies over the 10-fold crossvalidation procedure for increasing values of k, while Table 2 reports the statistics over the fixed training - validation - test partitioning of the data. The model with $k = 7$ was used as a reference for the other models (last column of Table 2) due to its yielding the highest accuracies overall. As usual, in Table 2 the terms specificity and sensitivity refer to the true negative and true positive rates, respectively. The acronym AUC stands for Area Under the ROC Curve. The p-value of AUC was computed in reference to the model with the highest cross-validated accuracy. It is seen that the results (all considered to be medically relevant) are not affected significantly by the choice of k, especially in the $k = 3, \ldots, 9$ range. Therefore, practitioners are recommend to follow the minimum description length principle and to apply $k = 3$ (the simplest and faster model among the best performing, yielding a p-values above 0.05 that retains the null hypothesis w.r.t. the reference $k = 7$).

Next, random forest (RF) models [8] were evaluated. We first studied the impact of varying values for `mtry` (i.e., the number of randomly drawn candidate variables in each split) and `ntree` (i.e., the number of trees in the forest) on a dataset with 16 predictor variables (Fig. 1). It is seen that the RF performed better when `mtry` was set to 4, 5, or 6. RF1 achieved the highest average accuracy and was selected as the reference model for further comparisons (Table 4). Results on the training - validation - test setup (Table 4) show no statically significant differences among the models (see, in particular, the corresponding p-values) (Table 3).

As for the DNNs, we selected 8 architectures and architecture-specific hyperparameters (Table 5) based on the 10-fold cross-validated accuracy. The DNNs

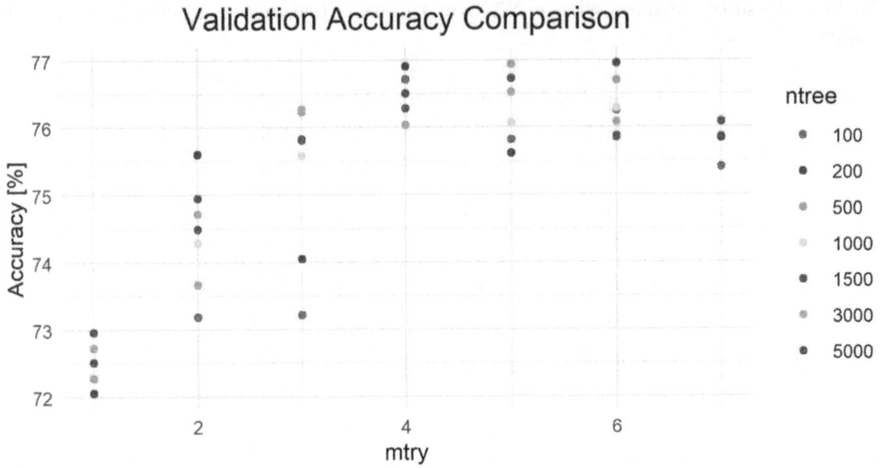

Fig. 1. Average performance of RFs for different values of `mtry` and `ntree` over the 10-fold cross-validation procedure. Distinct colors corresponds to different `ntree` values. (Color figure online)

Table 3. Average accuracy yielded by the 5-best RFs (RF1–RF5) over the 10-fold cross-validation procedure.

label	mtry	ntree	average accuracy (%)
RF1	6	200	76.95
RF2	5	500	76.94
RF3	4	200	76.91
RF4	5	5000	76.73
RF5	4	100	76.72

selected had a number of layers ranging between 3 and 6. For each number of layers, two different forms of activation functions were selected, namely logistic sigmoid (S) and leaky-ReLU (L). Training with early stopping was performed for a model-specific number of epochs ranging from 20 to 500. Table 5 reports the average accuracies yielded by the different DNNs over the 10-fold evaluation procedure. It is seen that, for any given number of layers, the L model improves systematically (yet, not significantly) over the corresponding S machine. There is a substantial (yet, not monotonically increasing) improvement in terms of accuracy when using deeper DNNs, the highest average accuracy being yielded by the DNN4L (6-layer DNN with leaky-ReLU activation functions). The DNNs were then evaluated in the usual fixed training - validation -test setup. The results are shown in Table 7, where DNN4L was used as the reference model for the computation of the p-value of the AUC. It is seen that no significant gaps emerged among the performance yielded by the different DNNs in terms of AUC, although the DNN4L and the (much simpler and faster) DNN1L achieved most of the best scores overall (Table 6).

Table 4. Results obtained with RFs over the fixed training - validation - test partitioning of the dataset.

label	specificity	sensitivity	AUC	p-value
RF1	94.4	63.1	84.2	reference
RF2	94.4	57.8	86.5	0.334
RF3	94.4	63.1	86.0	0.410
RF4	94.4	57.8	86.8	0.234
RF5	94.4	63.1	86.8	0.198

Table 5. Architectures and hyperparameters of the DNNs selected via 10-fold cross-validation. *Legend*: the writing $\langle n_1, \ldots, n_k \rangle$ refers to a DNN with $k+2$ layers (input layer, k hidden layers, output layer) whose i-th hidden layer (for $i = 1, \ldots, k$) contains n_i neurons. DNNs are labeled as "DNN<*k*><*AF*>" where <*k*> is the number of hidden layers and <*AF*> refers to the form of the activation function, i.e. either "S" (sigmoid) or "L" (leaky ReLU).

DNN label	nb. of hidden neurons	activation function	learning rate	mini-batch size
DNN1S	$\langle 5 \rangle$	sigmoid	0.01	16
DNN1L	$\langle 11 \rangle$	leaky ReLU	0.01	32
DNN2S	$\langle 12, 7 \rangle$	sigmoid	0.01	16
DNN2L	$\langle 12, 7 \rangle$	leaky ReLU	0.01	32
DNN3S	$\langle 11, 8, 5 \rangle$	sigmoid	0.01	16
DNN3L	$\langle 16, 13, 10 \rangle$	leaky ReLU	0.001	8
DNN4S	$\langle 12, 7, 6, 5 \rangle$	sigmoid	0.0001	8
DNN4L	$\langle 13, 12, 10, 8 \rangle$	leaky ReLU	0.01	32

A first ensemble [15] of classifiers (EC1) based on stacking [5] was built by combining the outputs of the aforementioned (and, already trained) k-NNs, RFs, and DNNs via a 3-layer combinator DNN, as in [17,19]. Grid-search was used for selecting the proper hyper-parameters, as we did for the DNNs earlier. During training, the latter is expected to learn an improved decision rule over current feature vector $\mathbf{x}$ (a generic patient in the dataset) based on the sole outputs yielded by k-NNs, RFs, and DNNs fed with $\mathbf{x}$, respectively. Since the RF is itself an ensemble method in nature, being an instance of bagging [2], the resulting machine can be seen as a meta-ensemble model. In the light of the results observed so far, the following base models were selected as building blocks for the present ensemble:

- k-NN model with `k`=3;
- RF model with `mtry`=4 and `ntree`=100;
- NN model *DNN1L* with following configuration:
  - 16 input neurons,
  - 1 hidden layer with 11 neurons and leaky ReLU activation function,

Table 6. Average accuracy yielded by the 8-best DNNs over the 10-fold cross-validation procedure.

DNN label	average accuracy (%)
DNN1S	79.00
DNN1L	79.08
DNN2S	78.96
DNN2L	82.92
DNN3S	79.62
DNN3L	80.25
DNN4S	81.54
DNN4L	83.54

Table 7. Results obtained with the DNNs over the fixed training - validation - test partitioning of the dataset.

DNN label	specificity	sensitivity	AUC	*p*-value
DNN1S	82.4	63.1	82.7	0.840
DNN1L	94.2	68.4	83.3	0.900
DNN2S	88.2	63.1	82.4	0.816
DNN2L	82.3	63.1	83.3	0.910
DNN3S	88.2	63.1	78.9	0.561
DNN3L	76.5	78.9	84.8	0.920
DNN4S	88.2	57.8	77.1	0.456
DNN4L	83.2	68.4	84.2	reference

- 1 output neuron with sigmoid activation function,
- trained with 20 epochs with a learning rate of 0.01 and batch size of 32.

Table 8 presents the average accuracy yielded by the EC1 over the 10-fold cross-validation procedure as a function of the number of hidden neurons in the combinator DNN. The EC1 achieved the highest average accuracy when the latter had 6 hidden neurons (therefore, this was selected as a reference model for computing the *p*-value in the following experimental stage). Table 9 reports the other evaluation metrics evaluated on the fixed training - validation -test partitioning of the dataset.

A second, enhanced version of the ensemble classifier (EC2) was trained and evaluated next. In the EC2, the combinator DNN was fed with the original feature vector $\mathbf{x}$ along with the corresponding outputs from the same base models (k-NN, RFs and DNN) selected in EC1. In so doing, the combinator could modulate the contribution from the base models to the final decision made by the ensemble on $\mathbf{x}$ according to the specific position of $\mathbf{x}$ itself in the feature space. A wider range of DNN architectures were evaluated, ranging from 3 to 6

Table 8. Average accuracy over the 10-fold cross-validation procedure yielded by the first ensemble classifier EC1 as a function of the number of hidden neurons in the combinator DNN.

nb. of hidden neurons	average accuracy (%)
1	95.37
2	97.33
3	97.37
4	97.33
5	95.37
6	97.41
7	95.33
8	95.33
9	95.41
10	95.33

Table 9. Results obtained with the first ensemble classifier EC1 over the fixed training - validation - test partitioning of the dataset.

nb. of hidden neurons	specificity	sensitivity	AUC	*p*-value
1	88.2	73.7	91.0	1.000
2	70.6	94.7	89.5	0.213
3	70.6	94.7	89.5	0.213
4	88.2	73.7	91.0	1.000
5	88.2	73.7	91.0	1.000
6	88.2	73.7	91.0	reference
7	88.2	73.7	91.0	1.000
8	70.6	94.7	89.5	0.213
9	70.6	94.7	89.5	0.213
10	88.2	73.7	91.0	1.000

layers. The results are reported in Table 10) in terms of average accuracy over the 10-fold cross-validation procedure. For each depth (e.g., number of layers) of the combinator DNN, two different types of activation functions were compared, namely logistic sigmoids ("S") and leaky ReLU ("L"), respectively. In the present experimental setup, all the combinator DNNs were trained for 100 epochs with learning rate set to 0.001 and mini-batch size equal to 16. Based on these results, the EC2-3L model was selected as a reference for computing the *p*-values in the next experimental round. Table 11 reports the other evaluation metrics for the EC2 evaluated over the fixed training - validation - test partitioning of the dataset. We found no significant differences in their performance

Table 10. Architectures of the combinator DNNs for the EC2 ensemble classifier, selected via 10-fold cross-validation. *Legend*: the writing $\langle n_1, \ldots, n_k \rangle$ refers to a DNN with $k+2$ layers (input layer, k hidden layers, output layer) whose i-th hidden layer (for $i = 1, \ldots, k$) contains n_i neurons. The models are labeled as "EC2-$<k><AF>$" where $<k>$ is the number of hidden layers used and $<AF>$ refers to the form of the activation function, i.e. either "S" (sigmoid) or "L" (leaky ReLU).

DNN label	nb. of hidden neurons	activation function	average accuracy (%)
EC2-1S	$\langle 17 \rangle$	sigmoid	92.75
EC2-1L	$\langle 17 \rangle$	leaky ReLU	98.66
EC2-2S	$\langle 13, 6 \rangle$	sigmoid	92.80
EC2-2L	$\langle 10, 5 \rangle$	leaky ReLU	99.33
EC2-3S	$\langle 16, 13, 10 \rangle$	sigmoid	96.75
EC2-3L	$\langle 17, 8, 7 \rangle$	leaky ReLU	100.00
EC2-4S	$\langle 15, 7, 5, 4 \rangle$	sigmoid	98.00
EC2-4L	$\langle 9, 8, 6, 5 \rangle$	leaky ReLU	100.00

Table 11. Results obtained with the second ensemble classifier EC2 over the fixed training - validation - test partitioning of the dataset.

DNN label	specificity	sensitivity	AUC	p-value
EC2-1S	82.4	94.7	91.6	0.508
EC2-1L	88.2	89.5	92.3	0.317
EC2-2S	82.4	94.7	91.0	0.627
EC2-2L	88.2	84.2	92.0	0.291
EC2-3S	88.2	78.9	90.4	0.635
EC2-3L	82.4	84.2	88.9	reference
EC2-4S	88.2	78.9	90.1	0.747
EC2-4L	88.2	78.9	88.2	0.768

based on AUC, although the EC2-1L turned out to be advantageous since it yielded a prominent AUC of 92.3% jointly with a simple architecture.

4 Conclusion

The paper contributed a detailed dataset that was analyzed, subjected to proper feature extraction and selection, and made available online to the Community via GitHub. A series of challenging machine learning-based benchmarks was presented within a clear cross-validated experimental evaluation design. These include baseline results yielded by traditional pattern recognition techniques (k-NN, RFs) and DNNs. Two ensemble methods based on stacking the aforementioned stand-alone classifiers were proposed, as well, that achieved the highest performance in terms of cross-validated accuracy and AUC.

From a general standpoint, the results presented confirm radiomic analysis and machine learning as a possible direction in differentiating hyperfunctional parathyroid glands from surrounding thyroid glands of PET/CT image with FCH and support their future clinical use. The proposed ensemble models (that achieved an AUC of up to 92.30%) could serve as a practical and effective tool for diagnostic support (in particular, as regards those patients for whom the diagnostic differentiation proves uncertain or problematic).

References

1. Bailly, C., et al.: Exploring tumor heterogeneity using PET imaging: the big picture. Cancers **11**(9), 1282 (2019)
2. Breiman, L.: Bagging Predictors. Mach. Learn. **24**(2), 123–140 (1996)
3. Broos, W.A.M., Wondergem, M., Knol, R.J.J., van der Zant, F.M.: Parathyroid imaging with 18F-fluorocholine PET/CT as a first-line imaging modality in primary hyperparathyroidism: a retrospective cohort study. EJNMMI Res. **9**(1), 72 (2019)
4. G. Chandrashekar and F. Sahin: a survey on feature selection methods. Comput. Electr. Eng. **40**(1), 1 (2014)
5. Clarke, B.: Comparing Bayes model averaging and stacking when model approximation error cannot be ignored. J. Mach. Learn. Res. **4**, 683–712 (2003)
6. Duda, R.O., Hart, P.E., Stork, D.G.: Pattern Classification, 2nd edn. Wiley, New York (2001)
7. Flury, B.: A First Course in Multivariate Statistics. Springer, New York (2013). https://doi.org/10.1007/978-1-4757-2765-4
8. Ho, T.K.: Random decision forests. In: Proceedings of 3rd International Conference on Document Analysis and Recognition, vol. 1, pp. 278–282 IEEE (1995)
9. Kursa, M., Jankowski, A., Rudnicki, W.: Boruta - a system for feature selection. Fund. Inform. **101**, 271–285 (2010)
10. Michaud, L., et al.: Is ^{18}F-fluorocholine-positron emission tomography/computerized tomography a new imaging tool for detecting hyperfunctioning parathyroid glands in primary or secondary hyperparathyroidism? J. Clin. Endocrinol. Metab. **99**(12), 4531–4536 (2014)
11. Michaud, L., et al.: A pilot comparison of ^{18}F-fluorocholine PET/CT, ultrasonography and 123I/99mTc-sestaMIBI dual-phase dual-isotope scintigraphy in the preoperative localization of hyperfunctioning parathyroid glands in primary or secondary hyperparathyroidism: influence of thyroid anomalies. Medicine **94**(41), e1701 (2015)
12. Nioche, C., et al.: LIFEx: a freeware for radiomic feature calculation in multimodality imaging to accelerate advances in the characterization of tumor heterogeneity. Can. Res. **78**(16), 4786–4789 (2018)
13. Petranović, P., et al.: The EANM practice guidelines for parathyroid imaging. Eur. J. Nucl. Med. Mol. Imaging **48**(9), 2801–2822 (2021)
14. Piciucchi, S., Barone, D., Gavelli, G., Dubini, A., Oboldi, D., Matteuci, F.: Primary hyperparathyroidism: imaging to pathology. J. Clin. Imaging Sci. **2**, 59 (2012)
15. Rokach, L.: Ensemble-based classifiers. Artif. Intell. Rev. **33**(1–2), 1–39 (2010)
16. Talbot, J.N., et al.: ^{18}F-fluorocholine PET/CT detects parathyroid gland hyperplasia as well as adenoma: 401 PET/CTs in one center. Q. J. Nucl. Med. Mol. Imaging **67**(2), 96–113 (2023)

17. Trentin, E.: Multivariate density estimation with deep neural mixture models. Neural Process. Lett. **55**(7), 9139–9154 (2023)
18. Trentin, E.: Soft-constrained neural networks for nonparametric density estimation. Neural Process. Lett. **48**(2), 915–932 (2018)
19. Trentin, E., Lusnig, L., Cavalli, F.: Parzen neural networks: fundamentals, properties, and an application to forensic anthropology. Neural Netw. **97**, 137–151 (2018)
20. Van Timmeren, J.E., Cester, D., Tanadini-Lang, S., Alkadhi, H., Baessler, B.: Radiomics in medical imaging- "how-to" guide and critical reflection. Insights Imaging **11**(1), 91 (2020)

A Hybrid Neuroevolutionary Approach to the Design of Convolutional Neural Networks for 2D and 3D Medical Image Segmentation

Nivedha Ramesh(✉), Tabish Ashfaq, and Nawwaf Kharma

Department of Electrical and Computer Engineering, Concordia University, Montreal, QC, Canada
nivedharamesh9351@gmail.com

Abstract. The evolution of Convolutional Neural Networks (CNNs) has revolutionized medical image segmentation, yet designing optimal architectures remains a challenge. In this paper, we introduce a hybrid evolutionary algorithm that advances the design of CNNs for medical image segmentation. By integrating Cartesian Genetic Programming with Simulated Annealing, our approach efficiently explores the architectural design space, yielding CNN architectures that perform robustly across several medical imaging tasks. Our model demonstrated competitive performance on the DRIVE and Medical Segmentation Decathlon datasets, achieving Dice Similarity Coefficients up to 0.828, and excelling in computational efficiency with reduced GFLOP values. Notably, our algorithm's design eliminates the need for extensive pre-training, reducing computational overhead and enhancing its suitability for resource-constrained environments.

Keywords: Medical Image Segmentation · Evolutionary Algorithms · Computational Efficiency · Neural Architecture Search (NAS)

1 Introduction

Automated segmentation of medical images plays a pivotal role in clinical applications ranging from anatomical analysis to disease diagnosis. Neural Architecture Search (NAS) is a technique to automate the design of neural network architectures. Among NAS methodologies, Evolutionary Algorithms (EAs) have emerged as promising architecture design and parameter optimization methods.

Adopting NAS-based methods for designing neural networks results in the training of a significant number of candidate models, presenting a considerable challenge due to the high dimensionality of medical images. For instance, C2FNAS [1] requires nearly a year of continuous GPU use to search for a single 3D segmentation architecture using an evolutionary algorithm. Another major

C. Y. Suen et al. (Eds.): ANNPR 2024, LNAI 15154, pp. 87–98, 2024.
https://doi.org/10.1007/978-3-031-71602-7_8

challenge in medical image segmentation is creating models that perform well across a variety of tasks, such as detecting tumors or delineating blood vessels.

In this work, we introduce a hybrid evolutionary algorithm specifically designed to evolve Convolutional Neural Network (CNN) architectures for different medical image segmentation tasks. Our approach uses an evolutionary framework to find a diverse set of sub-optimal solutions, which are then individually optimized using simulated annealing.

Our key contributions are:

1. A hybrid evolutionary algorithm that effectively evolves CNN architectures for medical image segmentation tasks, achieving competitive results with reduced computational demands compared to competing methods.
2. A new Graph Edit Distance (GED) based metric designed to quantify the structural differences between neural network architectures.

2 Literature Review

Contemporary approaches to medical image segmentation utilize fully convolutional neural networks (FCNNs) with a U-shaped architecture, including models such as 3D U-Net [2], Res-UNet [3], U-Net++ [4], and UNet3+ [5]. These models are renowned for their robust feature-learning capabilities, effectively tackling complex segmentation tasks across various medical imaging modalities [6–8].

Advanced NAS techniques, especially those using evolutionary algorithms, show real promise in outperforming manually-designed architectures in performance and efficiency. However, their full potential in complex medical image segmentation remains under-explored [1,9–11].

The application of evolutionary algorithms, as demonstrated by Hassanzadeh et al. [12], involves evolving deep convolutional neural networks (DCNNs), proving their ability to tailor network structures and optimize training parameters for specific tasks. DiNTS [9] introduced gradient-based optimization with NAS, which streamlined the search process, improving computational efficiency. However, this method faces challenges in generalizability and scalability. HyperSeg-NAS [13] uses a one-shot NAS method, enabling the discovery of optimal 3D segmentation architectures that balance computational efficiency with adaptability. Yet, this approach requires extensive (hence, lengthy) hyper-parameter tuning, which limits its practical application. AdwU-Net [14] presents an adaptive approach to optimizing U-Net architectures, which significantly reduces computational and memory demands. While this method offers tailored solutions for efficient architecture discovery, its application potential is limited by the complexity of the setup and the associated risk of task-specific overfitting.

Overall, while NAS methods have driven significant advancements in automating and optimizing network designs, they often face challenges with generalizability across different medical tasks and also with scalability. By integrating Cartesian Genetic Programming (CGP) [15] with Simulated Annealing [16], our hybrid EA methodology presents a new approach to the design of CNN architectures for medical image segmentation, one that addresses the central challenges of task-specific customization and tuning. The results below show that our method achieves competitive segmentation results while maintaining a high degree of computational efficiency.

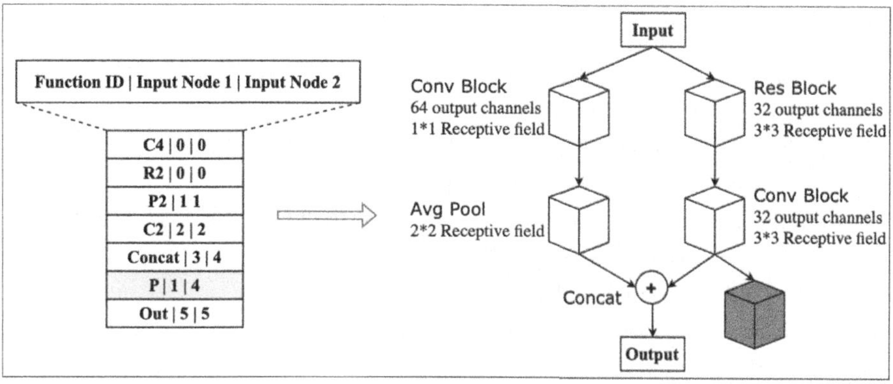

Fig. 1. Illustration of a typical genotype alongside its corresponding phenotype. Each cell within the genotype grid is specified by a function ID and two input connection IDs. This demonstrates how genetic encoding is mapped to a network architecture.

3 Methodology

3.1 GED Based Novelty Computation

In this work, the *novelty* of a network is calculated based on the average GED to its k-nearest neighbors within the population. GED measures the minimum number of edit operations—such as adding, deleting, or substituting vertices or edges—required to transform one graph into another. However, GED is an NP-hard problem, meaning that the time required to compute the exact distance increases exponentially with the size of the graphs. For example, even state-of-the-art algorithms can take excessively long (potentially days or longer) [17] to compute GED for graphs with more than 16 nodes.

Thus, we employ Graph Neural Networks, specifically SimGNN, implemented as outlined in the paper [17] to compute GED. SimGNN reduces the time complexity to a quadratic function of the number of nodes. Implementing SimGNN requires a time-consuming initial training phase, but once trained, it quickly computes graph similarities.

In evolutionary algorithms, maintaining diversity in the population is crucial for exploring a wide range of solutions and avoiding premature convergence to sub-optimal (or simply unacceptable) solutions. Our GED-based novelty computation enhances diversity by identifying and promoting unique architectures.

3.2 Evolutionary Algorithm

Table 1. Overview of the CGP Function Set with Variations in Parameters Used for Constructing the Genotype.

Node Operation	Variation
ConvBlock	Output Channels = {1, 2, 4, 8, 16, 32, 64, 128, 256, 512, 1024} Kernel Size = {1, 3, 5}
ResBlock	Output Channels = {1, 2, 4, 8, 16, 32, 64, 128, 256, 512, 1024} Kernel Size = {1, 3, 5}
DeConvBlock	Output Channels = {1, 2, 4, 8, 16, 32, 64, 128, 256, 512, 1024} Kernel Size = {1, 3, 5}
Pooling	Average and Max Pooling
MergeBlock	Concatenation and Summation

Representation. We encode the network architecture using CGP. Inspired by Suganuma et al. [18], CGP represents the architecture as a two-dimensional grid. As shown in Fig. 1, each cell in this grid contains a function gene and two connection genes, ensuring consistent representation across various models.

The chosen function operations in Table 1, include Convolutional Blocks (ConvBlocks), Residual Blocks (ResBlocks), and Deconvolutional Blocks (DeConvBlocks), pooling and fusion techniques. ConvBlocks extract features from the input images, ResBlocks help in preventing the vanishing gradient problem, and DeConvBlocks increase the spatial resolution of the feature maps. Pooling layers reduce the spatial dimensions of the data, and fusion techniques like concatenation and summation combine different feature maps to improve the network's performance through layers of batch normalization and ReLU activation functions.

This architectural blueprint is transformed into a functional CNN using a Directed Acyclic Graph (DAG), which is an intermediate phenotypic representation.

The actual CNN model, denoted as the phenotype, is implemented within the Keras framework as outlined by Chollet et al. [19]. This stage involves converting the abstract functions and connections depicted in the DAG into executable layers and linkages in a Keras-based CNN model. A final softmax activation layer is appended, to generate the output.

Rich Initialization. We initiate our evolutionary process by randomly generating a population of 20 individuals and optimizing it for novelty (metric developed in Sect. 3.1), using the Novelty Search algorithm [20]. To ensure consistent architectures and avoid unconnected paths, we verify node connectivity during initialization.

Parent Selection. For parent selection, we use a tournament selection algorithm with a window size of 5, without replacement. We end-up selecting 20% of the population as parents.

Mutation. Supported by findings from [21], crossover was not utilized in our methodology. Instead, mutation served as the primary genetic operator, introducing variability into the population by altering node functions and connections between nodes.

Fitness Evaluation. Fitness evaluation is conducted by training each model on a designated dataset for a set number of epochs (20 in our case) and then assessing performance on a held-out validation dataset. An epoch is one complete cycle of training with all training data. Fitness is computed as the sum of Dice Similarity Coefficient (DSC), calculated using Eq. 1 and Intersection over Union (IoU), calculated using Eq. 2. These metrics effectively handle class imbalance, accurately reflecting the segmentation of regions of interest. In Eq. 1 and Eq. 2, TP (True Positive): is the number of correctly predicted positive pixels, FP (False Positive): is the number of pixels incorrectly predicted as positive and FN (False Negative): is the number of actual positive pixels incorrectly predicted as negative.

$$DSC = \frac{2TP}{(2TP + FP + FN)} \tag{1}$$

$$IoU = \frac{TP}{TP + FP + FN} \tag{2}$$

Survivor Selection. For survivor selection, we use tournament selection without replacement, applied to parents and offspring, with a window size of 5. Individuals are selected based on fitness, novelty, and model size. The algorithm may select a less fit individual if its fitness is within 0.05 of the fitter one and it has fewer parameters or higher novelty score. This encourages architectural diversity and balances highly fit complex individuals with simpler ones.

Termination. The evolutionary process concludes after 200 generations, as average fitness improvement beyond this point was minimal, only about 1% over an additional 30 generations. Parameter values are listed in Table 2, and Algorithm 1 outlines the EA phase pseudocode.

Table 2. List of all the parameters, their ranges, and configurations used in the evolutionary process.

Parameter	Value
Population size	20
Tournament Size	5
Mutation Probability	0.1
Optimizer	Adam
Learning Rate	Uniform $(1e-4)$
Number of epochs trained during Fitness Evaluation	20
Loss Function	Weighted Cross-Entropy+Dice loss
Survivor Selection Fitness Tolerance	0.05
Number of Generations	200

Algorithm 1. Evolutionary Algorithm

procedure EVALUATEFITNESS(individual)
    Train individual for 20 epochs using Adam optimizer
    $Fitness \leftarrow \sum(\text{Dice Score}, \text{IoU})$
end procedure
procedure EVALUATENOVELTY(population, ind, k)
    Calculate GED of *ind* to every other individual in *population* using SimGNN
    $ind.novelty \leftarrow \frac{1}{k}\sum_{i=1}^{k} GED[i]$
end procedure
procedure TOURNAMENTSELECTION(population, size)
    $Selected \leftarrow \emptyset$
    for $i \leftarrow 1$ to size **do**
        $Tournament \leftarrow$ Randomly select 5 individuals from *population*
        $Selected \leftarrow Selected \cup$ Individual with highest fitness in Tournament
    end for
    return *Selected*
end procedure
procedure MUTATION(individual)
    if random() < 0.1 **then**
        Mutate a gene in the individual
    end if
end procedure
procedure SURVIVORSELECTION(parents, offspring, optimizeMode)
    $Combined \leftarrow parents \cup offspring$
    $NextGeneration \leftarrow \emptyset$
    for $i \leftarrow 1$ to Population size **do**
        $Tournament \leftarrow$ Randomly select 5 individuals from *Combined*
        if $optimizeMode =$ Fitness **then**
            $Best \leftarrow$ Individual with highest fitness in $Tournament$
            $RunnerUp \leftarrow$ Individual with second highest fitness in $Tournament$

```
            if (RunnerUp.fitness ≥ Best.fitness − 0.05) and (RunnerUp.params < Best.params) then
                NextGeneration ← NextGeneration ∪ {RunnerUp}
            else
                NextGeneration ← NextGeneration ∪ {Best}
            end if
        else if optimizeMode = Novelty then
            Best ← Individual with highest novelty in Tournament
            RunnerUp ← Individual with second highest novelty in Tournament
            if (RunnerUp.novelty ≥ Best.novelty − 0.05) and (RunnerUp.params < Best.params) then
                NextGeneration ← NextGeneration ∪ {RunnerUp}
            else
                NextGeneration ← NextGeneration ∪ {Best}
            end if
        end if
    end for
    return NextGeneration
end procedure
procedure Evolve(Population, maxGen, optimizeMode)
    for gen ← 1 to maxGen do
        Parents ← TournamentSelection(Population, 0.2× Population size)
        for parent ∈ Parents do
            child ← Mutation(parent)
            child.fitness ← EvaluateFitness(child)
            EvaluateNovelty(population, child, k)
            Offspring ← Offspring ∪ {child}
        end for
        Population ← SurvivorSelection(Parents, Offspring, optimizeMode)
    end for
    return Population
end procedure
procedure EA
    Population ← Randomly generate 20 individuals
    Rich Initialization Population ← Evolve(Population, 30, Novelty)
    Evolve(Rich Initialization Population, 200, Fitness)
end procedure
```

3.3 Local Optimization

In this phase, we identify individuals for further refinement by analyzing the mean novelty of individuals across several generations of the evolutionary algorithm (EA). Novelty helps identify different solutions that might not be the fittest but offer unique approaches to the problem. The variation of mean novelty over the first 100 generations for different datasets is shown in Fig. 2. We select the population from the generation with the highest mean novelty.

Next, we use the Non-dominated Sorting Genetic Algorithm II (NSGA-II) [22] to rank the individuals from the chosen generation based on both novelty and fitness.

The solutions found by NSGA-II are then optimized using Simulated Annealing (SA), a probabilistic local optimization algorithm. The algorithm starts with a high initial temperature, increasing the likelihood of accepting sub-optimal solutions. As the algorithm progresses, the temperature is gradually reduced according to a cooling schedule, and Simulated Annealing terminates after a predetermined number of iterations.

4 Results and Discussion

4.1 Experimental Setting

All experiments were conducted on two Intel Xeon Gold 6238 Cascade Lake 2.10 GHz CPUs with 44 cores, four NVIDIA T4 Turing GPUs, and 187 GB of RAM. The DRIVE dataset [23] includes 40 images, split equally into 20 for training and 20 for testing. Data augmentation techniques created overlapping and non-overlapping patches of 128 × 128 pixels. The Medical Segmentation Decathlon (MSD) dataset [24] includes multiple segmentation tasks across MRI and CT scans. Images were normalized using Z-score normalization, contrast-enhanced using CLAHE, and sliced to 2D images. CT volumes were clipped to isolate relevant Hounsfield Units (HU) ranges: −500 to 1500 for lungs and 40 to 75 for liver [25].

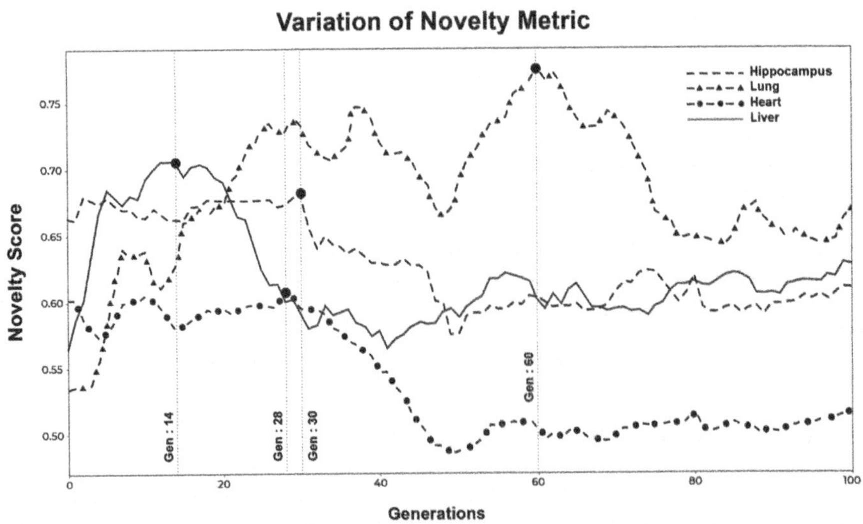

Fig. 2. Temporal progression of mean novelty across generations. Novelty scores are on the y-axis and generations on the x-axis. Distinct curves represent different models: Hippocampus peaks at the 30th generation, Heart at the 28th, Liver at the 14th, and Lung at the 60th. Vertical dotted lines indicate the generation of highest mean novelty observed for each model.

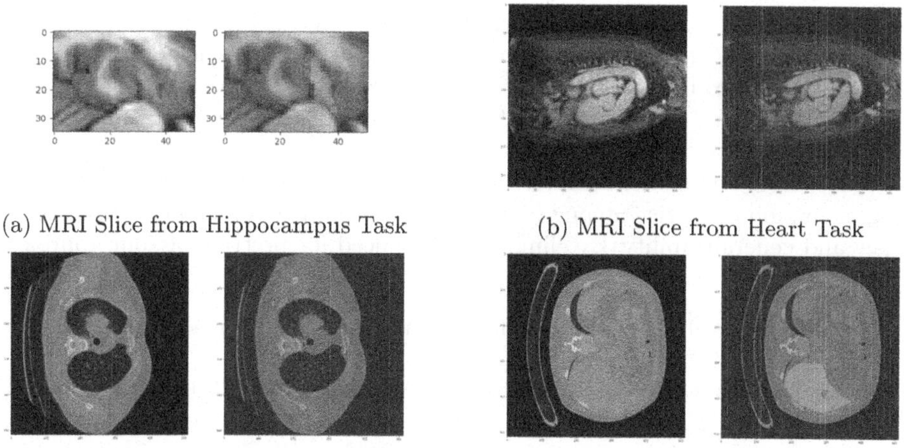

(a) MRI Slice from Hippocampus Task

(b) MRI Slice from Heart Task

(c) MRI Slice from Lung Task

(d) MRI Slice from Liver Task

Fig. 3. Visualization of (i) the original input slice on the left and (ii) its respective segmentation mask overlay on the right for the MSD dataset.

4.2 Segmentation Results

We analyzed the performance of several models on the MSD and DRIVE datasets, focusing on the DSC for segmentation tasks and GFLOPs as a measure of computational cost.

The top two models for each of the five segmentation tasks were trained extensively for 1,000 epochs using the complete dataset, with early stopping set to 30 epochs to prevent overfitting. The performance details are provided in Table 3, and the segmentation results are shown in Fig. 3. For retinal vessel segmentation, our model achieved a DSC of 82.8%, demonstrating efficient learning. In contrast, the lung dataset segmentation showed a lower DSC, mainly due to the small size of lung nodules worsening the existing foreground-background imbalance. Despite these challenges, our model achieved a DSC of 71.3%, outperforming the great majority (96%) of previously published architectures.

4.3 FLOPs and Comparison with Related Works

FLOPs (Floating Point Operations) measure a model's computational complexity, indicating the number of operations required to generate a prediction. In medical image segmentation, optimizing for lower FLOPs is crucial for deploying models in real-time applications, where computational resources are limited. Although our model did not always achieve the highest DSC, it demonstrated significant advantages in computational efficiency—a critical factor in real-world applications that require both accuracy and cost-effectiveness.

Our model's computational efficiency is a key highlight. With a GFLOPs value of 63.81, it outperforms top models like nnUNet [7] and Kim et al. [11], which have GFLOPs of 156.6 and 212.38, respectively. This efficiency leads to faster processing times and reduced energy consumption, demonstrating our model's ability to balance accuracy with computational demands. While models like DiNTS [9] and nnUNet deliver higher accuracy, their computational requirements are substantially greater.

Moreover, our model achieves this efficiency without the need for pretraining, a common practice among high-performing models to enhance initial accuracy and generalizability. By eliminating the need for pretraining, our approach

Table 3. Results of the best-performing models for each segmentation task.

Dataset	Segmentation Task	DSC 1	IoU 1	Accuracy 1	DSC 2	IoU 2	Accuracy 2
DRIVE	Retinal Vessel	0.828	0.716	0.981	0.814	0.736	0.978
MSD	Hippocampus	0.857	0.748	0.986	0.849	0.719	0.983
MSD	Heart	0.881	0.772	0.956	0.924	0.823	0.962
MSD	Lung	0.706	0.596	0.975	0.713	0.628	0.978
MSD	Liver	0.811	0.684	0.973	0.798	0.592	0.976

Table 4. Comparitive performance metrics of models on the **DRIVE** dataset, showing Accuracy (Acc), Dice Similarity Coefficient (DSC), Intersection over Union (IoU), and GFLOPs, sorted by GFLOPs. Key results are highlighted with dashes indicating unpublished GFLOPs.

Model	Acc	DSC	IoU	GFLOPs (GMACS)	Pretraining
RV-GAN	97.90	**86.90**	**97.62**	48.36	**No**
FR-UNet	97.05	83.16	71.20	53.19	No
Our Model	**98.10**	82.80	71.60	**43.41**	**No**
MERIT-CASCADE	96.89	82.21	69.08	33.31	Yes
UNet	96.78	81.41	68.64	–	No

Table 5. Comparative performance metrics of models for segmentation tasks in the **MSD** dataset, sorted by GFLOPs. Key results are highlighted with dashes indicating unpublished GFLOPs.

Model	Liver	Lung	Hippocampus	Heart	GFLOPs (GMACS)	Pretraining
Kim et al.	0.836	0.625	**0.894**	0.964	–	**No**
nnUNet	**0.858**	0.760	0.894	**0.967**	156.6	**No**
DiNTS	0.849	0.772	0.891	0.963	146	Yes
Our Model	0.811	0.713	0.857	0.924	**63.81**	**No**
Swin UNETR	0.855	**0.774**	–	–	25.06	**Yes**
C2FNAS	0.839	0.704	0.879	0.925	–	Yes
Models Gen.	0.866	0.745	–	–	–	No

significantly reduces computational overhead and simplifies the training process, enhancing the model's applicability in resource-constrained environments. Comparative results, including performance metrics and computational demands derived from our own experimental testing, are tabulated in Tables 4 and 5.

5 Summary, Conclusion and Future Work

Our work addresses the challenge of optimizing neural network architectures using NAS for medical image segmentation, focusing on balancing performance and computational efficiency. By integrating evolutionary algorithms with NSGA-II and Simulated Annealing, we have created a comprehensive optimization framework. Novelty is measured using a GED-based metric, which computes the average distance to the k-nearest neighbors, ensuring a diverse population and facilitating the discovery of innovative solutions. The evolutionary algorithm starts with a rich initialization, evolves over 200 generations, and selects candidates from the generation with the highest novelty scores for further refinement using NSGA-II. The top solutions are then optimized with Simulated Annealing.

Our experimental results on the DRIVE and MSD datasets demonstrate that our models achieve competitive DSC scores while maintaining lower computational costs. Specifically, our model achieved a DSC of 82.8% on the retinal vessel segmentation task and a GFLOPs value of 43.41, outperforming most published architectures. It demonstrated superior efficiency with a GFLOPs value of 63.81 on the MSD dataset.

Future research will focus on enhancing model performance through advanced data preprocessing and augmentation techniques to improve robustness and accuracy. Expanding the CGP function set with more complex layers can further improve segmentation outcomes. Additionally, integrating transfer learning strategies within the local optimization phase could address data scarcity issues, making the models more efficient and applicable across various medical imaging tasks.

References

1. Yu, Q., et al.: C2FNAS: coarse-to-fine neural architecture search for 3D medical image segmentation. arXiv (2020)
2. Çiçek, Ö., Abdulkadir, A., Lienkamp, S.S., Brox, T., Ronneberger, O.: 3D U-net: learning dense volumetric segmentation from sparse annotation. arXiv preprint arXiv:1606.06650 (2016)
3. Xiao, X., Shen, L., Luo, Z., Li, S.: Weighted Res-UNet for high-quality retina vessel segmentation. In: Proceedings, pp. 327–331 (2018)
4. Zhou, Z., Rahman Siddiquee, M.M., Tajbakhsh, N., Liang, J.: UNet++: a nested U-net architecture for medical image segmentation. In: Stoyanov, D., et al. (eds.) DLMIA/ML-CDS -2018. LNCS, vol. 11045, pp. 3–11. Springer, Cham (2018). https://doi.org/10.1007/978-3-030-00889-5_1
5. Huang, H., et al.: UNet 3+: a full-scale connected UNet for medical image segmentation. arXiv preprint arXiv:2004.08790 (2020)

6. Ronneberger, O., Fischer, P., Brox, T.: U-net: convolutional networks for biomedical image segmentation. arXiv preprint arXiv:1505.04597 (2015)
7. Isensee, F., Jaeger, P.F., Kohl, S.A.A., Petersen, J., Maier-Hein, K.H.: nnU-Net: a self-configuring method for deep learning-based biomedical image segmentation. Nat. Methods **18**(2), 203–211 (2021)
8. Jin, Q., Meng, Z., Sun, C., Cui, H., Su, R.: RA-UNet: a hybrid deep attention-aware network to extract liver and tumor in CT scans. Front. Bioeng. Biotechnol. **8** (2020)
9. He, Y., Yang, D., Roth, H., Zhao, C., Xu, D.: DiNTS: differentiable neural network topology search for 3D medical image segmentation (2021)
10. Weng, Y., Zhou, T., Li, Y., Qiu, X.: NAS-Unet: neural architecture search for medical image segmentation. IEEE Access **7**, 44247–44257 (2019)
11. Kim, S., et al.: Scalable neural architecture search for 3D medical image segmentation. In: Shen, D., et al. (eds.) MICCAI 2019. LNCS, vol. 11766, pp. 220–228. Springer, Cham (2019). https://doi.org/10.1007/978-3-030-32248-9_25
12. Hassanzadeh, T., Essam, D., Sarker, R.: 2D to 3D evolutionary deep convolutional neural networks for medical image segmentation. IEEE Trans. Med. Imaging (2020)
13. Peng, C., et al.: HyperSegNAS: bridging one-shot neural architecture search with 3D medical image segmentation using HyperNet. arXiv preprint arXiv:2112.10652 (2022)
14. Huang, Z., Wang, Z., Yang, Z., Gu, L.: AdwU-net: adaptive depth and width U-net for medical image segmentation by differentiable neural architecture search. In: Medical Imaging with Deep Learning (2022)
15. Miller, J.: Cartesian Genetic Programming, vol. 43. Springer, Heidelberg (2003). https://doi.org/10.1007/978-3-642-17310-3
16. Kirkpatrick, S., Gelatt, C.D., Vecchi, M.P.: Optimization by simulated annealing. Science **220**(4598), 671–680 (1983). https://doi.org/10.1126/science.220.4598.671
17. Bai, Y., Ding, H., Bian, S., Chen, T., Sun, Y., Wang, W.: SimGNN: a neural network approach to fast graph similarity computation. arXiv preprint arXiv:1808.05689 (2020)
18. Suganuma, M., Shirakawa, S., Nagao, T.: A genetic programming approach to designing convolutional neural network architectures. arXiv preprint arXiv:1704.00764 (2017)
19. Chollet, F., et al.: Keras. GitHub (2015). https://github.com/fchollet/keras
20. Lehman, J., Stanley, K.: Novelty search and the problem with objectives. In: Riolo, R., Vladislavleva, E., Moore, J. (eds.) Genetic Programming Theory and Practice IX. Genetic and Evolutionary Computation, pp. 37–56. Springer, New York (2011). https://doi.org/10.1007/978-1-4614-1770-5_3
21. Real, E., et al.: Large-scale evolution of image classifiers. arXiv (2017)
22. D'Souza, R., Sekaran, C., Kandasamy, A.: Improved NSGA-II based on a novel ranking scheme. J. Comput. **2** (2010)
23. Staal, J., Abramoff, M.D., Niemeijer, M., Viergever, M.A., van Ginneken, B.: Ridge-based vessel segmentation in color images of the retina. IEEE Trans. Med. Imaging **23**(4), 501–509 (2004). https://doi.org/10.1109/TMI.2004.825627
24. Simpson, A.L., et al.: A large annotated medical image dataset for the development and evaluation of segmentation algorithms. CoRR abs/1902.09063 (2019). http://arxiv.org/abs/1902.09063
25. Radiopaedia. Windowing (CT). https://radiopaedia.org/articles/windowing-ct?lang=us

An Improved Pix2Pix GAN for Medical Image Generation

Yanlin Deng[1], Jingwen Ling[1], Xiaqing Rao[1], Jun Tan[1,2], Xiaoyong Fu[1], and Sheng Li[3](✉)

[1] School of Mathematics, Sun Yat-Sen University, Guangzhou 510275, People's Republic of China
[2] Guangdong Province Key Laboratory of Computational Science, Sun Yat-Sen University, Guangzhou 510275, People's Republic of China
[3] State Key Laboratory of Oncology in South China, Guangdong Provincial Clinical Research Center for Cancer, Department of Radiology, Sun Yat-sen University Cancer Center, Guangzhou 510060, People's Republic of China
lisheng@sysucc.org.cn

Abstract. Generating missing modality medical images using deep learning is crucial for disease diagnosis, treatment planning, and medical education. With the rapid development of deep learning, Generative Adversarial Networks (GANs) have shown significant potential by producing realistic, high-quality images. However, GANs face challenges such as blurred organ edges, diffuse tissue boundaries, and failure to maintain the target modality's style.

This study improves the classic GANs model by introducing the Channel Attention Module (CBAM) into the generator's Residual Block, enhancing feature extraction and encoding capabilities. We integrate the VGG16 network to enhance the depiction of image edge intricacies and introduce a histogram chi-square comparison loss to tackle the challenges associated with bone structure generation, stemming from data imbalance. To enhance image clarity, we include a gradient loss term in the loss function. Additionally, real CT images are input into the generator to help the model learn the target modality's style features. Using Conditional GANs, mismatched data pairs are input into the discriminator with false labels to improve content matching sensitivity. Extensive experiments on public brain datasets demonstrate that our method significantly enhances GANs's image generation capabilities, producing images that excel in both visual quality and quantitative metrics.

Keywords: Medical Images · Cross-modal Image Generation · GANs

1 Introduction

1.1 Research Background and Significance

As an important tool to reflect the information of human tissues and organs, medical image processing tasks cover high-definition [1], semantic segmenta-

C. Y. Suen et al. (Eds.): ANNPR 2024, LNAI 15154, pp. 99–110, 2024.
https://doi.org/10.1007/978-3-031-71602-7_9

tion [2], image generation [3], image registration [4], and are applied to different scenes. This paper focuses on cross-modal image generation.

Magnetic resonance imaging (MRI) and computed tomography (CT) are two imaging modes commonly used in clinical diagnosis. MRI has superior soft tissue contrast and no radiation hazard. CT is better than MRI In bone structure imaging speed and effect, but the radiation dose is large and the soft tissue imaging is poor. Therefore, the cross-modal image conversion between MRI And CT is both challenging and of great medical value.

Benefit from the depth development of learning technology, GANs [5] technology in the field of image generation show excellent capability. The task of cross-modal medical image generation can be roughly divided into paired image generation and unpaired image generation. Pix2Pix [6] is considered the baseline model in the field of paired image generation for one-to-one image generation. CycleGAN(cyclic consistent Adversarial Network) [7] is considered the baseline model in the field of unpaired image generation to produce a generation effect that competes with Pix2Pix on unpaired datasets.

Since GANs was proposed in 2014, several hundred variants have emerged, and image translation based on GANs has achieved impressive results. In 2014, Mirza et al. proposed conditional GAN(C-GAN) [8] to control the model of sample generation by introducing conditional information such as labels or categories. However, CGAN has strong dependence on labels and high cost of image annotation, which requires a lot of manpower and computing power. RegGAN [9] treats an incomplete pixel-by-pixel aligned image as a noise label and then registers the resulting image extra. However, its performance may depend on the registered network. For medical images in complex scenes, registration is also a difficult problem. Pix2Pix, proposed by Isola et al. in 2017, applies GANs to networks with supervised image translation. It introduces image-to-image conversion tasks on the basis of CGAN. By taking the input image as conditional information, the generator is able to convert the input image to the associated output image. This improvement makes Pix2Pix show better performance and effect in image conversion tasks, however, Pix2Pix requires pair matching of training data, which has higher requirements for training data. CycleGAN is relatively less rigorous and works well on unpaired or misaligned images. However, CycleGAN does not align the source image with the content of the corresponding generated image. To solve this problem, some researchers have proposed edge constraints using Sobel features, shape constraints based on semantic label information, and deep semantic constraints based on pre-trained networks. Yang combined with MIND(Modal independent neighborhood descriptor) [10] for structural constraints. Although MIND can extract cross-modal commonalities, this method can only extract specific feature commonalities, and the computational cost is high. In this paper, we propose the Chi-square comparison loss of histogram, which is combined with VGG16 network to better solve the problem of rib generation caused by data imbalance.

Our main contributions can be summarized as follows:

(a) The real image information from the target modality is incorporated in this approach to enhance the alignment between the generated CT images and the corresponding recCT images at each bottleneck layer and decoder layer. Additionally, real images from different anatomical levels are introduced to the discriminator, which are deliberately assigned false labels. This strategy aims to improve the discriminator's ability to discern and match content accurately.
(b) In this study, we inserts a CBAM attention mechanism into the residual Block module, which bottleneck stage of the generator network, to improve the generator's ability to capture image channels and locations.
(c) To address the challenge of rib generation resulting from data imbalance, this paper introduces a histogram chi-square comparison loss for approximating image features. Building upon this approach, the shallow output of the VGG16 network is employed to capture additional discrepancies in tissue edge details between real CT and generated CT images, while a gradient loss is incorporated to tackle the blurriness issue arising from the L_1 loss. These techniques contribute to enhancing the fidelity and sharpness of the generated images.

2 Methods

2.1 Network Structure

The improved Pix2Pix generator network structure and data flow process are shown below. The generator presents a U-shaped encoderbottleneck-decoder structure. It consists of the basic Convolution layer (conv) and ConvTranspose deconvolution layer, InstanceNorm normalization layer, relu activation and tanh activation. In this network, in order to ensure that the generated data conforms to the image data distribution, the tanh activation function is used in the last layer. The encoder converts the (1, 256, 256) dimension image into (256, 64, 64) dimension. In this stage, a series of convolution and pooling layers are used to improve the abstraction and richness of the data, and the feature information of the image is extracted and encoded. bottleneck does not do any change in image dimensions. It is mainly responsible for feature conversion and key feature extraction. The decoder stage is responsible for feature decoding, which takes a high-dimensional representation of the output of the bottleneck portion and renders it into fake data that resembles the real data through a series of deconvolution operations. During training, realMR Images (realMR) are fed into the generator to generate fakeCT images, and this generation capability is continuously improved in subsequent training. At the same time, the realCT image (realCT) is also fed into the generator to obtain a reconstructed CT (recCT), which is also compared with the realCT to help the generator learn more about the target mode and stabilize its generation performance. In order to ensure that the generator can gradually render the image of the target mode style during the decoder stage, a strategy is adopted in this study: each layer of the decoder is output. In the training process, the output of fakeCT and recCT in these stages is close to ensure that the decoding of each layer in the decoder stage is moving towards the target mode, so as to stabilize the style rendering ability of the generator and generate images with more style of the target mode (Fig. 1).

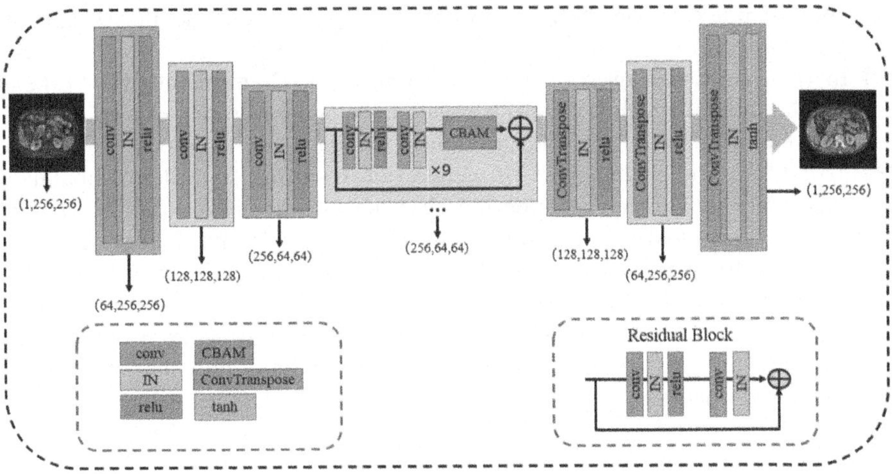

Fig. 1. Generator Network Architecture Diagram.

The discriminator adopts the conditional GAN mode, and the input contains not only the image that needs to be judged, but also the source mode image as the conditional information. The input to the discriminator comprises three distinct sets of data, $(realMR_i, realCT_i)$, $(realMR_i, fakeCT_i)$ and $(realMR_i, realCT_j)$, Among them $(realMR_i, realCT_i)$ data on the label is true, $(realMR_i, fakeCT_i)$ tags to false, $(realMR_i, realCT_j)$ the corresponding labels to false, it is because we hope that discrimination is not only have the ability to identify data authenticity, more has the ability to judge whether match on both in the content. By inputting mismatched MR And CT sections into the discriminator and giving false labels, the discriminator's perception of content matching is strengthened, and the content matching capability of the generator is improved by generating an adversarial learning process.

2.2 Image Feature Approximation

In order to solve the two problems of unclear organ structure and blurred image in the generated image, this paper attempts to capture features of multiple dimensions, from low to high dimensions: pixel-by-pixel feature, edge feature with structure, and content feature of the image. In the training process, the generated image approximates the real image in these dimensions, so as to improve

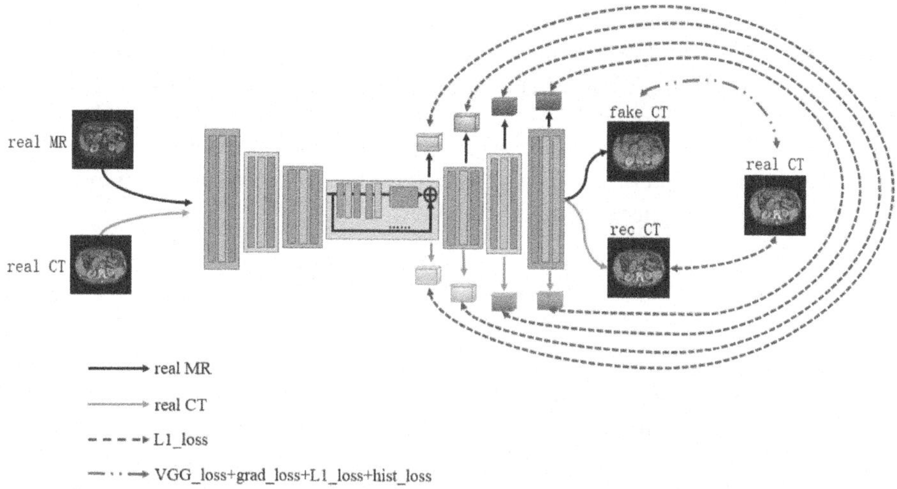

Fig. 2. Generator Data Flow Diagram.

the generation performance of the generator. For this reason, this paper tries to use VGG16 [14] to extract the advanced features of the image. VGG16 has 16 layers, including 13 convolutional layers and 3 fully connected layers. In the convolutional neural network, shallow convolution pays more attention to the low-level features of the image, while deep convolution pays more attention to the high-level features of the image. After several experiments, this paper selects the feature map after the fourth convolutional relu as the feature extraction value of the texture edge details of the image. Input both the generated image and the real image into VGG16, then take the output of layer 4, take L_1 loss for both tensors, and incorporate it into the generator's loss function. This allows the generator to enhance the learning of organ tissue edges. The average value of the channel layer C is calculated in order to visualize the feature map. Feature map visualization is shown in the following figure (Figs. 2 and 3).

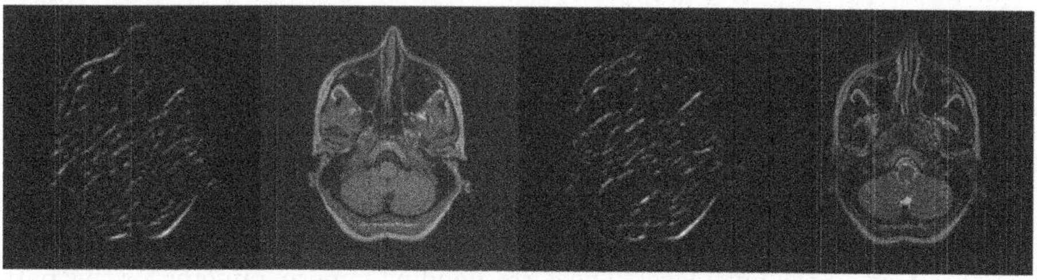

Fig. 3. VGG feature maps of T1 and T2.

The L_1 loss treats all pixel differences equally, without considering the importance or distribution of pixels. In order to evaluate the difference between images more comprehensively, this paper measures the difference of data distribution between two images by calculating the similarity of gray histogram. In the actual comparison, the Chi-square comparison method is chosen. As shown in the formula, $i$represents a bin of the histogram, $H_1(i)$ and $H_2(i)$ respectively represent the values of the two histograms in this bin.

$$d(H_1, H_2) = \sum_i \frac{(H_1(i) - H_2(i))^2}{H_1(i)} \tag{1}$$

Since the comparison of each interval can be weighted, the distribution of data in different intervals is taken into account. This is different from simply calculating the pixel difference between the generated image and the real image. The method of Chi-square comparison can better deal with the problem of data imbalance, so that the comparison results can be balanced to guide the generation of generators in each organization.

2.3 Loss Function

The loss function in this paper is greatly improved on the basis of pix2pix. As shown in the following formula, the first term is the L1 loss; Meanwhile, we hope that the data generated by the generator can fool the discriminator, that is, $D(G(MR))$ is as close to 1 as possible, and the corresponding loss function is in the second term of the formula with hyperparameter λ_{adv}.

In order to stabilize the performance of the generator, recCT is expected to be as close to the real CT as possible during training. The corresponding hyperparameter is λ_{rec}, which corresponds to the third item in the formula.

In order to solve the problem of image blurring caused by L_1 loss, the gradient loss L_{grad} is added.

In order to strengthen the generator's ability of information compression and feature extraction, and to render the image style of target modes stably at decoder stage, this study restricts the output that bottleneck stage ends, and the output of real MR And real CT at decoder stage of the generator. Bottleneck output can be regarded as the 0th layer of decoder, which has three layers of output, so there are a total of four layers of output. The corresponding hyperparameter of these four layers of output is λ_{cf}.

To solve the problem of imbalance in abdominal image data and difficulty in generating bone regions, L_{hist} loss is introduced in this study, which uses chi-square distance to measure the difference of gray histogram between generated image and real image, thus providing a more comprehensive image similarity measure beyond the pixel level. The corresponding hyperparameter for this part is λ_{hist}; In order to make the generated CT close to the real CT in the feature level. In this study, the output of generated CT and real CT after convolution at the fourth layer of VGG network is extracted, and L_1 loss is taken for the two tensors. The corresponding hyperparameter is λ_{VGG}.

$$L_G = \lambda_1 \times L_1(G(MR), CT) + \lambda_{adv} \times E[\log(1 - D(G(MR)))] \\ + \lambda_{rec} \times L_1(G(CT), CT) + L_{grad} + L_{hist} + L_{cf} + L_{VGG} \quad (2)$$

$$L_{cf} = \lambda_{cf} \sum_{i=0}^{3} L_1(decoder_i G(MR), decoder_i G(CT)) \quad (3)$$

$$L_D = E[\log(D(CTi|MRi))] + E[\log(1 - D(CTi|MRi))] \\ + E[\log(1 - D(CTj|MRi))] \quad (4)$$

The loss function of the discriminator is shown in Formulas (4), and each training of the discriminator consists of the calculation of three pairs of data, (realMRi, realCTi, True), (realMRi, fakeCTi, False) and (realMRi, realCTj, False). i refers to the number of the image currently being trained, and j is taken by randomly selecting one of the first 100 trained CT sections. The overall training objectives are shown in Formula (5):

$$minL_D, minL_G \quad (5)$$

2.4 Dateset and Data Preprocessing

In the research of the medical imaging data collection from March 1, 2020 to December 1, 2021 use 3T mri scanner (SignaExcite GEHealthcare, Milwaukee, Wisconsin). The imaging regimen consisted of acquiring a pre-gadolinium T1-weighted image with a repeat time (TR) of 450–600 ms-s and an echo time (TE) of 12–16 ms, followed by a gadolinium enhanced T1-weighted image us- at a dose of 0.1–0.2 mmol/kg. The SYSU abdominal dataset consisted of 24 patients with MRT1-weighted images and corresponding CT images. For clinical diagnosis, we selected CT plain and enhanced scans, as well as T1-weighted images. We use advanced to a chemical (AdvancedNormalizationTools, ANTs) in each of the subjects between MR and CT images registration, then the application N4BiasFieldCorrection30 MR images. In order to eliminate fixed blocked areas in CT images, we establish thresholds at specific locations to remove them. All CT and MR Images are then normalized to the range [−1, 1] by min-max normalization. The pre-processed training set contained 1729 slices and the test set contained 299 slices. In addition, we utilized the IXI brain dataset, which was previously processed by Dar, and included 40 patients. In our study, we selected 30 patients for training (2,730 pieces) and 10 patients for testing (910 pieces), ensuring a comprehensive dataset for our study.

3 Results

3.1 Quantitative Comparisons

To test the model improvements described in this paper, we used a publicly available brain dataset, IXI, to generate internal image generation of MR Modes from T1 to T2. The model compared involves Pix2Pix, CycleGAN, MunitGAN [11], Nice-GAN [12], Unit [13], RegGAN [24].

In order to make a more comprehensive comparison, this paper from the qualitative and quantitative perspectives. From the qualitative point of view, this paper will discuss in detail from each tissue and organ. Because the structure of the brain is so simple, the brain only conducts discussions at the edges of the organization. From the quantitative point of view, this paper selects PSNR, SSIM and LPIPS, which are commonly used in the field of image generation, for comprehensive comparison.

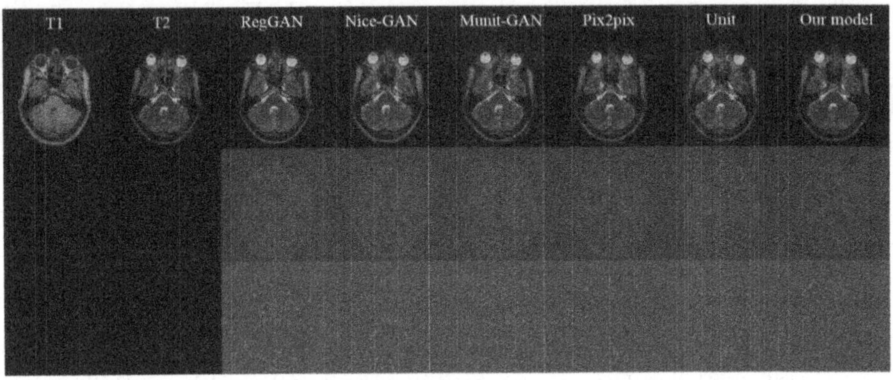

Fig. 4. Experimental model comparison on the IXI dataset.

Figure 4 show a cross section of the human brain. In this case, the first line shows the T2 images synthesized by each model, while the second line shows the error graphs of the model synthesis compared with the real T2 images. Further, the third line measures the synthesis ability of the model at the edge details through the error graph between the Canny edge detection results of the synthesized T2 image and the Canny edge detection results of the real T2 image. The brain synthesis task is relatively simple, so all models can achieve satisfactory synthesis accuracy, and their error graphs are not obvious. Ourmodel, in particular, showed the smallest difference in the error plot and the edge Canny error plot of the synthesized image, indicating that it performed best in maintaining image authenticity and edge detail, with greater accuracy and fidelity in processing image detail and edge information.

Table 1 compare the performance indicators of RegGAN, Nice-GAN, Munit-GAN, Pix2Pix, Unit and Ourmodel in the experiment. The results show that in IXI data set, from the perspective of SSIM index and PSNR index, Ourmodel reaches the optimal level; in LPIPS index, the index of IXI data set is better reflected, and Ourmodel reaches the optimal level of 0.062. Compared with other models, Ourmodel's superiority in experimental indicators reflects its excellent performance in image generation tasks. Benefiting from Ourmodel's effectiveness in learning data set features and capturing image details, Ourmodel's generation effect on IXI data sets is outstanding, which may have positive significance for the practical application of image generation tasks and medical image processing.

Table 1. Quantitative comparison of various models on the IXI dataset.

Model	SSIM↑	PSNR↑	Ipips↓
RegGAN [9]	0.824	24.706	0.079
Nice-GAN [12]	0.807	23.811	0.091
Munit-GAN [10]	0.837	25.502	0.081
Pix2Pix [6]	0.867	26.664	0.066
Unit [13]	0.844	25.615	0.077
Our model	**0.875**	**27.084**	**0.062**

3.2 Ablation Study

In order to explore the effectiveness of the series of losses proposed in this paper, this paper takes Pix2Pix+CBAM+VGG16 as the baseline to verify the hist-gram histogram loss (hist) and gradient gradient loss (grad) proposed later. The validity of cf(contentfeature) feature graph is lost, and the validity of df(discriminatorflow) discriminator data flow is increased (Table 2).

Table 2. Ablation study of models.

Model	SSIM↑	PSNR↑	Ipips↓
baseline	0.869	26.590	0.064
baseline+hist	0.868	26.684	0.065
baseline+hist+cf	0.825	24.318	0.099
baseline+hist+cf+grad	0.833	24.655	0.099
baseline+hist+cf+grad+df	**0.875**	**27.084**	**0.062**

As shown in the ablation experiment, baseline can reach a good generation after the addition of CBAM and VGG. After that, with the increase of a series of loss functions, SSIM value first decreased and then increased, PSNR fluctuation increased, and LPIPS gradually increased. After the last discriminator data flow is increased, the model achieves the optimal result under the parameter adjustment.

As mentioned above, in order to better compare the histogram losses of the two graphs, this paper divides the graphs first and then compares them. First, the picture is divided into K × K, and then the corresponding region is compared by chi-square. Table 4–6 lists the attempts for parameter K. In theory, the more blocks, the more accurate the comparison. In the actual training test, it is found that with the increase of K, the increase of the index is not very obvious, and even the Ipips index has a downward trend. This may be because the position of the tissue in the MR and CT images does not correspond exactly one to one. As a result, when the image is segmented more finely, the chi-square comparison

Table 3. Chi-square comparison of histogram with parameter K attempt.

Model	SSIM↑	PSNR↑	Ipips↓
Pix2Pix	0.587	16.512	0.218
Pix2Pix+$hist_{loss}$ (2 × 2)	0.589	16.308	**0.213**
Pix2Pix+$hist_{loss}$ (3 × 3)	**0.590**	**16.539**	0.219

of the gray histogram is more inaccurate. Therefore, in the actual training, the block parameter K used in the experiment is 2 (Table 3).

4 Conclusion

Cross-modal medical image generation technology holds great potential for applications in the healthcare field, offering valuable support for clinical diagnosis, comprehensive understanding of pathological processes, and various characteristic analyses. Since 2010, the field of deep learning has witnessed remarkable advancements, with techniques such as transfer learning, GANs, and reinforcement learning emerging to tackle complex problems. Against this backdrop, the utilization of generative adversarial networks for medical image generation has yielded significant achievements. Building upon prior research, this study enhances the network architecture and feature extraction of the Pix2Pix model, enabling it to effectively capture modal-specific characteristics in medical images.

In this study, the proposed approach focuses on enhancing the Pix2Pix network by incorporating the CBAM into its bottleneck. This integration aims to improve the model's capability to capture both channel and spatial features, thereby playing a crucial role in optimizing the generator's performance. To achieve more accurate feature representation of real CT images, a specific strategy is employed in this paper. Specifically, at the final bottleneck and decoder stages of the generator, the real CT images are progressively compared with the MR images produced at the same stages. The generator's outputs at these stages are gradually adjusted to resemble the feature expression of real CT images. This contrast mechanism facilitates better learning and rendering of the intricate details and structural characteristics present in real CT images, ultimately enhancing the overall quality and accuracy of the generated results.

In the context of abdominal image generation, this study specifically addresses the challenge of rib generation in CT images. To tackle this issue, the paper introduces a histogram loss. By comparing the gray histograms of the generated and real images, the importance of each intensity interval is effectively balanced, mitigating data imbalance problems and enhancing the accuracy of rib generation. Furthermore, to improve the fidelity of the generated images to their real counterparts, this work focuses on capturing differences between low-dimensional and high-dimensional features of the real and generated images. Firstly, the shallow outputs of the VGG16 network are leveraged to extract additional texture information from the images. Secondly, a gradient loss is introduced alongside the L1 loss to comprehensively capture detailed image features. These techniques significantly enhance the model's performance in the image generation task while enriching the edge details of the generated CT images.

To comprehensively assess the effectiveness of the introduced improvements, comparative experiments were conducted on the publicly available brain dataset IXI. The results clearly demonstrate that the proposed model outperforms other models in terms of visual perception and evaluation metrics. Subsequently, a series of ablation experiments were performed to validate the efficacy of the enhancements. Additionally, diverse training strategies were explored to explore further possibilities for cross-modal generation of medical images. These findings contribute valuable insights to the field and offer potential directions for advancing the field of medical image generation.

Acknowledgements. This work was supported by Guangdong Province Key Laboratory of Computational Science at the Sun Yat-sen University (2020B1212060032), and the National Science Foundation of China (11971491, 12171490, 62376291),and the Foundation of Guangdong-Hong Kong-Macao National Center for Applied Mathematics (2021B1515310002).

Disclosure of Interests. The authors have no competing interests to declare.

References

1. Yuan, G., Zhi, L., Pinle, Q., et al.: Medical image super resolution algorithm based on deep residual generation adversarial network. J. Comput. Appl. **38**(09), 2689–2695 (2018)
2. Yu, D.D., Chen, T., Zhang, J.P., et al.: A review of MRI stroke lesion segmentation methods based on deep learning. J. Intell. Sci. Technol. **5**(03), 293–312 (2023)
3. Chen, F., Zhu, F., Wu, Q.-X., et al.: Review of generative adversarial networks and their applications in image generation. Chin. J. Comput. **44**(02), 347–369 (2021)
4. Jincai, C., Zhang, X., Weina, W., et al.: Multi-modal medical image registration algorithm based on CT and MRI. J. North China Univ. Sci. Technol. (Nat. Sci. Edn.) **43**(04), 108–118 (2021)
5. Goodfellow, I., Pouget-Abadie, J., Mirza, M., et al.: Generative adversarial nets. In: Advances in Neural Information Processing Systems, vol. 27 (2014)
6. Isola, P., Zhu, J.Y., Zhou, T., et al.: Image-to-image translation with conditional adversarial networks. In: Proceedings of the IEEE Conference on Computer Vision and Pattern Recognition, pp. 1125–1134 (2017)
7. Zhu, J.Y., Park, T., Isola, P., et al.: Unpaired image-to-image translation using cycle-consistent adversarial networks. In: Proceedings of the IEEE International Conference on Computer Vision, pp. 2223–2232 (2017)
8. Mirza, M., Osindero, S.: Conditional generative adversarial nets. arXiv preprint arXiv:1411.1784 (2014)
9. Zheng, Y., Zhang, L., Ding, Y., Wang, L.: RegGAN: a regularization-based generative adversarial network for image-to-image translation. In: Proceedings of the IEEE International Conference on Computer Vision (ICCV), pp. 1234–1243 (2021)

10. Yang, X., He, X., Zhang, H., Ren, B.: MIND: a framework for efficient image-to-image translation. In: Proceedings of the IEEE Conference on Computer Vision and Pattern Recognition (CVPR), pp. 6789–6798 (2020)
11. Zhang, C., Xi, W., Liu, X., et al.: Unsupervised multimodal image-to-image translation: generate what you want. In: 2022 International Joint Conference on Neural Networks (IJCNN), pp. 1–8. IEEE (2022)
12. Chen, R., Huang, W., Huang, B., et al.: Reusing discriminators for encoding: towards unsupervised image-to-image translation. In: Proceedings of the IEEE/CVF Conference on Computer Vision and Pattern Recognition, pp. 8168–8177 (2020)
13. Liu, M.Y., Breuel, T., Kautz, J.: Unsupervised image-to-image translation networks. In: Advances in Neural Information Processing Systems, vol. 30 (2017)
14. Simonyan, K., Zisserman, A.: Very deep convolutional networks for large-scale image recognition. arXiv preprint arXiv:1409.1556 (2014)

Vision Transformer Features-Based Leukemia Classification

Karima Ben-Suliman and Adam Krzyżak(✉)

Department of Computer Science and Software Engineering, Concordia University, Montrèal, QC H3G 1M8, Canada
{k_bensul,Krzyzak}@cse.concordia.ca

Abstract. Acute Lymphoblastic Leukemia (ALL) is a disease that is caused by the uncontrollable growth of immature and malignant White Blood Cells (WBCs) which are called lymphoblasts. It occurs when the bone marrow contains 20% or more lymphoblasts. Therefore, Leukemia is diagnosed by counting White Blood Cells (WBCs) in the microscopic smears of bone marrow and blood. There are several attempts for effective leukemia classification with the help of computer aided-based methods by analyzing microscopic smear images. However, how to further improve the classification accuracy is still challenging. Therefore, in this paper, we propose an end-to-end framework to classify ALL. First, we train a more robust specific pre-trained Vision Transformer (ViT) trained on the same type of data, ALL-IDB1. Then, we examine the extracted features of Deep ViT across layers and utilize these features as dense descriptors to make predictions. Based on the observation in this paper, we find and empirically demonstrate that such features, when extracted from the ViT model, consistently improve the classification performance. Moreover, we observe that the pre-trained ViT which is fine-tuned on specific dataset is more effective than the general images dataset. We show that our proposed method achieves an average accuracy of 98.75% which is a competitive result with recent state-of-the-art by a large margin.

1 Introduction

Pretrained-based Deep Features represent the extracted features from deep layers of pre-trained models on large data sets, e.g., ImageNet [26]. These visual descriptors have been used in a wide range of visual tasks [29,30]. For example, pre-trained-CNN-based deep features can be utilized directly as a generic representation in various visual tasks such as object detection, image classification, image manipulation, and generation(e.g., VGG [28,31]). Furthermore, some related studies have shown that visual features extracted from the middle layers of CNN can facilitate image classification and complement semantic features missing [27].

Recently, transformer-based Deep Features have emerged and have been used successfully in a variety of tasks such as image generation [20], image classification [16,21], segmentation [22], and object detection [23,24]. Moreover, pre-

C. Y. Suen et al. (Eds.): ANNPR 2024, LNAI 15154, pp. 111–120, 2024.
https://doi.org/10.1007/978-3-031-71602-7_10

trained-transformers-based deep features have been employed for image pre-processing tasks including super-resolution, denoising, and deblurring [25]. For example, Vision Transformers (ViTs) achieve impressive results in various visual tasks e.g., Dosovitskiy et al. [16] proposed a vision transformer (ViT) for the first time to be applied to image classification. ViTs have the advantage of robustness to occlusions, and texture bias adversarial attacks [17]. Moreover, ViTs provide residual links to feed the extracted features directly to a networked multilayer perceptron head unaffected by its depth. Also, it uses a self-attention mechanism to identify important parts of the image and extract features. This makes ViTs learn complex relationships between different regions of the image and capture local and global features such as color and overall shapes [11,18,19]. This raises the following question: Do the aforementioned properties of ViTs reflect on the representations learned by internal layers of pre-training ViTs and help to improve microscopic images of leukemia? To answer this question, we explore the utilization of internal features of pre-trained ViT as dense descriptors and their effectiveness in a leukemia classification.

The main contributions of this paper can be summarized as follows.

- We are the first to propose an approach aims to design an end-to-end framework so that it can help hematologists in their diagnosis of ALL.
- The proposed approach is inspired by clinical procedures capable of accurately classifying leukemia using a pre-trained Vision Transformer (ViT).
- We investigate the effectiveness of pre-trained ViT features from middle layers in leukemia classification, and our new observations find that the middle layer features are useful in classifying WBCs as lymphocytes or blast cells from peripheral blood smear images.
- We conclude that fine-tuning of pre-trained ViT for task-related images can improve the classification performance significantly.
- We validate the effectiveness of our proposed method with a benchmark dataset. The experiments show it can effectively enhance the accuracy of leukemia classification.

2 Related Work

Deep Features Based on CNN. Rastogi et al. [2] introduced Leukocyte Feature-Extraction model (LeuFeatx) which is a customized pre-trained VGG16 model. The authors applied an upsampling technique for balancing the data. Also, the images were pre-processed by normalization and downsizing. Finally, the training set was augmented using translations, random rotations, and random flipping. They obtained an accuracy of 96.15% using the extra trees classifier (ETC) for ALL-IBD classification. In [3], the authors proposed a new transfer-learning method (SqueezNet). More specifically, ShuffleNet was used to classify benign and malignant. They applied data argumentation to enhance classification performance. The proposed approach was validated on the ALL-IDB and achieved accuracy of 96.97%. In the work of Sahlol et al. [4], pre-trained VGGNet19 was applied to extract features from WBC images. Then, the extracted features were reduced using a statistically enhanced Salp Swarm Algorithm (SESSA) where they reached an accuracy of 96.11% on ALL-IDB. Shanmugam et al. [15] proposed an approach for various classes of Alzheimer disease (AD) detection using pre-trained deep learning models, GoogLeNet, AlexNet, and ResNet-18. The images were resized, pre-processed, and augmented. The overall accuracy of the AlexNet, GoogleNet, and ResNet-18 in AD detection were obtained as 94.08%, 96.39%, and 97.51% respectively. Gokulkrishnan et al. [7] applied pre-trained Convolutional Neural Networks (CNNs): ResNet-101 and ResNet-50 to classify between three stages of malignant ALL lymphoblast and benign cells that were segmented using color thresholding. The authors achieved an accuracy of 98.62% using the Kaggle dataset. In [8], Claro et al. proposed ensemble and multilevel approaches for leukemia detection using convolutional neural networks (CNNs). In the ensemble approach, the votes of a set of pre-trained models are merged resulting in a final classification. However, in a multilevel framework, the features extracted from each used CNN were concatenated. The proposed methodologies achieved accuracies of 94.59% and 94.73% obtained using ensemble and multilevel techniques. Jawahar et al. [9] suggested a deep neural network-based model (ALNett). Moreover, four pre-trained models were used to classify normal and ALL images: AlexNet, VGG-16, ResNet-50, and GoogleNet. The images were pre-processed by resizing and cropping. Finally, the training set was augmented using random contrast enhancement, horizontal flipping, zooming, and 15-degree rotation. The proposed ALNett model achieved the highest accuracy of 91.13% using C-NMC-2019.

Deep Features Based on Transformers. Choa et al. [5] adopted the Vision Transformer (ViT) model to classify white blood cells. They achieved an accuracy of 88.4% on the C-NMC dataset. In [11], the authors performed an empirical study to classify histology images using CNN and Vision transformer models pre-trained on ImageNet. The experiment was performed on the LC25000 dataset which consists of five classes. Selvano et al. [12] proposed a pre-trained ViT model for lung disease detection. The results of the proposed approach demonstrated excellent results compared to state-of-the-art. In the work of [13], five pre-trained vision transformer (ViT) models were employed for brain tumor classification. The obtained results outperformed the existing approaches showcasing the proposed approach efficacy. Yang et al. [14] proposed a novel method (ViTfSCD) for skin cancer classification. The authors' method consists of four blocks which were rebalancing, preprocessing, transformer encoding, and a classification block. To develop the proposed mode, a pre-trained ViT model on ImageNet was implemented. The obtained results achieved an accuracy of 94.1% outperforming the state-of-the-art. Although all the related work aforementioned achieved a satisfactory results for medical image classification, they only considered the features that are extracted from the last layer of the pre-trained model and utilized these features as global representations. In contrast, our approach examines the intermediate features of pre-trained ViT and leverages these internal representations for the classification of microscopic blood smear images.

3 Proposed Method

ALL is diagnosed by detecting at least 20% of blasts in the peripheral blood or bone marrow. Most literature relies on microscopic blood smear datasets that have already individual blood cells or microscopic blood smears containing many blasts and detecting at least one blast, which is inconvenient in medical laboratories. To the best of our knowledge, we are the first to handle such a challenging case. Inspired by the long-range dependence of the Transformers and the homogeneity of the features of WBCs images, we can capture more abundant WBCs feature information. Moreover, the positional information decreases gradually with an increase in the number of network layers. Therefore, we explore the effectiveness of pr-trained ViT by considering its middle layers features and their use in leukemia classification improvement.

3.1 Dataset

To validate the performance of the proposed method, we employ the benchmark dataset: Acute Lymphoblastic Leukemia Image Database (ALL-IDB) [1]. It is a public dataset proposed by Donida Labati. All-IDB includes microscopic images of peripheral blood samples of healthy individuals and unhealthy patients suffering from leukemia. The ALL-IDB dataset is subdivided into two versions. The first version is ALL-IDB1 which has multiple leukocytes per image, as shown in Fig. 1. It contains 108 microscopic images, 59 healthy and 49 unhealthy images that are in full size of 1712 × 1368. The second version, ALL-IDB2, contains 130 normal and 130 single lymphoblasts in size 257 × 257. The images in both versions are manually labeled by expert oncologists to be used as ground truth. In this study, images belonging to ALL-IDB1 are used. To assess the proposed system for ALL classification, 90% of the images are used for training, and 10% are used to test our model.

3.2 Data Pre-processing and Augmentation

The diagnosis of ALL requires detecting at least 20% of lymphoblasts. Therefore, to obtain individual WBCs for classifying it as lymphocyte or a blast cell, circular Hough transform (CHT) is applied, and a single blood cell appears in each image as shown in Fig. 2.

After individual WBCs are obtained, these images are cropped to 120 × 120. The dataset is split into training and testing sets. Then, data augmentation methods, such as random rotation random and horizontal inversion are applied to the training set.

3.3 Methodology and Implementation Details

Figure 3 illustrates the pipeline framework of the proposed method. The overall framework is divided into three parts: data pre-processing, middle layer feature

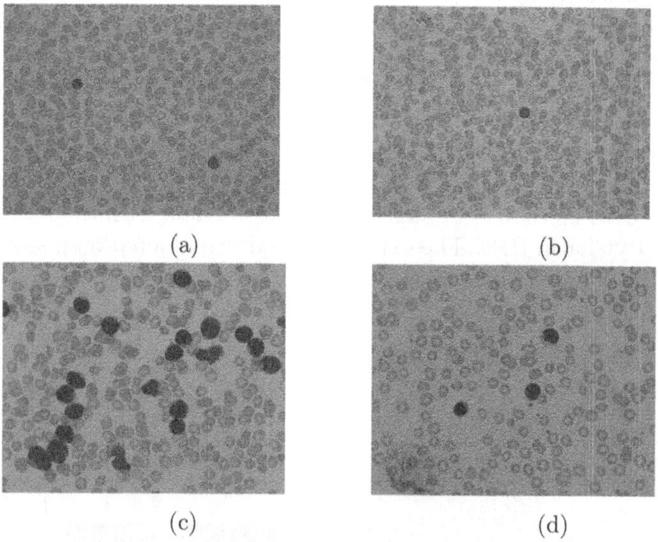

Fig. 1. Samples from ALL-IDB1 Samples from ALL-IDB1. The first row (a) and (b) represent healthy patients, and the second row (c) and (d) for unhealthy patients.

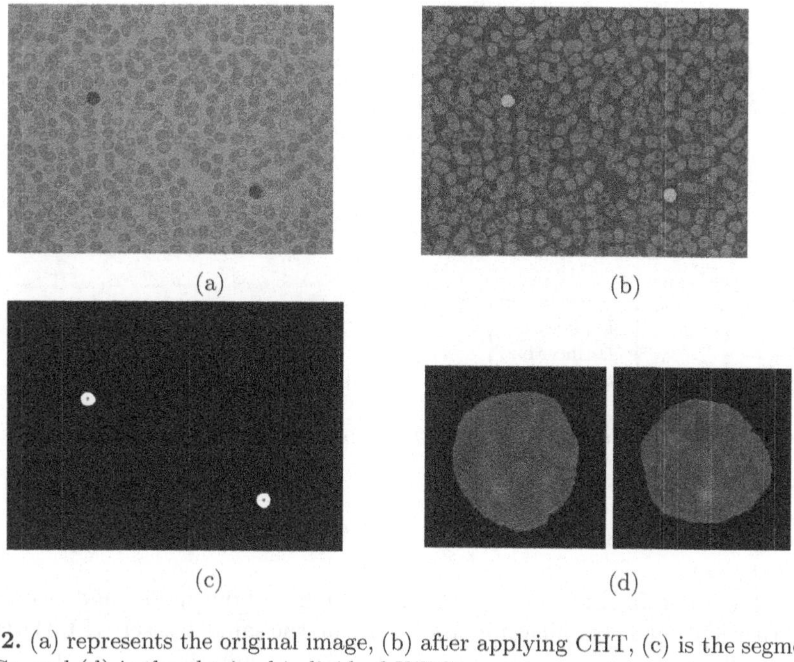

Fig. 2. (a) represents the original image, (b) after applying CHT, (c) is the segmented WBCs, and (d) is the obtained individual WBCs.

extraction, and classification. firstly, microscopic blood smear images are preprocessed to obtain single WBCs. Subsequently, the individual WBCs are fed into a pre-trained ViT-B32 [16] to fine-tune the model. So, we retain all the structure of the pre-trained ViT-B32 model except the last five encoder layers and the final classification head. Then, the output of the seventh transformer block is passed through a customized MLP. This customized MLP consists of two fully connected layers for WBCs classification.

The proposed method is implemented using Python with the PyTorch framework in the PyCharm IDE. The experiments are conducted on a single NVIDIA GeForce RTX 3080 Ti GPU, using the Ubuntu 18.04 operating system.

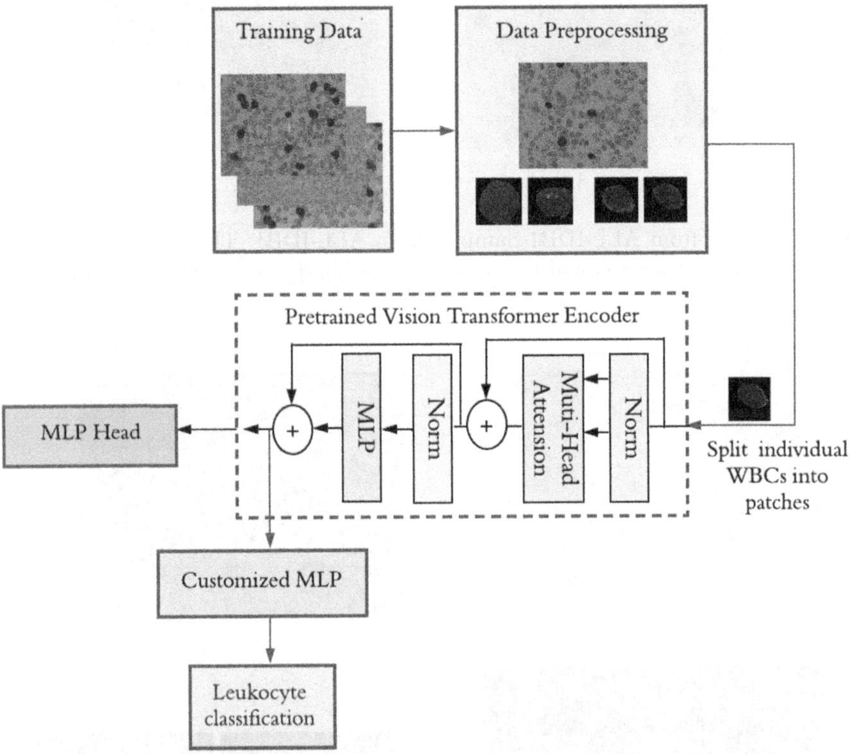

Fig. 3. Workflow of the proposed approach.

4 Experiments Evaluation and Comparisons

We conduct experiments to assess the performance of our model and evaluate it with other leukemia classification approaches on the same dataset ALL-IDB1. As shown in Table 1, our proposed approach is better, exhibiting at least 1.78% and at most 2.64% increase in accuracy. Additionally, it achieves better performance, with at least 4.37% and at most 6.57% increase in precision. Also, our approach is better, obtaining at least 1.93% and at most 2.33% increase in F1-score. We show that our proposed approach achieves competitive results with state-of-the-art by a large margin.

Table 1. Comparison of performance results obtained by the proposed method and different existing methods on ALLIDB1 dataset.

Method	Accuracy (%)	Precision (%)	Recall (%)	F1-score (%)
Rastogi et al. [2]	96.15	94.28	98.69	96.4
Sahlol et al. [4]	96.11	93.43	99.55	96.22
Das et al. [3]	96.97	95.63	98.0	96.80
Ours	**98.75**	**100.0**	97.5	**98.73**

4.1 Effectiveness of Internal Feature

We compare the performances by utilizing separately the internal features and dense features as discriminative features. The experimental results are shown in Fig. 4. The results show that by considering the internal features as dense descriptors, the accuracy is significantly improved by 5% indicating its benefit for the model's performance.

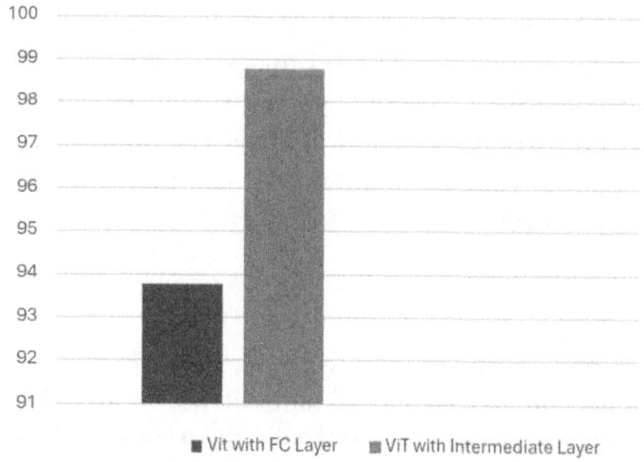

Fig. 4. Classification accuracy of the proposed approach. Intermediate features-based model performance vs. Dense features-based model performance.

4.2 Effectiveness of Task-Related Dataset

We show the comparison of the performances of the pre-trained Vit with/without fine-tuning on our dataset. The accuracy of the classification increased by 3.75% which demonstrates that the result is improved after incorporating specific pre-training ViT significantly, which is trained on the same kind of dataset, as shown in Fig. 5. This observation indicates the superiority of utilizing specialized pre-trained network for achieving improved performance.

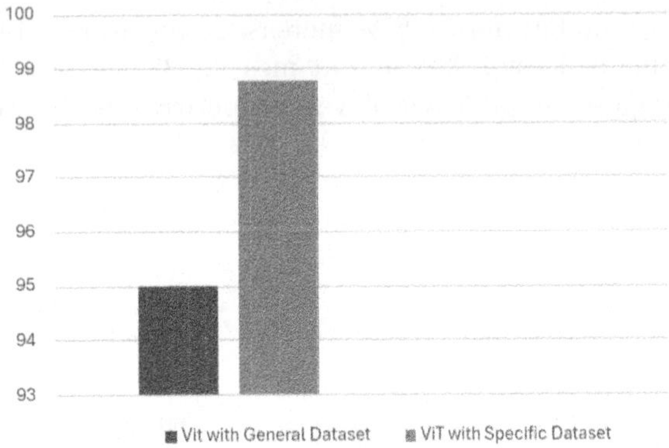

Fig. 5. Classification accuracy of the proposed approach. General dataset-based pre-training model performance vs. task-related dataset-based pre-training model performance.

5 Conclusions

We have provided an end-to-end framework for more effective ALL classification which can be harnessed to aid hematologists in the diagnosis of the patients who may suffer from ALL. We demonstrate that by considering the internal features learned by ViT has a crucial role in holding great promise for our outstanding and competitive performance. We also show the benefit of the specific pre-trained network, which is trained on task-related images, achieves higher accuracy than the general image recognition pre-trained network. We would like to point out that our proposed end-to-end methodology is the first study of its kind that considers the procedures that are followed in the medical laboratories for ALL classification of microscopic images.

References

1. ALL-IDB Homepage.https://homes.di.unimi.it/scotti/all/. Accessed 25 Oct 2023
2. Rastogi, P., Khanna, K., Singh, V.: LeuFeatx: deep learning-based feature extractor for the diagnosis of acute leukemia from microscopic images of peripheral blood smear. Comput. Biol. Med. **142**, 105236 (2022)
3. Das, P., Meher, S.: An efficient deep convolutional neural network based detection and classification of acute lymphoblastic leukemia. Expert Syst. Appl. **183**, 115311 (2021)
4. Sahlol, A., Kollmannsberger, P., Ewees, A.: Efficient classification of white blood cell leukemia with improved swarm optimization of deep features. Sci. Rep. **10**(1), 2536 (2020)
5. Cho, P., Dash, S., Tsaris, A., Yoon, H.J.: Image transformers for classifying acute lymphoblastic leukemia. In: Medical Imaging 2022: Computer-Aided Diagnosis, vol. 12033, pp. 647–653. SPIE (2022)

6. Khandekar, R., Shastry, P., Jaishankar, S., Faust, O., Sampathila, N.: Automated blast cell detection for Acute Lymphoblastic Leukemia diagnosis. Biomed. Signal Process. Control **68**, 102690 (2021)
7. Gokulkrishnan, N., Nayak, T., Sampathila, N.: Deep learning-based analysis of blood smear images for detection of acute lymphoblastic leukemia. In 2023 IEEE International Conference on Electronics, Computing and Communication Technologies (CONECCT), pp. 1–5. IEEE (2023)
8. Claro, M., et al.: Assessing the impact of data augmentation and a combination of CNNs on leukemia classification. Inf. Sci. **609**, 1010–1029 (2022)
9. Jawahar, M., Sharen, H., Jani, L., Gandomi, A.: ALNett: a cluster layer deep convolutional neural network for acute lymphoblastic leukemia classification. Comput. Biol. Med. **148**, 105894 (2022)
10. Zeng, W., et al.: Microscopic hyperspectral image classification based on fusion transformer with parallel CNN. IEEE J. Biomed. Health Inform. (2023)
11. Aitazaz, T., Tubaishat, A., Al-Obeidat, F., Shah, B., Zia, T., Tariq, T.: Transfer learning for histopathology images: an empirical study. Neural Comput. Appl. **35**(11), 7963–7974 (2023)
12. Selvano, E., Paulindino, A., Elwirehardja, G., Pardamean, B.: Evaluating self-supervised pre-trained vision transformer on imbalanced data for lung disease classification. ICIC Express Lett. Part B: Appl. (2023)
13. Asiri, A., et al.: Advancing brain tumor classification through fine-tuned vision transformers: a comparative study of pre-trained models. Sensors **23**(18), 7913 (2023)
14. Yang, G., Luo, S., Greer, P.: A novel vision transformer model for skin cancer classification. Neural Process. Lett. **55**(7), 9335–9351 (2023)
15. Shanmugam, J., Duraisamy, B., Simon, B., Bhaskaran, P.: Alzheimer's disease classification using pre-trained deep networks. Biomed. Signal Process. Control **71**, 103217 (2022)
16. Dosovitskiy, A., et al.: An image is worth 16×16 words: transformers for image recognition at scale. arXiv preprint arXiv: pp2010.11929 (2020)
17. Naseer, M., Ranasinghe, K., Khan, S., Hayat, M., Khan, F., Yang, M.: Intriguing properties of vision transformers. Adv. Neural. Inf. Process. Syst. **34**, 23296–23308 (2021)
18. Puls, E., Todescato, M., Carbonera, J.: An evaluation of pre-trained models for feature extraction in image classification. arXiv preprint arXiv:2310.02037 (2023)
19. Maurya, R., Pandey, N., Dutta, M.: VisionCervix: papanicolaou cervical smears classification using novel CNN-Vision ensemble approach. Biomed. Signal Process. Control **79**, 104156 (2023)
20. Chen, H., et al.: Pre-trained image processing transformer. In: Proceedings of the IEEE/CVF Conference on Computer Vision and Pattern Recognition, pp. 12299–12310 (2021)
21. Touvron, H., Cord, M., Douze, M., Massa, F., Sablayrolles, A., J'egou, H.: Training data efficient image transformers and distillation through attention. arXiv preprint arXiv:2012.12877 (2020)
22. Ye, L., Rochan, M., Liu, Z., Wang, Y.: Cross-modal self-attention network for referring image segmentation. In: CVPR (2019)
23. Carion, N., Massa, F., Synnaeve, G., Usunier, N., Kirillov, A., Zagoruyko, S.: End-to-end object detection with transformers. arXiv preprint arXiv:2005.12872 (2020)
24. Zhu, X., Su, W., Lu, L., Li, B., Wang, X., Dai, J.: Deformable DETR: deformable transformers for end-to-end object detection. arXiv preprint arXiv:2010.04159 (2020)

25. Deng, J., Dong, W., Socher, R., Li, Li-J., Li, K., Fei-Fei, L.: ImageNet: a large-scale hierarchical image database. In: 2009 IEEE Conference on Computer Vision and Pattern Recognition, pp. 248–255, 5206848 (2009)
26. Amir, S., Gandelsman, Y., Bagon, S., Dekel, T.: On the effectiveness of ViT features as local semantic descriptors. In: Karlinsky, L., Michaeli, T., Nishino, K. (eds.) ECCV 2022. LNCS, vol. 13804, pp. 39–55. Springer, Cham (2022). https://doi.org/10.1007/978-3-031-25069-9_3
27. Kataoka, H., Iwata, K., Satoh, Y.: Feature evaluation of deep convolutional neural networks for object recognition and detection. arXiv:1509.07627 (2015)
28. Razavian, A., Azizpour, H., Sullivan, J., Carlsson, S.: CNN features off-the-shelf: an astounding baseline for recognition. In: IEEE Conference on Computer Vision and Pattern Recognition Workshops, pp. 512–519 (2014)
29. Oquab, M., Bottou, L., Laptev, I., Sivic, J.: Learning and transferring mid-level image representations using convolutional neural networks. In: Computer Vision and Pattern Recognition, pp. 1717–1724 (2014)
30. Wei, Y., et al.: CNN: single-label to multi-label. Comput. Sci. (2014)
31. Simonyan, K., Zisserman, A.: Very deep convolutional networks for large-scale image recognition. In: ICLR (2015)

Comparative Study of Deep Learning Models in Melanoma Detection

Farnaz Haghshenas[1], Adam Krzyżak[1(✉)], and Stanislaw Osowski[2]

[1] Department of Computer Science and Software Engineering, Concordia University, Montreal H3G 1M8, Canada
f_haghsh@encs.concordia.ca, krzyzak@cs.concordia.ca

[2] Faculty of Electrical Engineering, Warsaw University of Technology, pl. Politechniki 1, 00-661 Warsaw, Poland
stanislaw.osowski@ee.pw.edu.pl

Abstract. The increasing number of skin cancers underscores the critical importance of early detection and accurate classification to improve treatment outcomes. Melanoma, a malignant skin cancer, has the highest mortality rate among all skin cancer types. Early detection of melanoma significantly enhances the chances of effective treatment and survival rates. This research evaluates advanced deep learning techniques in medical imaging, specifically Vision Transformers (ViT) and Convolutional Neural Networks (CNNs), for detecting melanoma. In this study, we used an annotated dataset of melanoma dermoscopic images. In addition, we employed the k-fold cross-validation technique to evaluate the reliability of our models. Our experimental results indicate that pre-trained Vision Transformers achieved an exceptional diagnostic accuracy of 97.97%, outperforming other models, specifically the pre-trained CNNs models.

Keywords: Medical image analysis · Deep learning · Vision transformer · Melanoma detection · Skin cancer detection

1 Introduction

In recent years, the increasing number of cancer cases has highlighted the critical importance of early detection and effective treatment strategies. The year 2020 alone recorded nearly 10 million deaths worldwide due to various forms of cancer, prompting the global health community to escalate its efforts in utilizing advanced technologies to improve diagnostic and treatment protocols [19]. Among the various types of cancer, skin cancer is notably widespread. Melanoma, in particular, is a life-threatening form of skin cancer, yet it is treatable when detected and treated in its early stages [12].

The last decade has marked a significant transformation in the landscape of cancer diagnosis, primarily driven by advancements in the integration of Artificial Intelligence (AI) in medical imaging analysis. The introduction of Convolutional Neural Networks (CNNs) has been a cornerstone of this evolution by

C. Y. Suen et al. (Eds.): ANNPR 2024, LNAI 15154, pp. 121–131, 2024.
https://doi.org/10.1007/978-3-031-71602-7_11

automating the detection process and achieving more accurate results. CNNs have been particularly effective in analyzing medical images for the detection of various cancers, including skin, prostate, and breast cancers [1,4,7].

More recently, Vision Transformers (ViT) have introduced a new paradigm by adapting the transformer model, initially developed for natural language processing, to computer vision tasks [10]. ViT employs a self-attention mechanism that allows it to weigh the importance of various parts of an image. This approach has shown promising results in different image classification tasks, including skin cancer detection [3,21].

In this research, we evaluate the performance of the state-of-the-art CNNs and ViT models on dermoscopic images for melanoma detection. This evaluation utilizes transfer learning, with each model being initially pre-trained on the ImageNet dataset [9]. This study aims to shed light on the strengths and weaknesses of these advanced deep-learning models and improve the accuracy of melanoma detection. To the best of the authors' knowledge, this study represents the first comparative analysis of CNNs and ViT models using the specified annotated dermoscopic image dataset for melanoma detection.

2 Related Work

The timely and accurate detection and classification of melanoma are critical for effective treatment and improving patient survival outcomes. In response to this need, the field has seen significant advancements in Computer-Aided Diagnosis (CAD) systems. These systems, which are powered by AI, analyze skin images to identify unusual tissue patterns and differentiate between malignant and benign skin lesions. Such AI-empowered technologies have markedly improved the precision of diagnosis, thereby facilitating the early detection and subsequent treatment of melanoma [5].

CNNs, with their deep learning capabilities, have been instrumental in automating the detection and classification of skin cancers. Their architecture, designed to mimic the human visual system, excels in extracting hierarchical features from dermoscopic images, offering substantial improvements over manual diagnostic methods. In the evolving landscape of melanoma diagnosis utilizing deep learning, Belattar et al. (2022) [6] conducted a comprehensive comparison of seven CNN models, including Baseline CNN, InceptionV3, ResNet50, VGG16, Xception, MobileNetV2, and DenseNet201, for melanoma versus nevus classification using a balanced subset of the ISIC 2019 dataset. The study is notable for its thorough evaluation of different architectures, showing that models like Baseline CNN, DenseNet201, and Xception outperformed others. Interestingly, both simpler models like the Baseline CNN and more complex ones such as DenseNet201 achieved high performance, suggesting that model effectiveness is not solely dependent on architectural complexity. In [14], Kassani et al. (2019) presented a comprehensive analysis comparing several advanced CNNs for melanoma detection using dermoscopic images. By employing diverse preprocessing steps and data augmentation techniques like horizontal and vertical flipping to address class imbalance, this study sought to enhance image quality and model accuracy. Through their experimental evaluation, they assessed the performance of state-of-the-art CNN architectures. The study demonstrated that preprocessing and data augmentation could significantly improve detection accuracy, underscoring the potential of deep learning in early and accurate melanoma diagnosis. Kim et al. (2024) [16] developed a comprehensive melanoma classification framework that integrates 54 combinations of 11 datasets and 24 advanced deep learning models, resulting in 1,296 experimental comparisons. Their study introduces Mela-D, a lightweight model optimized for web deployment, which achieves a 33x increase in processing speed and a 24x reduction in parameters while maintaining an accuracy of 88.8%, comparable to ResNet50. This model is designed for efficient, real-time melanoma detection on consumer-grade hardware. The study by Faghihi et al. (2024) [11] presents an innovative approach to skin cancer detection using customized transfer learning models based on VGG16 and VGG19 architectures. The researchers trained their models on 2,541 skin lesion images, employing dropout and early stopping techniques to prevent overfitting. Notably, using k-fold cross-validation, the models achieved average accuracies of 97.51% for VGG16 and 98.18% for VGG19, demonstrating significant improvement over existing approaches without resorting to data augmentation techniques.

Despite the excellent performance of CNNs, the advent of Vision Transformers has introduced a paradigm shift in the field, particularly in the context of distinguishing between closely resembling skin lesions. Unlike traditional CNNs, which rely on local receptive fields and hierarchical feature extraction, ViTs utilize self-attention mechanisms to capture global contextual information within images. This ability to process global relationships among visual features is crucial for tasks that require a comprehensive understanding of the entire image, such as differentiating between skin lesions that may appear similar [8,18]. Due to this fact, several studies have explored the use of ViT for melanoma skin cancer detection. The study by Arshed et al. (2023) [2] underscores the growing importance of ViT models in skin cancer classification. Their research highlights the superiority of pre-trained vision transformers in achieving an accuracy of 92.14%, surpassing CNN-based transfer learning models across several evaluation metrics for skin cancer diagnosis. This comparative analysis suggests that while traditional CNNs remain valuable, ViTs offer a compelling alternative that could redefine the landscape of medical image classification, especially in the context of skin cancer detection.

3 Dataset Description and Pre-processing

In this section, we explore the dataset and the preprocessing strategies we utilized for this research.

3.1 Melanoma Dataset

The melanoma dataset was developed at the Warsaw Maria Skłodowska-Curiebreak National Research Institute of Oncology, Department of Soft Tissue/Bone Sarcoma and Melanoma [13]. It features 112 RGB images of verruca seborrhoica as non-melanoma examples and 134 images of basal cell carcinoma categorized as melanoma. These images were captured at a 20× magnification using dermoscopy from various body areas. Expert dermatologists annotated the collection using the ABCDE criteria, with confirmation by precise pathomorphological analysis. The images, stored in JPEG format, were taken at different times and with various resolutions, leading to a range in image sizes from 767 × 576 to 4273 × 2848 pixels. Figure 1 shows some examples of melanoma and non-melanoma cases in the target dataset.

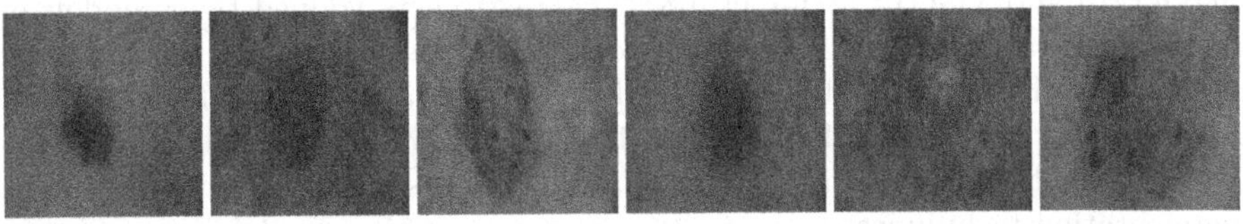

Fig. 1. Sample images from the melanoma dataset. From left to right, the first three samples (a, b, and c) belong to non-melanoma cases, and the next three samples belong to melanoma cases.

3.2 Pre-processing

For this study, three preprocessing steps were taken to improve the models' performances and generalization ability. Subsequent sections explain these three steps in more detail.

Normalization. In our approach, we used normalization for our dataset. We applied normalization to adjust the pixel values of images to a range of [0, 1], which is crucial for facilitating model convergence during the training phase. This process can be mathematically represented as follows

$$x' = \frac{x - x_{\min}}{x_{\max} - x_{\min}} \tag{1}$$

where x' represents the normalized value, x represents the original pixel value, $x_{\min}$ represents the minimum pixel value in the dataset, and $x_{\max}$ represents the maximum pixel value in the dataset.

Augmentation. Recognizing the importance of model robustness and generalization, we utilized a data augmentation step exclusively for the training data in our target dataset. We first split the data using a stratified sampling approach to ensure that each subset maintained the same class distribution as the original dataset. Specifically, we divided the dataset into training, validation, and test sets. Initially, we allocated 20% of the data to the test set. The remaining 80% of the data was then split again, with 20% of this subset assigned to the validation set and the remaining 80% used for training. After splitting the data, we applied data augmentation techniques exclusively to the training set to enhance the model's generalization capabilities. For each training example, we generated three augmented examples using a variety of transformations. These transformations included rotation, width and height shifts, shear, zoom, and horizontal flips. These augmentations artificially expand the variety of training data, simulating real-world variations in image orientation, scale, and perspective. This approach significantly enhances the model's ability to generalize from the training data to unseen images.

Resizing. One of the common challenges in medical image analysis is diversity in image sizes. This is particularly evident in our target melanoma dataset, which contains dermatoscopic images of varying dimensions. To address this issue, a critical preprocessing step involves resizing the images to ensure they meet our models' input requirements. For this purpose, all images are standardized to a resolution of 224×224 pixels. This step is vital for facilitating batch processing and ensuring that the network is trained on a consistent input format, thereby reducing the likelihood of bias towards specific image sizes.

4 Methodology

This research provides an in-depth evaluation and comparison of four distinct models used for the binary classification of melanoma. We utilized two well-known CNN architectures, Inception-V3 and MobileNet, along with two versions of ViT models, ViT-b16 and ViT-b32, to conduct a comparative analysis for melanoma detection. Each model is initially pre-trained on the ImageNet dataset and then fine-tuned using our target annotated melanoma dataset. All models were developed using Python with the Keras and TensorFlow libraries. We will provide detailed descriptions of each model in subsequent sections. Additionally, we will explain the performance metrics used in this study.

4.1 Inception-V3

Inception-V3, introduced by Szegedy et al. [20], marked a significant advancement in the series of Inception networks. It is recognized as a significant architecture due to its innovative design that enhances computational efficiency while maintaining high accuracy in image classification tasks. This model is distinguished by the integration of Inception modules, which consist of parallel convolutional layers with filters of varying sizes, such as 1×1, 3×3, and 5×5, alongside max-pooling layers. These modules facilitate the network's ability to execute various convolution operations efficiently, optimizing parameter usage and minimizing memory consumption.

4.2 MobileNet

MobileNet, introduced by Howard et al. [15] in their 2017 paper titled 'MobileNets: Efficient Convolutional Neural Networks for Mobile Vision Applications', marks a notable advancement in neural network design tailored for mobile and embedded devices. This architecture is distinguished by its use of depthwise separable convolutions, which significantly decrease the model's size and computational demands while maintaining strong performance. These depthwise convolutions divide the convolution process into two stages: a depthwise convolution that applies a single filter per input channel, followed by a pointwise convolution that merges these outputs. This approach substantially reduces computational load, making MobileNet highly effective for applications where efficiency is important.

4.3 Vision Transformers (ViT)

In our study, we employed the ViT architecture, a leading innovation which represents a significant shift from CNNs. ViT introduces a novel approach by applying self-attention mechanisms, a technique originally designed for natural language processing, to the domain of image classification. By breaking down images into smaller pieces called patches, the ViT can focus on and evaluate the significance of various parts of the image. This approach allows for a more detailed and context-aware examination of visual information, enabling the model to capture complex relationships within the visual data [10]. In our research methodology, we implemented two different configurations of the Vision Transformer: ViT-b16 and ViT-b32. The ViT-b16 variant divides images into smaller 16×16 pixel patches, while the ViT-b32 variant processes images using larger 32×32 pixel patches.

4.4 Performance Metrics

In evaluating the effectiveness of our models, we rigorously employ a suite of performance metrics that are standard in medical image analysis. These metrics include:

Accuracy: This metric reflects the ratio of correctly identified cases to the total cases evaluated, offering a general overview of the model's effectiveness.

$$\text{Accuracy} = \frac{TP + TN}{TP + TN + FP + FN} \tag{2}$$

Area Under the Curve (AUC): This metric assesses the model's ability to discriminate between classes, with an AUC of 1.0 indicating perfect classification and 0.5 suggesting no discriminative power.

Precision: Precision calculates the proportion of true positive findings out of all positives predicted by the model, indicating the reliability of positive classifications. The precision score ranges from 0 to 1, where a score of 1 indicates that all positive predictions were correct, and a score of 0 indicates that all samples labeled as positive are negative.

$$\text{Precision} = \frac{TP}{TP + FP} \tag{3}$$

Recall: Recall evaluates the percentage of true positives, highlighting its ability to detect all true instances. The range of recall is from 0 to 1. A recall score of 1 means the model correctly identified all positive cases. In contrast, a score of 0 means the model did not correctly identify any positive cases.

$$\text{Recall} = \frac{TP}{TP + FN} \tag{4}$$

5 Results

In this section, we analyze the outcomes of melanoma detection using four widely recognized deep learning architectures: Inception-V3, MobileNet, and two versions of the Vision Transformer, ViT-b16 and ViT-b32. To ensure a rigorous evaluation, we employed a 5-fold cross-validation technique. This method allowed us to use every part of the dataset for training, validation, and testing, providing a comprehensive assessment of the model's ability to perform consistently across different data segments and demonstrating their capacity to generalize. Additionally, details on the experimental configurations and training parameters

Table 1. The experimental settings employed.

Parameters	Values
Optimizer	Adam
Learning Rate	10^{-5}
Loss Function	Binary Cross-Entropy
Batch Size	32
Epochs	20
Early Stopping Patience	10
Execution Environment	GPU

for this study are outlined in Table 1. Throughout this research, we maintained a uniform experimental setup across all models to ensure the consistency and reliability of our results.

Table 2 showcases the performance outcomes from our evaluation of four advanced deep learning models. Notably, the Vision Transformer variant ViT-b16 achieved a notable accuracy of 97.97%. Additionally, it scored an AUC of 0.9925, reflecting its superior capability in accurately classifying different classes. Table 3 presents a comparison between the findings of this study and those of previous research that utilized the same dataset. Our proposed method's accuracy, while slightly lower than the 98.6% achieved by the deep ensemble method of F. Gil et al. [13], is remarkable considering it utilizes a single model rather than an ensemble of nine complex CNN models. This demonstrates the efficiency of our approach, which requires significantly fewer computational resources, making it highly practical for real-world applications. Figure 2 shows our models' accuracy and loss curves with respect to the number of epochs.

Table 2. Classification Results Obtained from the Models Using 5-Fold Cross-Validation.

Classifier	Accuracy	AUC	Precision	Recall	Loss	k-Fold Accuracy Variance
ViT-b16	**97.97%**	**0.9925**	**0.9857**	**0.9772**	**0.0976**	**0.0002**
ViT-b32	95.76%	0.9974	0.9575	0.9653	0.1623	0.0002
Inception-V3	87.39%	0.9469	0.8689	0.9100	0.2796	0.0013
MobileNet	85.49%	0.9492	0.9058	0.8212	0.3132	0.0022

Table 3. Comparison of this study with previous research using the same dataset.

Author	Classification Method	Evaluation Method	Best Results (ACC.)
Kruk et al. [17]	SVM and RF	K-fold CV	93.8 %
F. Gil et al. [13]	Shallow Ensemble (SVM, RF, Softmax) and deep ensemble including nine deep CNN models	Mean of 10 experiments	98.6%
Proposed	ViT, Inception-V3, and MobileNet	K-fold CV	97.97%

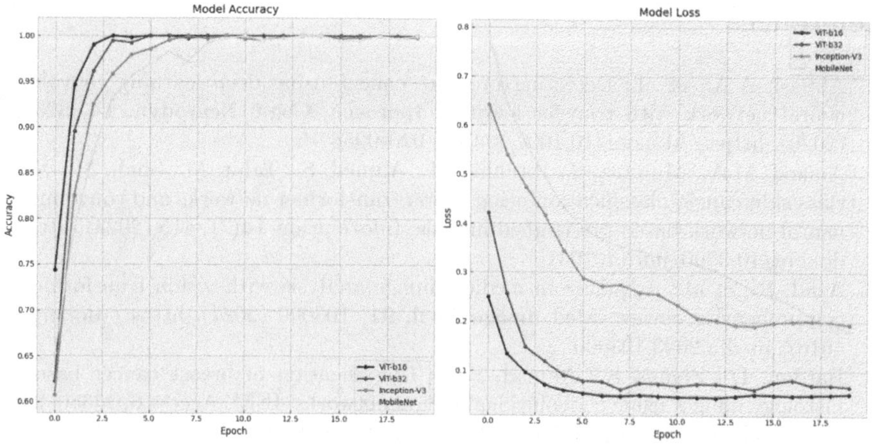

Fig. 2. Accuracy and loss curves with respect to the number of epochs.

6 Conclusions and Future Work

The early and precise detection of melanoma is crucial for improving patient survival rates. In this research, we evaluated the effectiveness of four deep-learning models for melanoma detection using a dataset of melanoma dermoscopic images. We prepared the dataset for this study by meticulously performing three preprocessing steps: normalization, resizing, and augmentation. Additionally, we used the k-fold cross-validation methodology to ensure the robustness of our models' performance. Our results reveal that the Vision Transformer model, ViT-b16, has shown exceptional performance, achieving an accuracy of 97.97% and an AUC of 0.9925. Another variation of Vision Transformer, the ViT-b32 model, also showed acceptable results, with an accuracy of 95.76%, outperforming the traditional CNN models, Inception-V3 and MobileNet, in terms of accuracy and AUC. The accuracy variance of both ViT-b16 and ViT-b32 was notably low, indicating a highly stable and reliable performance. Inception-V3 and MobileNet exhibited slightly higher variances but still demonstrated reasonably consistent performance. These outcomes underscore the potential of Vision Transformers to enhance melanoma detection, surpassing the performance of traditional CNNs, which generally showed lower accuracy and higher loss rates in this study. Future research should focus on further refining ViT models by optimizing and fine-tuning their hyperparameters to enhance precision in melanoma detection. Additionally, an ablation study on the influence of image resizing on model performance would provide valuable insights for optimizing preprocessing steps and improving accuracy. Integrating these advanced models into clinical settings has the potential to revolutionize melanoma diagnostics, thereby improving patient outcomes and care.

References

1. Abbasi, A.A., et al.: Detecting prostate cancer using deep learning convolution neural network with transfer learning approach. Cogn. Neurodyn. **14**, 523–533 (2020). https://doi.org/10.1007/s11571-020-09587-5
2. Arshed, M.A., Mumtaz, S., Ibrahim, M., Ahmed, S., Tahir, M., Shafi, M.: Multi-class skin cancer classification using vision transformer networks and convolutional neural network-based pre-trained models. Information **14**(7), 415 (2023). https://doi.org/10.3390/info14070415
3. Azad, R., et al.: Advances in medical image analysis with vision transformers: a comprehensive review. Med. Image Anal. **91**, 103000 (2024). https://doi.org/10.1016/j.media.2023.103000
4. Bardou, D., Zhang, K., Ahmad, S.M.: Classification of breast cancer based on histology images using convolutional neural networks. IEEE Access **6**, 24680–24693 (2018). https://doi.org/10.1109/ACCESS.2018.2831280
5. Barros, W., Morais, D., Fernandes Lopes, F., Torquato, M., De Melo Barbosa, R., Fernandes, M.: Proposal of the cad system for melanoma detection using reconfigurable computing. Sensors **20**, 3168 (2020). https://doi.org/10.3390/s20113168
6. Belattar, K., Adjadj, M., Bakir, M., Ait Mehdi, M.: A comparative study of CNN architectures for melanoma skin cancer classification. In: ICT Innovations, pp. 74–89 (2022)
7. Brinker, T.J., et al.: Skin cancer classification using convolutional neural networks: systematic review. J. Med. Internet Res. **20**(10), 365–372 (2018). https://doi.org/10.2196/11936
8. Deininger, L., et al.: A comparative study between vision transformers and CNNs in digital pathology. arXiv preprint arXiv:2206.00389 (2022). https://doi.org/10.48550/arXiv.2206.00389
9. Deng, J., Dong, W., Socher, R., Li, L.J., Li, K., Li, F.F.: ImageNet: a large-scale hierarchical image database. In: IEEE Conference on Computer Vision and Pattern Recognition, pp. 248–255 (2009). https://doi.org/10.1109/CVPR.2009.5206848
10. Dosovitskiy, A., et al.: An image is worth 16×16 words: transformers for image recognition at scale. arXiv preprint arXiv:2010.11929 (2020). https://doi.org/10.48550/arXiv.2010.11929
11. Faghihi, A., Fathollahi, M., Rajabi, R.: Diagnosis of skin cancer using VGG16 and VGG19 based transfer learning models. arXiv preprint arXiv:2404.01160 (2024). https://doi.org/10.48550/arXiv.2404.01160
12. Geller, A.C., Swetter, S.M., Brooks, K., Demierre, M.F., Yaroch, A.L.: Screening, early detection, and trends for melanoma: current status (2000–2006) and future directions. J. Am. Acad. Dermatol. **57**(4), 555–572 (2007). https://doi.org/10.1016/j.jaad.2007.06.032
13. Gil, F., Osowski, S., Swiderski, B., Słowińska, M.: Ensemble of classifiers based on deep learning for medical image recognition. Metrol. Meas. Syst. **30**(1), 139–156 (2023). https://doi.org/10.24425/mms.2023.144400
14. Hosseinzadeh Kassani, S., Hosseinzadeh Kassani, P.: A comparative study of deep learning architectures on melanoma detection. Tissue Cell **58**, 76–83 (2019). https://doi.org/10.1016/j.tice.2019.04.009
15. Howard, A., et al.: MobileNets: efficient convolutional neural networks for mobile vision applications. arXiv preprint arXiv:1704.04861 (2017). https://doi.org/10.48550/arXiv.1704.04861

16. Kim, S., Gaibor, E., Haehn, D.: Web-based melanoma detection. arXiv preprint arXiv:2403.14898 (2024). https://doi.org/10.48550/arXiv.2403.14898
17. Kruk, M., Świderski, B., Osowski, S., Kurek, J., Słowińska, M., Walecka, I.: Melanoma recognition using extended set of descriptors and classifiers. EURASIP J. Image Video Process. **2015**(43), 1–10 (2015). https://doi.org/10.1186/s13640-015-0099-9
18. Pu, Q., Xi, Z., Yin, S., Zhao, Z., Zhao, L.: Advantages of transformer and its application for medical image segmentation: a survey. Biomed. Eng. Online **23**, 14 (2024). https://doi.org/10.1186/s12938-024-01212-4
19. Sung, H., et al.: Global cancer statistics 2020: globocan estimates of incidence and mortality worldwide for 36 cancers in 185 countries. CA: Cancer J. Clin. **71**(3), 209–249 (2021). https://doi.org/10.3322/caac.21660
20. Szegedy, C., Vanhoucke, V., Ioffe, S., Shlens, J., Wojna, Z.: Rethinking the inception architecture for computer vision. In: 2016 IEEE Conference on Computer Vision and Pattern Recognition (CVPR), pp. 2818–2826 (2016). https://doi.org/10.1109/CVPR.2016.308
21. Xin, C., et al.: An improved transformer network for skin cancer classification. Comput. Biol. Med. **149**, 105939 (2022). https://doi.org/10.1016/j.compbiomed.2022.105939

A Metaheuristic Optimization Based Deep Feature Selection for Oral Cancer Classification

Akash Halder[1], Sugata Laha[1], Saptarshi Bandyopadhyay[1], Friedhelm Schwenker[2], and Ram Sarkar[1(✉)]

[1] Department of Computer Science and Engineering, Jadavpur University, Kolkata, India
ram.sarkar@jadavpuruniversity.in
[2] Institute of Neural Information Processing, Ulm University, Ulm, Germany

Abstract. Oral cancer is a serious hazard to world health, with many new cases recorded each year. Researchers have been concentrating on developing medical image analysis-based computer-aided diagnostic (CAD) systems for oral cancer. To this end, we propose a novel model that we name Gray Wolf Optimization (GWO) based deep Feature Selection Network (GFS-Net). Initially, we use an attention-aided NASNet Mobile, a convolutional neural network (CNN) architecture, to extract features from the input images. Next, we use a metaheuristic-based optimization algorithm, called GWO, to get rid of the extraneous features obtained from the CNN model. For the final classification task, the K-Nearest-Neighbours (KNN) classifier is applied with this optimal feature set. Our model is evaluated on two publicly accessible oral cancer datasets, histopathologic oral cancer identification dataset and oral cancer (lips and tongue) dataset, that yields classification accuracies of 92.86% and 93.94%, respectively. The code and additional results are available at https://github.com/stellarsb7/GFS-Net.

Keywords: Oral Cancer · Deep Learning · Feature Selection · Gray Wolf Optimization · Attention Mechanism · Medical Image

1 Introduction

Oral cancer represents a significant global health burden, with over 350,000 new cases reported annually worldwide. It encompasses malignant neoplasms that occur in the oral cavity, including the lips, tongue, gums, floor of the mouth, and other intraoral sites. The majority of oral cancers are squamous cell carcinomas, although other histological types such as adenocarcinomas and melanomas can also occur. Risk factors include tobacco use, alcohol consumption, betel quid chewing, human papillomavirus infection [1], and genetic predisposition.

Early detection and accurate diagnosis are crucial for improving patient outcomes in oral cancer. Traditional diagnostic methods often rely on visual examination, biopsy, and histopathological analysis, which can be time-consuming and

C. Y. Suen et al. (Eds.): ANNPR 2024, LNAI 15154, pp. 132–143, 2024.
https://doi.org/10.1007/978-3-031-71602-7_12

subjective. However, advancements in medical imaging technologies and machine learning offer promising avenues for enhancing oral cancer diagnosis and classification. The transition from machine learning to deep learning methods marks a paradigm shift from traditional algorithms that rely on manual feature extraction to neural networks capable of automatic feature learning. This change is driven by advancements in computational power, larger datasets, and improvements in algorithms, enabling deep learning to outperform machine learning in complex tasks. As a result, deep learning has become the dominant approach in many computer vision applications including medical image analysis due to its ability to model intricate patterns and representations directly from raw data. On the other hand, deep feature selection involves identifying and utilizing the most relevant features extracted by deep learning models for a specific task, enhancing interpretability, efficiency, and clinical applicability by focusing on key indicators rather than black-box predictions.Thus to reduce compuation expenses, one can leverage the deep feature selection for deep learning tasks. To this end, we have proposed Gray Wolf Optimization (GWO) based on deep Feature Selection Network (GFS-Net) for oral cancer classification. After feature extraction using a convolutional neural network (CNN) model, GWO is utilized feature selection, ensuring that the most relevant and significant ones are retained for the improved model performance. This combination leverages the deep learning model's ability to capture complex data representations and GWO's efficient search capability to optimize the feature subset. To the best of our knowledge, there is no work done in the field of oral cancer classification that leverages the advantage of metaheuristic algorithms post feature extraction using a CNN.

The **highlighting points** of this paper are as follows:

1. Our proposed GFS-Net uses the NASNet Mobile architecture as the base CNN model for feature extraction from input images. This architecture is combined with a squeeze-and-excitation (SE) based attention module to better focus on the important regions of an input image.
2. To eliminate the redundant features obtained from attention-aided CNN model, the proposed model applies a popular metaheuristic based optimization algorithm, called GWO. This also helps to reduce the computation time and memory usage. The optimized feature set is fed to the K-Nearest Neighbors (KNN) classifier for the final classification task.
3. The proposed model is evaluated on the two oral cancer datasets, namely histopathologic oral cancer identification dataset and oral cancer (lips and tongue) dataset, and yields 92.86% and 93.94% classification accuracies, respectively.

2 Literature Review

Amidst the intricate web of oral cancer's diagnostic challenges, deep learning emerges as a beacon of hope, promising enhanced detection and prognosis capabilities. Shephard et al. [2] introduce a deep learning framework to simultaneously segment and classify nuclei and (intra-)epithelial layers from oral cancer-based H & E stained slides, consisting of an encoder branch and four decoder branches for simultaneous instance segmentation of nuclei and semantic segmentation of the epithelial layers. Dos et al. [3] propose a method that uses color features in the HSV color model to identify tissue regions in a pre-processing step to remove background and non-relevant areas. The identified tissue regions are then transformed into the Commission Internationale de l'Eclairage $l \times \alpha \times \beta$ color model and split into image-patches. Bansal et al. [4] use different pre-trained backbones like VGG16, MobileNetV2, DenseNet, ResNet50, DenseNet and VGG19 and optimizers like Adam, SGD and RMSprop. Welikala et al. [5] combine bounding box annotations. Then Faster R-CNN is used for detection and ResNet101 is used for object classification improving detection accuracy and automation in oral cancer diagnostics.

Matias et al. [6] compare three different deep learning approaches: segmentation, object detection, and image classification. The results show that the binary object detection with Faster R-CNN is the best approach for nuclei detection and localization,enhancing accuracy in oral cancer diagnostics. Folmsbee et al. [7] use active learning, a process for iterative interactive training, to create a modified U-Net classifier on the region of interest scale, which is then compared with random learning, where images for addition to the dataset for retraining are randomly selected.The significance of this work lies in its potential to improve the accuracy and efficiency of oral cancer detection. By utilizing active learning, the classifier can intelligently select the most informative images for training, thereby reducing the need for manual annotation and potentially enhancing the model's performance.

GWO is a popular metaheuristic algorithm, which is computationally efficient due to its simple structure, and it requires fewer control parameters compared to other optimization algorithms. GWO demonstrates strong exploration and exploitation capabilities, effectively balancing global and local searches to avoid premature convergence and improve solution accuracy. Additionally, it has been successfully applied across various complex optimization problems, showcasing its robustness and versatility in different fields such as medical imaging segmentation, classification, stock market prediction, feature selection etc. Albadr et al. [8] propose a hybrid method combining GWO and extreme learning machine (ELM) to improve the classification accuracy of diabetic retinopathy from fundus images in the field of medical imaging.

Sujithra et al. [9] propose an innovative method for detecting glaucoma in fundus images, employing GWO to enhance the performance of the ResNet50 model. Zhao et al. [10] integrates Enhanced Nonlinear Bayesian Learning System (ENBLS) with GWO, the method aims to enhance the accuracy and efficiency of performance trend prediction, potentially offering significant improvements in forecasting capabilities for various applications. Jordehi et al. [11] employs GWO to find the most efficient schedule for operating household appliances, minimizing energy costs or maximizing energy efficiency while considering user preferences and constraints.

3 Datasets

Here, we describe the macroscopic and microscopic datasets used for training and evaluating our model. The histopathologic oral cancer identification dataset [12] comprises high-resolution images of oral cancer lesions obtained through microscopy.

The oral cancer (lips and tongue) dataset [13] consists of images of oral cancer lesions captured through macroscopic imaging techniques. Comprising of 87 cancerous and 44 non-cancerous images, a small dataset size posed a significant challenge. To address this limitation, a 75:25 split is applied, augmenting the training set to increase its sample count. Post-augmentation, the training set is expanded to 4528 images, with 2273 categorized as cancerous and 2255 as non-cancerous, while the test set remains at 33 images, comprising 22 cancerous and 11 non-cancerous images. The distribution of the samples within the dataset is shown in Table 1.

Table 1. Distribution of the images in the microscopic and macroscopic dataset

Dataset type	Image type	Training set	Validation set	Test set
Microscopic	Normal	2435	28	31
	Cancerous	2511	92	95
	Total	4946	123	126
Macroscopic	Normal	2255	–	11
	Cancerous	2273	–	22
	Total	4528	–	33

4 Methodology

Here, we describe the model architecture and GWO algorithm we use. The overall pipeline of our proposed method is shown in Fig. 1.

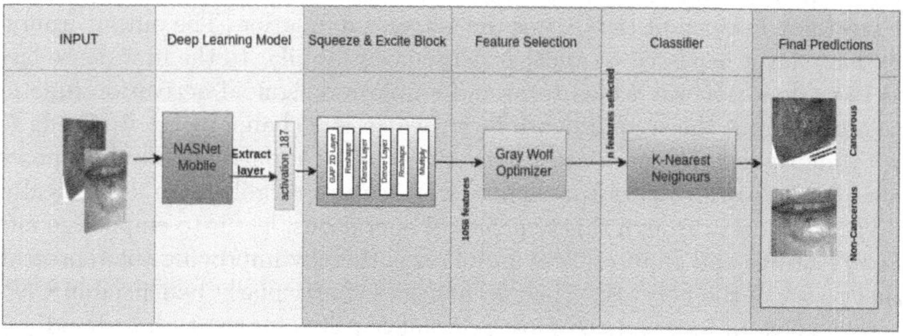

Fig. 1. Block diagram of the proposed GFS-Net used for oral cancer classification

4.1 NASNet Mobile Architecture

Neural Architecture Search (NAS) [14] is a recent technique in the field of DL. It was proposed by the Google brain team in 2016 and has three constituents: search space, search strategy and performance estimation. Search space involves searching for convoluted performances, fully-connected, max-pooling, etc., and then checking the connection between the layers through which complete feasible network architectures are formed. The search strategy involves the use of random search and reinforced learning to sample the population of network architecture candidates by receiving child model performance rewards (maximum accuracy, time management). The architecture of the NASNet Mobile is shown in Fig. 2.

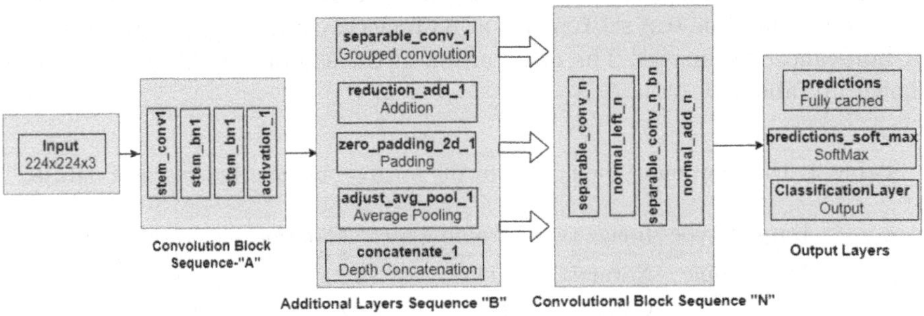

Fig. 2. An illustration of the NASNet Mobile model [15]

4.2 Squeeze-and-Excitation Block

The SE block describes a squeeze and excitation block, designed to enhance the representational capacity of the network. The feature map is first processed through a GlobalAveragePooling2D (GAP) layer. This operation computes the spatial average for each channel in the input feature map, effectively squeezing the spatial dimensions to 1×1 while retaining the depth (channel-wise information). The squeezed output is then fed into a dense (fully connected) layer. This layer reduces the channel dimensionality. This layer does not use a bias term and uses rectified linear unit (ReLU) as the activation function. The output from the first dense layer is passed to another dense layer. Similar to the first dense layer, this layer does not use a bias term and employs a sigmoid activation function. The output from the second dense layer, which is a channel-wise weighting factor, is multiplied (element-wise) with the original input. This scaling process allows the network to perform channel-wise feature recalibration dynamically.

The SE block, through this sequence of operations, learns to emphasize informative features and suppress less useful ones, thereby improving the representational power of the network. This mechanism is particularly beneficial in CNNs, where it helps to focus on the most relevant features for the task at hand.

$$x_{final} = \sigma(Dense(ReLU(Dense(GAP(x_{input})))) \times x_{input} \tag{1}$$

4.3 Gray Wolf Optimizer

GWO [16–19] is a metaheuristic optimization algorithm that emulates the natural social order and hunting behaviour observed in gray wolves. The algorithm is imported from the Py_FS library [20] This technique uses four classifications of a pack of wolves - alpha, beta, gamma and delta wolves to realize the social order. Let the position vector of wolf w_1 at time t be $\overrightarrow{d}_{w_1}(t)$ If we knew the position of the prey in space at time t (which in this context denotes the vector with the best fitness), let us call it say $\overrightarrow{d_p}(t)$, then the distance vector of the wolf w_1 from the prey with respect to coefficient vector $\overrightarrow{m}$ is given by

$$\overrightarrow{D}_{w_1,p} = |\overrightarrow{m}.\overrightarrow{d}_p(t) - \overrightarrow{d}_{w_1}(t)|$$

and the position vector of wolf w_1 at the next instant of time t+1, $\overrightarrow{d}_{w_1}(t+1)$ will be calculated as

$$\overrightarrow{d}_{w_1}(t+1) = \overrightarrow{d}_p(t) - \overrightarrow{n}.\overrightarrow{D}_{w_1,p}$$

Here, $\overrightarrow{m}$ and $\overrightarrow{n}$ are coefficient vectors for updating the position of the wolf, and are given by

$$\overrightarrow{m} = 2.\overrightarrow{p_1}$$
$$\overrightarrow{n} = 2.\overrightarrow{z}.\overrightarrow{p_2}$$

where $\overrightarrow{p_1}$ and $\overrightarrow{p_2}$ are vectors chosen randomly (with magnitudes lying between 0 and 1) to widen the search space of *exploring* for better fitting result and $\overrightarrow{z}$ is a vector with coefficients that follow a linear diminution from 2 to 0 to reduce the search space locally and *exploit* the region and maximise fitness there.

However, here we do not know a pre-determined location of the prey to approach it and encircle it. Instead, we want to find the vectors (wolves) with the best fitness for which we use the current position of the alpha, beta and delta wolves to update the positions of all the wolves in the pack. Now, among the wolves in its new positions, choose the wolf having the best fitted position vector as the alpha of the pack and the next best fitted wolf as the beta and the next as the delta.

Say we take a pack of 4 wolves- alpha, beta, delta and another subordinate wolf. At time t say i^{th} member of the pack is at position $\overrightarrow{d}_{w_i}(t)$ then the distance of the ith wolf from the alpha, $\overrightarrow{D}_{w_1,\alpha}$, from the beta $\overrightarrow{D}_{w_1,\beta}$, from the delta $\overrightarrow{D}_{w_1,\delta}$ are given by

$$\overrightarrow{D}_{w_i,\alpha} = |\overrightarrow{m}_{w_i,\alpha}.\overrightarrow{d}_\alpha(t) - \overrightarrow{d}_{w_i}(t)|$$
$$\overrightarrow{D}_{w_i,\beta} = |\overrightarrow{m}_{w_i,\beta}.\overrightarrow{d}_\beta(t) - \overrightarrow{d}_{w_i}(t)|$$
$$\overrightarrow{D}_{w_i,\delta} = |\overrightarrow{m}_{w_i,\delta}.\overrightarrow{d}_\delta(t) - \overrightarrow{d}_{w_i}(t)|$$

where $\overrightarrow{m}_{w_i,x}$ denotes the coefficient vector $\overrightarrow{m}$ of the ith wolf with respect to wolf x and $\overrightarrow{d}_x(t)$ denotes the position vector of wolf x at time instant t.

Now the position of the ith wolf to be updated according to the alpha, $\vec{d}_{w_i,\alpha}$, according to beta, $\vec{d}_{w_i,\beta}$, and according to delta $\vec{d}_{w_i,\delta}$, are given as

$$\vec{d}_{w_i,\alpha} = \vec{d_\alpha}(t) - \vec{n}_{w_i,\alpha}.\vec{D}_{w_i,\alpha}$$
$$\vec{d}_{w_i,\beta} = \vec{d_\beta}(t) - \vec{n}_{w_i,\beta}.\vec{D}_{w_i,\beta}$$
$$\vec{d}_{w_i,\delta} = \vec{d_\delta}(t) - \vec{n}_{w_i,\delta}.\vec{D}_{w_i,\delta}$$

where $\vec{n}_{w_i,x}$ denotes the other coefficient vector $\vec{n}$ of the ith wolf with respect to wolf x.

Now the position of the ith wolf at time instant t+1 is updated as the mean of the position vectors as calculated to be updated according to the alpha, beta and delta wolves.

$$\vec{d}_{w_i}(t+1) = \frac{\vec{d}_{w_i,\alpha} + \vec{d}_{w_i,\beta} + \vec{d}_{w_i,\delta}}{3}$$

This is done for the entire pack of wolves and the best three solutions with highest fitness are set as alpha, beta and delta, in decreasing order of fitness, are to be used for calculating the position vectors of the wolves in the next iteration (instant of time). After a fixed number of iterations, the best wolf is considered as solution i.e., the optimal subset of features. As feature selection is a binary optimization problem, we use a sigmoid transfer function to convert these continuous values to the discrete values, where 1 represents a selected feature and 0 represents a discarded feature.

4.4 K-Nearest Neighbors

After selecting the features required for classification of the images, the binary string thus generated is passed to the KNN [21] classifier. The KNN algorithm is a type of instance-based learning, where the algorithm does not explicitly learn a model but instead makes predictions based on the stored training data. It being a non-parametric algorithm, stores the feature vectors and their labels only. While encountering a new test data, it calculates distances between new data point and all other points already classified. After calculating the distances, k training samples closest to the data point are found out and the point is classified under the label which is of highest frequency among all those neighbors.

5 Results and Discussion

The results presented in this section highlight the efficacy of the proposed model in oral cancer classification from microscopic and macroscopic images.

5.1 System Configuration

A Jupyter Notebook with a 12 GB NVIDIA Tesla T4 GPU in Google's collaborative environment is used to run the whole set of experiments. Tensorflow, Keras, Matplotlib, Scikit, Numpy, and Pandas are the main open-source modules that have been used for implementing the suggested method in Python.

5.2 Evaluation Metrics

In order to make a quantitative evaluation of our results, we have considered standard metrics like precision, recall, accuracy and F1 score [4].

5.3 Results

We have trained and evaluated our model on the histopathologic oral cancer identification dataset and the oral cancer (lips and tongue) dataset. We have experimented with and without SE attention module as well as with and without GWO. Results on the histopathologic oral cancer identification dataset and the oral cancer (lips and tongue) dataset is shown in Table 2. The performance graphs for the microscopic and macroscopic datasets are shown in Fig. 3. Confusion matrices obtained are shown in Fig. 4.

Table 2. Results on histopathologic oral cancer identification and oral cancer (lips and tongue) datasets

Dataset	GWO used	Attention	Accuracy	Recall	Precision	F1 Score	Features
Microscopic	No	No	0.8889	0.8889	0.8889	0.8469	1056
Microscopic	No	Yes	0.8968	0.8968	0.8968	0.8526	1056
Microscopic	Yes	No	0.8968	0.9474	0.9184	0.9326	**606**
Microscopic	Yes	Yes	**0.9286**	**0.9684**	**0.9387**	**0.9534**	613
Macroscopic	No	No	0.9091	0.9394	0.9118	0.9283	1056
Macroscopic	No	Yes	0.9394	0.9394	**0.9394**	**0.9394**	1056
Macroscopic	Yes	No	0.9091	0.9017	0.8968	0.8986	608
Macroscopic	Yes	Yes	**0.9394**	0.9324	0.9286	0.9298	**601**

From Table 2, it is clear that the accuracy, precision, recall and F1 Score have improved compared to NASNet Mobile model after applying the SE attention module at the last Global Average Pooling (GAP) layer. Also after applying GWO, it has been observed that although the accuracy remains almost same, the number of features has reduced significantly as most of the features are redundant which have not impact the accuracy.

We have also implemented different optimization algorithms apart from GWO like Binary Bat Algorithm (BBA), Genetic Algorithm (GA), Cuckoo Search (CS), and Gravitational Search Algorithm (GSA). Table 4 is shown the results (Table 3).

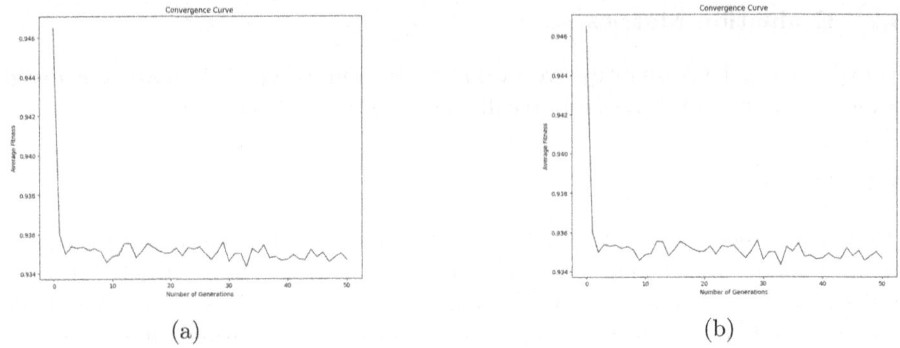

Fig. 3. Convergence graphs of GWO on (a) macroscopic and (b) microscopic datasets.

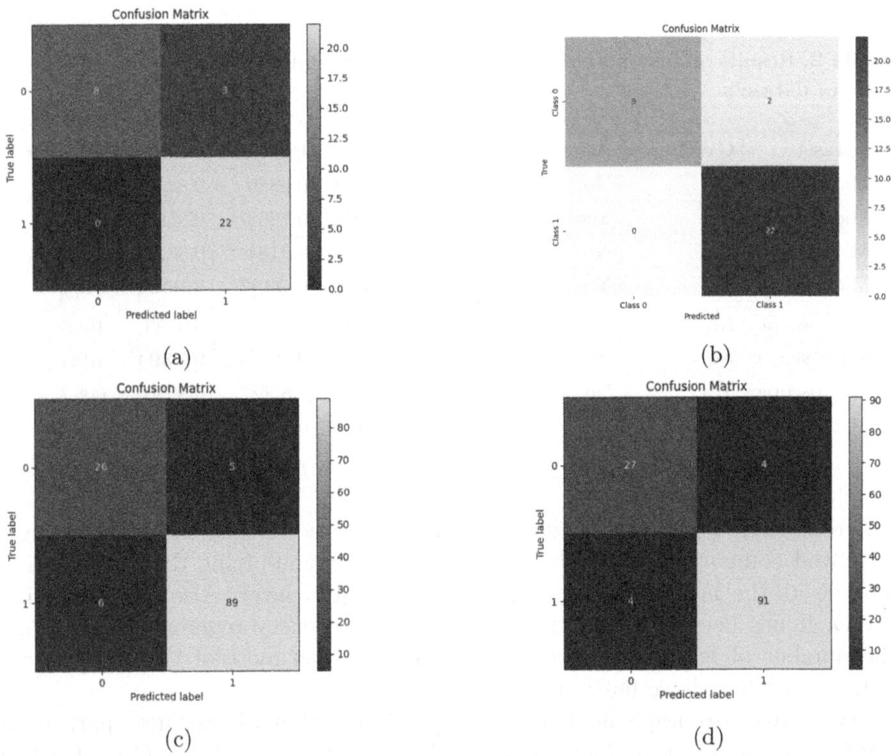

Fig. 4. Confusion matrix for the macroscopic dataset (a) without SE attention and (b) with SE attention, and for the microscopic dataset (c) without SE attention and (d) with SE attention.

Table 3. Comparison with other optimization algorithms on the histopathologic oral cancer identification dataset

Algorithm	Accuracy	F1 Score	No. of Features
BBA	0.8810	0.9223	325
GA	0.8968	0.9319	330
CS	0.9127	0.9436	330
GSA	0.8968	0.9319	542
GWO	0.9286	0.9534	613

Table 4. Comparison with other optimization algorithms on the oral cancer (lips and tongue) dataset

Algorithm	Accuracy	F1 Score	No. of Features
BBA	**0.9394**	**0.9565**	332
GA	**0.9394**	0.9565	**330**
GSA	0.8788	0.8754	542
CS	0.8788	0.8754	**330**
GWO	**0.9394**	0.9298	601

5.4 GradCAM Analysis

GradCAM (Gradient-weighted Class Activation Mapping) is a popular technique used to visualize the regions of input data that are most relevant to a model's decision-making process. In the context of oral histopathology and in oral cancer detection from macroscopic images of lips and tongue, where models are used to identify and classify carcinogenic conditions from medical images and macroscopic images, understanding the model's focus can provide critical insights into its reliability and decision-making process. As seen from Fig. 5, by comparing GradCAM heatmaps from models with and without attention mechanisms on both the datasets, it is observed that attention mechanism helps in highlighting regions that are more diagnostically relevant, whereas models without attention might focus on less pertinent regions, potentially leading to false positives.

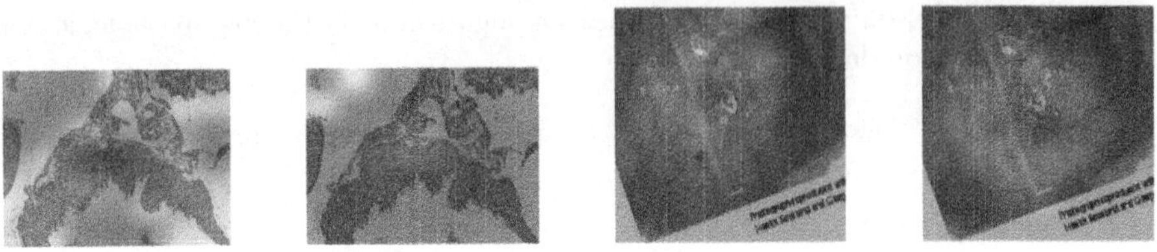

Fig. 5. GradCAM visualization of images for oral cancer detection by our model for microscopic dataset (a) with attention and (b) without attention, and for macroscopic dataset (c) with attention, and (d) without attention.

6 Conclusions and Future Work

In this study, we have put forth a new model for oral cancer classification, called GFS-Net. Initially, to extract features from the images, we have employed NAS-Net Mobile architecture combined with a SE attention module. Next, we have excluded the unimportant features generated from the CNN model using a meta-heuristic algorithm called GWO, and fed this optimal feature set to the KNN classifier for classification. Our model achieves 92.86% and 93.94% accuracies on histopathologic oral cancer identification and oral cancer (lips and tongue) datasets, respectively. Though the results obtained by our model are satisfactory, in order to make this model usable in a practical scenario, we have to reduce the false positive and false negative cases. In the future, we may use better attention module with explainable-AI to extract more informative features from the inputs. We will also try to apply some lightweight base CNN model in order to deploy the final model in a resource-constrained environment.

References

1. Münger, K., Howley, P.M.: Human papillomavirus immortalization and transformation functions. Virus Res. **89**(2), 213–228 (2002)
2. Shephard, A.J., et al.: Simultaneous nuclear instance and layer segmentation in oral epithelial dysplasia. In: Proceedings of the IEEE/CVF International Conference on Computer Vision, pp. 552–561 (2021)
3. dos Santos, D.F., de Faria, P.R., Travencolo, B.A., do Nascimento, M.Z.: Automated detection of tumor regions from oral histological whole slide images using fully convolutional neural networks. Biomed. Signal Process. Control **69**, 102921 (2021)
4. Bansal, K., Bathla, R.K., Kumar, Y.: Deep transfer learning techniques with hybrid optimization in early prediction and diagnosis of different types of oral cancer. Soft Comput. **26**(21), 11153–11184 (2022)
5. Welikala, R.A., et al.: Automated detection and classification of oral lesions using deep learning for early detection of oral cancer. IEEE Access **8**, 132677–132693 (2020)
6. Matias, A.V., Cerentini, A., Macarini, L.A.B., Amorim, J.G.A., Daltoé, F.P., von Wangenheim, A.: Segmentation, detection, and classification of cell nuclei on oral cytology samples stained with papanicolaou. SN Comput. Sci. **2**(4), 285 (2021)

7. Folmsbee, J., et al.: Histology segmentation using active learning on regions of interest in oral cavity squamous cell carcinoma. J. Pathol. Inform. **13**, 100146 (2022)
8. Albadr, M.A.A., Ayob, M., Tiun, S., Al-Dhief, F.T., Hasan, M.K.: Gray wolf optimization-extreme learning machine approach for diabetic retinopathy detection. Front. Public Health **10**, 925901 (2022)
9. Sujithra, B.S., Jerome, S.A..: Identification of glaucoma in fundus images utilizing gray wolf optimization with deep convolutional neural network-based resnet50 model. Multimed. Tools Appl. 1–19 (2023)
10. Zhao, H., Zhang, P., Zhang, R., Yao, R., Deng, W.: A novel performance trend prediction approach using ENBLS with GWO. Meas. Sci. Technol. **34**(2), 025018 (2022)
11. Jordehi, A.R.: Optimal scheduling of home appliances in home energy management systems using grey wolf optimisation (GWO) algorithm. In: 2019 IEEE Milan PowerTech, pp. 1–6. IEEE (2019)
12. Histopathologic oral cancer identification dataset, 2021. https://www.kaggle.com/datasets/ashenafifasilkebede/dataset
13. Barot, S.: Oral cancer (lips and tongue) dataset, 2020. https://www.kaggle.com/datasets/shivam17299/oral-cancer-lips-and-tongue-images
14. Zoph, B., Vasudevan, V., Shlens, J., Le, Q.V.: Learning transferable architectures for scalable image recognition. In: Proceedings of the IEEE Conference on Computer Vision and Pattern Recognition, pp. 8697–8710 (2018)
15. Shah, F.A., et al.: A cascaded design of best features selection for fruit diseases recognition. Comput. Mater. Contin. **70**(1), 1491–1507 (2022)
16. Mirjalili, S., Mirjalili, S.M., Lewis, A.: Grey wolf optimizer. Adv. Eng. Softw. **69**, 46–61 (2014)
17. Dhargupta, S., Ghosh, M., Mirjalili, S., Sarkar, R.: Selective opposition based grey wolf optimization. Expert Syst. Appl. **151**, 113389 (2020)
18. Malakar, S., Ghosh, M., Chaterjee, A., Bhowmik, S., Sarkar, R.: Offline music symbol recognition using daisy feature and quantum grey wolf optimization based feature selection. Multimed. Tools Appl. **79**, 32011–32036 (2020)
19. Pramanik, R., Pramanik, P., Sarkar, R.: Breast cancer detection in thermograms using a hybrid of GA and GWO based deep feature selection method. Expert Syst. Appl. **219**, 119643 (2023)
20. Guha, R., Chatterjee, B., Khalid Hassan, S.K., Ahmed, S., Bhattacharyya, T., Sarkar, R.: Py_FS: a python package for feature selection using meta-heuristic optimization algorithms. In: Das, A.K., Nayak, J., Naik, B., Dutta, S., Pelusi, D. (eds.) Computational Intelligence in Pattern Recognition. AISC, vol. 1349, pp. 495–504. Springer, Singapore (2022). https://doi.org/10.1007/978-981-16-2543-5_42
21. Fix, E.: Discriminatory analysis: nonparametric discrimination, consistency properties, vol. 1. USAF school of Aviation Medicine, 1985

Machine Learning for Clinical Score Prediction from Longitudinal Dataset: A Case Study on Parkinson's Disease

Nourin Ahmed(✉) and Ziad Kobti

School of Computer Science, University of Windsor, Windsor, ON N9B 3P4, Canada
{ahmed7k,kobti}@uwindsor.ca

Abstract. Accurate prediction of Parkinson's disease (PD) progression is vital for personalized treatment and effective clinical trials. This study presents a machine learning approach to predict the Movement Disorder Society-Unified Parkinson's Disease Rating Scale (MDS-UPDRS) Part III scores, quantifying motor symptom progression in PD patients. Using the longitudinal Parkinson's Progression Markers Initiative (PPMI) dataset, we examined the impact of dataset format (wide vs. cross-sectional), dimensionality reduction techniques (PCA, NMF), and regression models (Linear Regression, Random Forest, XGBoost, SVR) on prediction performance. Our findings indicate that models trained on wide-format datasets consistently outperformed those on cross-sectional data. The combination of Nonnegative Matrix Factorization (NMF) and Support Vector Regression (SVR) achieved the best performance, with a mean absolute error (MAE) of 1.91 and R^2 of 0.83. These results underscore the importance of data arrangement and highlight NMF's effectiveness in feature extraction for longitudinal datasets.

Keywords: Parkinson's Disease · Machine Learning · Longitudinal Data · Dimensionality Reduction · NMF · SVR · Prediction Models · MDS-UPDRS III

1 Introduction

Parkinson's disease (PD) is the second most common neurodegenerative disorder primarily affecting the elderly, typically presenting symptoms around age 60 [1]. It is characterized by neuronal loss in the substantia nigra, leading to motor and non-motor symptoms such as tremors, stiffness, bradykinesia, balance issues, and autonomic dysfunction [3–6].

Currently, there is no cure for PD, but early detection and treatment with levodopa [7] or dopaminergic agonists [8] can alleviate symptoms and improve quality of life [9,10]. The Unified Parkinson's Disease Rating Scale (UPDRS) is a key tool for assessing PD severity and progression [11]. It includes four parts: non-motor experiences of daily living, motor experiences of daily living, motor

C. Y. Suen et al. (Eds.): ANNPR 2024, LNAI 15154, pp. 144–155, 2024.
https://doi.org/10.1007/978-3-031-71602-7_13

examination (MDS-UPDRS III), and motor complications [12,13]. The MDS-UPDRS III is particularly reliable for measuring motor function and symptom development in early, untreated PD patients [14,15]. Higher MDS-UPDRS III scores correlate with lower quality of life and greater motor function decline [16].

The Parkinson's Progression Markers Initiative (PPMI) offers a comprehensive longitudinal dataset with clinical, imaging, and biological data from 1,500 participants over a decade [17,19]. This dataset is rich in potential but presents challenges such as high dimensionality, which can lead to overfitting and data sparsity [20]. Traditional statistical models often struggle with this complexity, whereas ML models are better suited to handle it and are more robust to missing data [23].

This study proposes a machine learning methodology to predict MDS-UPDRS III scores using the PPMI dataset. The remainder of this paper is structured as follows. Section 2 reviews related research. Section 3 describes the proposed framework. Section 4 details the experimental setup and results. Section 5 discusses the conclusion and future work.

2 Related Work

In PD patients, machine learning (ML) and deep learning (DL) algorithms have been increasingly employed to predict clinical scores, such as the UPDRS. Exley et al. (2022) [1] utilized Random Forest (RF) to predict UPDRS-III subscores related to the intensity of movement disorders, focusing on feature importance. However, they did not use ensemble models or compare their results with other models. Nilashi et al. (2022) [24] employed Support Vector Regression (SVR) ensembles and various clustering techniques to predict Motor-UPDRS and Total-UPDRS scores, demonstrating that the combination of expectation-maximization and SVR ensembles outperformed other methods. However, they did not propose any novel methods or use feature engineering techniques. Arias-Londoño and Gómez-García (2020) [11] investigated DNN and CNN models with transfer learning (TL) methodologies, using a multisensory technique that integrated speech signals and modulation spectra transformations, showing the potential of deep learning and transfer learning in predicting UPDRS scores.

Several studies have utilized ML models such as RF, k-Nearest Neighbors (kNN), Logistic Regression (LR), Support Vector Machines (SVM), and Classification and Regression Trees (CART) for UPDRS score prediction [25–31], as well as DL models such as Deep Belief Networks (DBN), Deep Neural Networks (DNN), Convolutional Neural Networks (CNN), and Long Short-Term Memory (LSTM) networks [32–37]. Deep learning-based approaches have demonstrated significant improvements in outcome prediction by discovering patterns in images [39].

In addition to identifying individual novel PD biomarkers, using machine learning and data mining to build predictive models of PD progression with multiple biomarkers as inputs are promising [18]. Feature selection, performed manually or automatically using feature subset selector algorithms (FSSAs), can improve prediction accuracy and enable algorithms to work with many input features [38]. Building on previous findings, this study aims to:

- Perform comprehensive experiments on longitudinal dataset arrangements' influence on MDS-UPDRS III score prediction performance.
- Propose a robust machine learning pipeline that prevents data leakage and employs rigorous validation techniques.
- Compare the performance of NMF and PCA feature reduction techniques in combination with Linear Regression, Random Forest, SVR, and XGBoost models using metrics such as MAE, RMSE, and coefficient of determination (R^2).

3 Proposed Method

3.1 Methodology Overview

This section outlines our comprehensive framework, as detailed in 1, for predicting MDS-UPDRS-III scores in Parkinson's (PD) patients using machine learning techniques. Each step is detailed in the next subsections to ensure the study is reproducible.

3.2 Dataset

We utilized data from the Parkinson's Progression Markers Initiative (PPMI), which includes clinical, imaging, and biological data from 423 de novo PD patients and 196 healthy controls, followed for up to five years. PD patients were included if they were diagnosed with PD for two years or less at the time of enrollment, were not on PD medication, and showed reduced dopamine transporter function as indicated by DaTscan imaging.

After ensuring complete data for 60 months and comprehensive phenotypic information, the final dataset included 294 PD cases and 154 controls. The mean age at screening was 61 ± 9.7 years for PD cases and 60.3 ± 11 years for controls. We excluded participants from other concurrent studies to avoid confounding variables. We extracted a total of 122 longitudinally collected clinical data elements from six visits, including: (i) MDS-UPDRS Part III, (ii) Cranial Nerve Examination, (iii) Montreal Cognitive Assessment, (iv) Hopkins Verbal Learning Test, (v) Semantic Fluency test, (vi) WAIS-III Letter-Number Sequencing Test, (vii) Judgment of Line Orientation Test, (viii) Symbol Digit Modalities Test, (ix) SCOPA-AUT, (x) State-Trait Anxiety Inventory for Adults, (xi) Geriatric Depression Scale, (xii) Questionnaire for Impulsive-Compulsive Disorders in Parkinson's Disease, (xiii) REM-Sleep Behavior Disorder Screening Questionnaire, (xiv) Epworth Sleepiness Scale, (xv) Biological and Genetic Features: Genotypes using imputed Illumina NeuroX array, vital signs, serum, CSF, and urine measurements, (xvi) Patient Characteristics: Height, Weight, Blood pressure, Demographic details, and (xvii) Biological Biomarkers: Alpha-synuclein, Total tau protein, β-amyloid 1–42 (Aβ42), phospho-tau181 (p-Tau181) in CSF, serum neurofilament light (NfL), and urine levels.

Algorithm 1 MDS-UPDRS-III Prediction

```
procedure ParkinsonsScorePrediction(data)
    Step 1: Data Processing
    data ← VectorizeData(ImputeMissingValues(TransformToUniformDirection(data)))
    Step 2: Dataset Arrangements
    wide_format_datasets ← CreateWideFormats(data)    ▷ Returns datasets like BL, BL+V04, etc.
    cross_sectional_datasets ← CreateCrossSectionalFormats(data)    ▷ Returns datasets like BL, V04, etc.
    all_datasets ← wide_format_datasets + cross_sectional_datasets
    for each dataset in all_datasets do
        Step 3: Target Variable Separation and Preprocessing
        (X, y) ← SeparateTargetVariable(dataset, "MDS-UPDRS-III")
        (X_train, X_test, y_train, y_test) ← TrainTestSplit(X, y)
        (X_train, X_test) ← MinMaxNormalize(X_train, X_test)
        Step 4: Dimensionality Reduction
        reduced_datasets ← ApplyDimensionalityReductions(X_train, X_test)
        for each (X_train_reduced, X_test_reduced) in reduced_datasets do
            Step 5: Model Training and Evaluation
            models ← {LinearRegression(), RandomForest(), XGBoost(), SVM()}
            for each model in models do
                tuned_model ← HyperParamTune(model, X_train_reduced, y_train)
                results ← EvaluateModel(tuned_model, X_test_reduced, y_test, metrics=[MAE, RMSE, R^2])
                ReportResults(results)
            end for
        end for
    end for
end procedure
```

3.3 Data Processing

We started by standardizing the directionality of clinical features using two-sample, one-tailed t-tests ($p < 0.05$), and manual confirmation to ensure that higher values uniformly indicated poorer outcomes. Next, we clipped feature values to the 2nd and 98th percentiles to mitigate the effects of extreme outliers. We then applied linear interpolation to impute missing values for features with less than 5% missing data, ensuring continuity across longitudinal data points. Finally, the dataset was transformed into a vectorized format, with each clinical feature represented as a column vector.

3.4 Dataset Arrangements

For comprehensive analysis and machine learning modeling, we organized the PPMI dataset into two distinct formats:

Wide Format Datasets: To illustrate disease progression, we constructed sequential wide-format datasets incorporating increasingly comprehensive visit data, starting with baseline (BL) and adding successive visits (V04, V06, V08, V10, V12).

Cross-Sectional Format: This format groups data by unique patients, with columns representing features from specific visits (BL, V04, V06, V08, V10, V12). It is tailored for analyses focusing on individual time points.

3.5 Target Variable and Dataset Splitting

The target MDS-UPDRS-III score was computed by summing up the relevant features of the score as available in the dataset. Following this, we employed 5-fold cross-validation to ensure robust model evaluation and prevent overfitting. Within each cross-validation fold, the data was split into training (80%) and testing (20%) sets using stratified sampling to maintain the distribution of the target variable.

We applied min-max normalization to scale the feature values to a [0, 1] range, which is vital for effective model performance, especially for algorithms like Non-negative Matrix Factorization (NMF). Normalization parameters were determined from the training set to avoid data leakage and were applied consistently across training and test datasets.

3.6 Dimensionality Reduction Algorithms (DRA)

Principal Component Analysis (PCA) reduces a high-dimensional dataset to a lower-dimensional subspace while preserving maximum variance. It constructs a projection matrix $\mathbf{W}$ that maps data matrix $\mathbf{X}$ onto a k-dimensional subspace ($k < p$), minimizing reconstruction error.

Non-negative Matrix Factorization (NMF) decomposes a non-negative matrix X into two smaller non-negative matrices U and V, such that:

$$X \approx U \times V$$

Here, U contains latent features, and V represents coefficients or activations. NMF minimizes the difference between X and $U \times V$ through iterative updates of U and V.

The transformed data X_{nmf} is given by:

$$X_{nmf} = U \times V$$

3.7 Machine Learning Algorithms

Linear regression models the relationship between the input features and the target variable as a linear equation. The algorithm finds the best-fitting line by minimizing the sum of squared residuals between the predicted values $\hat{y}_i$ and the actual values y_i. Although it cannot capture the complex non-linear relationships in datasets, but it is a good baseline to compare to while developing more complex models.

The Random Forest Regression algorithm is one of the robust machine-learning regression algorithms. It is an ensemble of multiple decision trees where each of the decision trees is independently trained on a random subset of training data and features. Each tree model $f(x_i)$ is defined as:

$$\hat{y}_i = f(x_i) + \epsilon_i \tag{1}$$

The final prediction is calculated by averaging the outputs of all the trees. Random forest algorithm is suitable for high-dimensional datasets because it can capture non-linear relationships, handle high-dimensional data, and offer feature importance measures that aid in selecting the most informative features.

The Support Vector Regression (SVR) tries to find a hyperplane that fits the data points with the smallest margin of error. Given the training data (x_i, y_i), where x_i are the independent variables and y_i is the dependent variable, SVR aims to find an optimal function $f(x)$ such that:

$$f(x) = \sum_{i=1}^{N} (\alpha_i - \alpha_i^*) K(x_i, x_j) + b \tag{2}$$

where α_i and α_i^* are the Lagrange multipliers, b is a constant, and K is the kernel function. The ϵ-insensitive loss function is used, which ignores errors within a certain threshold. SVR can handle nonlinear relationships by employing kernel functions to transform the input space into a higher-dimensional one where a linear separation becomes possible.

Extreme Gradient Boosting or XGBoost sequentially builds an ensemble of decision trees where each subsequent trees minimize the errors made by the previous trees. The model can be expressed as:

$$\hat{y}_i = \sum_{j=1}^{N} f_j(x_i) \tag{3}$$

XGBoost is known for its excellent predictive accuracy, speed, scalability, and ability to handle high-dimensional datasets efficiently.

The details of all hyperparameters we used for each of the machine learning models are listed in Table 1 to facilitate the reproducing of our work.

Table 1. ML models and their hyperparameters used for tuning.

Model	Hyperparameters
Support Vector Regression (SVR)	C: {0.1, 1.0, 10.0} kernel: {'linear', 'rbf'}
Random Forest (RF)	n_estimators: {50, 100, 200, 300, 500} max_features: {'auto', 'sqrt', 'log2'} max_depth: {None, 10, 20, 30, 40, 50}
XGBoost (XGBOOST)	n_estimators: {50, 100, 200} learning_rate: {0.01, 0.1, 0.2}

3.8 Evaluation Metrics

We evaluated model performance using three metrics: mean absolute error (MAE), root mean squared error (RMSE), and coefficient of determination (R^2). All experiments were conducted 10 times, and average values were reported to account for variability.

MAE measures the average absolute difference between predicted and actual values. Lower MAE indicates better accuracy.

RMSE is the square root of the average of squared differences between predicted and actual values. RMSE penalizes larger errors more heavily. Lower RMSE indicates better performance.

$\mathbf{R^2}$ represents the proportion of variance in the dependent variable predictable from the independent variables. R^2 ranges from 0 to 1, with 1 indicating a perfect fit.

Together, MAE, RMSE, and R^2 provide a comprehensive understanding of model performance, with MAE and RMSE quantifying prediction accuracy and R^2 assessing goodness of fit.

4 Results and Discussion

4.1 MAE Scores Across Wide Format and Cross-Sectional Datasets

From Fig. 1, we observe that the MAE scores of MDS-UPDRS III prediction are significantly lower on wide-format datasets than cross-sectional datasets. In wide format datasets, the combination of NMF with SVR consistently achieves the lowest MAE scores, ranging from 1.68 in the baseline dataset to 1.91 in the dataset comprising all time points. In cross-sectional format datasets, PCA+NMF with SVR performs best or close to best for most datasets, with MAE scores ranging from 3.40 in the baseline dataset to 4.90 in the V12 dataset.

Ultimately, NMF with SVR performed consistently better in the prediction.

4.2 RMSE Scores Across Wide Format and Cross-Sectional Datasets

In Fig. 2, we demonstrate that the RMSE scores of MDS-UPDRS III prediction on both wide format and cross-sectional datasets follow a similar trend to the MAE scores. The NMF, SVR combination trained on wide-format datasets consistently achieves lower RMSE scores than the cross-sectional datasets. In wide-format datasets, the RMSE scores range from 3.98 in the baseline dataset (BL) to 2.70 in the dataset comprising all time points. In cross-sectional format datasets, PCA+NMF with linear regression performs best or close to best for most cross-sectional datasets, with RMSE scores ranging from 4.00 in the baseline dataset (BL) to 5.96 in the V12 dataset.

These findings emphasize the importance of considering dataset format in developing machine learning methodology to predict clinical scores with superior accuracy.

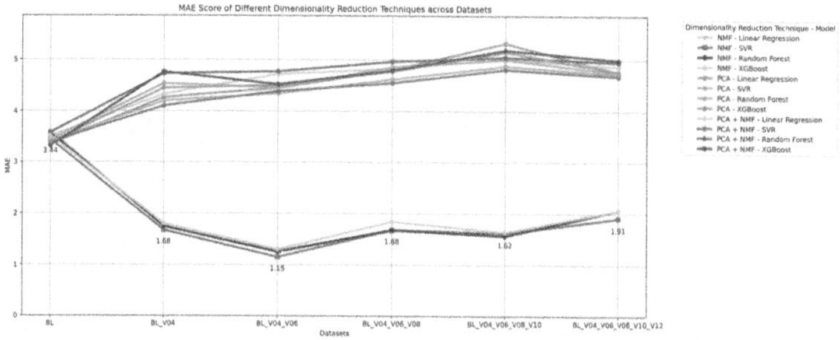

(a) MAE scores of machine learning models across wide format datasets

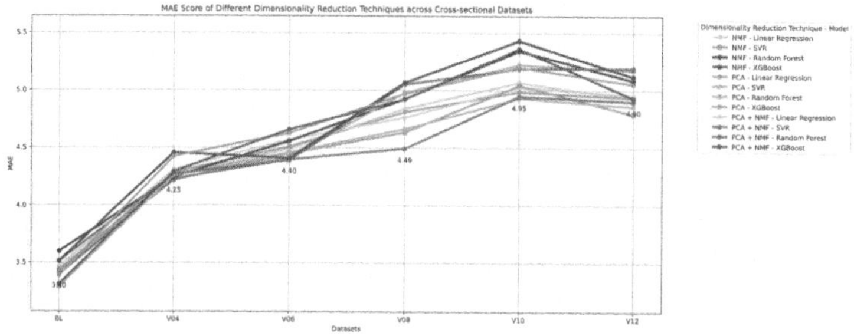

(b) MAE scores of machine learning models across cross-sectional datasets

Fig. 1. MAE scores

4.3 R^2 Scores Across Wide Format and Cross-Sectional Datasets

Figure 3 shows that in the wide format datasets, NMF with SVR consistently achieves the highest R^2 scores, ranging from 0.81 (baseline dataset) to 0.83 (all time points dataset), indicating that NMF effectively captures the most relevant features for prediction.

However, the cross-sectional format datasets show significantly low R^2 scores, suggesting that the models struggle to explain a meaningful portion of the variance in the MDS-UPDRS III scores. This could be due to the lack of information from past visits in cross-sectional datasets, making it challenging for dimensionality reduction algorithms and machine learning models to capture generalized patterns for predicting the target variable on unseen data.

The stark contrast between the R^2 scores in the wide format and cross-sectional datasets highlights the limitations of using cross-sectional data for this specific task.

From performing post-hoc Tukey's HSD tests across all three metrics, we found that the confidence intervals for the mean difference between dataset arrangements did not include zero, and the p-values were less than 0.05, indicating statistically significant differences and proving that wide-format arrangements result in better prediction performance.

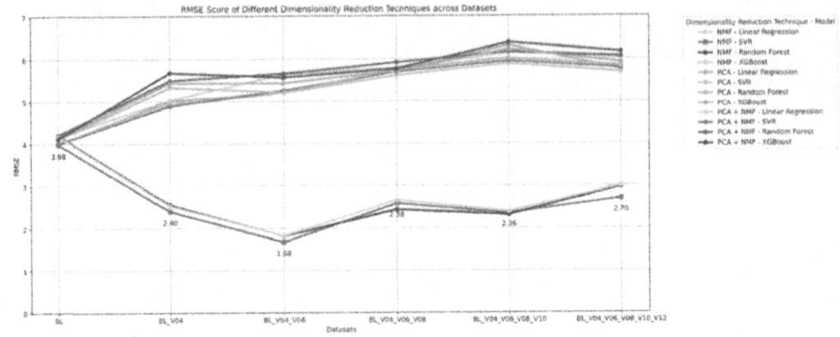

(a) RMSE scores of machine learning models across wide format datasets

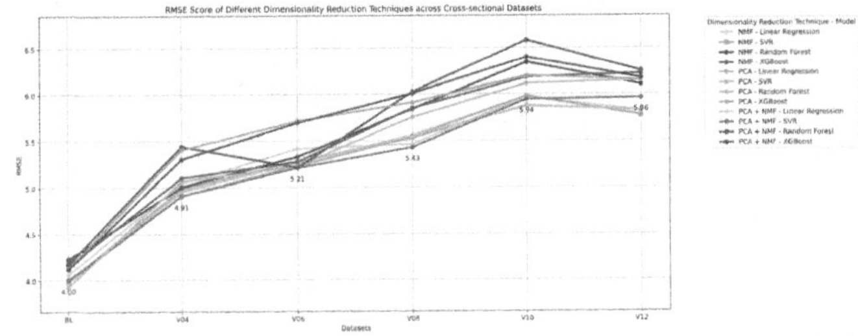

(b) RMSE scores of machine learning models across cross-sectional datasets

Fig. 2. MAE scores

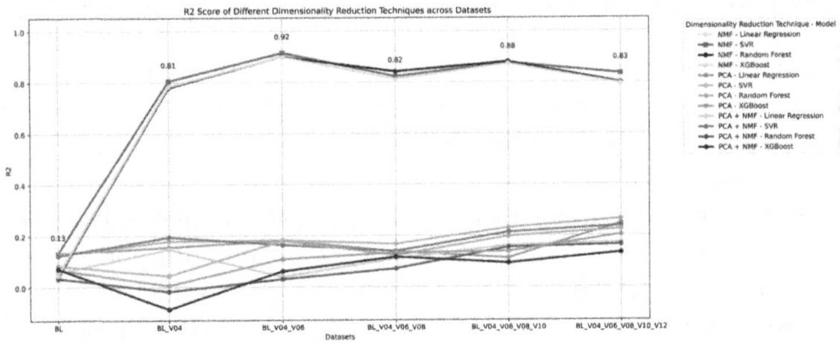

(a) R^2 scores of machine learning models across wide format datasets

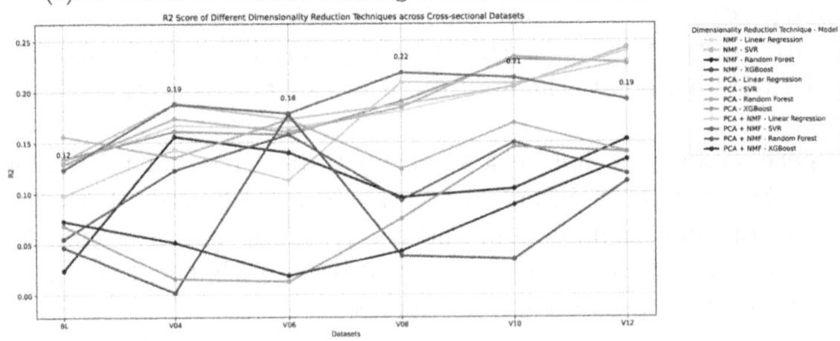

(b) R^2 scores of machine learning models across cross-sectional datasets

Fig. 3. R^2 scores

5 Conclusion and Future Work

In this study, we proposed a robust machine learning methodology to predict MDS-UPDRS III scores in Parkinson's disease patients using the longitudinal PPMI dataset. Our comprehensive approach involved data preprocessing, arranging data in wide and cross-sectional formats, dimensionality reduction using PCA, NMF, and a combination of both, and training various regression models.

Our results demonstrated that wide-format datasets consistently outperformed cross-sectional datasets, with the combination of NMF and SVR achieving the best performance in terms of MAE, RMSE, and R^2 scores.

The superior performance of wide-format datasets suggests that incorporating temporal information from longitudinal data leads to better prediction accuracy than cross-sectional data alone. NMF with SVR's ability to capture the most relevant features and explain a higher proportion of the variance in the target variable makes it a promising approach for future studies in this domain.

Despite the comprehensive nature of our study, it has some limitations. First, external validation of the findings of datasets like the Parkinson's Disease Biomarkers Program is necessary. Second, the study should be extended to develop prediction models for other parts of MDS-UPDRS score. Third, integrating imaging data or genetic information has been shown to achieve promising results in unsupervised clustering tasks; future studies should explore incorporating multimodal data to develop more accurate and robust predictive models for Parkinson's disease progression.

References

1. Aggrawal, R., Pal, S.: Elimination and backward selection of features (P-value technique) in prediction of heart disease by using machine learning algorithms. Turkish J. Comput. Math. Educ. (TURCOMAT) **12**, 2650–2665 (2021)
2. Shi, C., Zheng, Z., et al.: Exploring the effects of genetic variants on clinical profiles of Parkinson's disease assessed by the unified Parkinson's disease rating scale and the Hoehn-Yahr Stage. PLoS ONE **2016**, 1–11 (2016)
3. Brooks, D., Ibanez, V., et al.: Differing patterns of striatal F-18 Dopa uptake in Parkinsons-disease, multiple system atrophy, and progressive supranuclear palsy. Ann. Neurol. **28**, 547–555 (1990)
4. Garnett, E., Lang, A., et al.: A rostrocaudal gradient for aromatic acid decarboxylase in the human striatum. Can. J. Neurol. Sci. **14**, 444–447 (1987)
5. Stoessl, A., Martin, W., et al.: Advances in imaging in Parkinson's disease. Lancet Neurol. **10**, 987–1001 (2011)
6. Jankovic, J.: Parkinson's disease: clinical features and diagnosis. J. Neurol. Neurosurg. Psychiatry **79**(4), 368–376 (2008)
7. Antonini, A., Isaias, I., et al.: Duodenal levodopa infusion for advanced Parkinson's disease: 12-month treatment outcome. Int. Parkinson Mov. Disord. Soc. **22**(8), 1145–1149 (2007)
8. Tippmann-Peikert, M., Park, G., et al.: Pathologic gambling in patients with restless legs syndrome treated with dopaminergic agonists. Neurology **68**, 301–303 (2007)

9. Savitt, M., Dawson, V., et al.: Diagnosis and treatment of Parkinson disease: molecules to medicine. JCI **7**, 1744–1754 (2006)
10. Lang, A., Lozano, A.: Parkinson's disease. Second of two parts. N. Engl. J. Med. **1998**, 1130–1143 (1998)
11. Arias-Londoño, J.D., Gómez-García, J.A.: Predicting UPDRS scores in Parkinson's disease using voice signals: a deep learning/transfer-learning-based approach. In: Godino-Llorente, J.I. (ed.) AAPS 2019. CCIS, vol. 1295, pp. 100–123. Springer, Cham (2020). https://doi.org/10.1007/978-3-030-65654-6_6
12. Mischley, L., Lau, R., et al.: Use of a self-rating scale of the nature and severity of symptoms in Parkinson's Disease (PRO-PD): correlation with quality of life and existing scales of disease severity. NPJ Parkinson's Dis. **3**, 1–20 (2017)
13. Goetz, C., Tilley, B., et al.: Movement disorder society-sponsored revision of the unified Parkinson's disease rating scale: scale presentation and clinimetric testing results. Mov. Disord. **23**, 2129–2170 (2008)
14. Parashos, S., Luo, S., et al.: Measuring disease progression in early Parkinson disease: the national institutes of health exploratory trials in Parkinson Disease experience. JAMA Neurol. **6**, 710–716 (2014)
15. Movement disorder society task force on rating scales for Parkinson's disease.: the unified Parkinson's disease rating scale: status and recommendations. Movement **18**(7), 738–750 (2003)
16. Noyce, A., Schrag, A., et al.: Subtle motor disturbances in PREDICT-PD participants. J. Neurol. Neurosurg. Psychiatry **88**, 212–217 (2017)
17. Parkinson Progression Marker Initiative (PPMI). Prog. Neurobiol. **95**, 629–635 (2011)
18. Singh, Y., Bhatia, P.K., Sangwan, O.: A review OF studies ON machine learning techniques. Int. J. Comput. Sci. Secur. **1**(1), 70–84 (2007)
19. Gerraty, R.T., Provost, A., Li, L., Wagner, E., Haas, M., Lancashire, L.: Machine learning within the PPMI: review of the current state of affairs. Front. Aging Neurosci. **15**, 1076657 (2023)
20. Chen, S., Grant, E., Wu, T.T., Bowman, F.D.: Some recent statistical learning methods for longitudinal high-dimensional data. Wiley Interdiscip. Rev. Comput. Stat. **6**(1), 10–18 (2014)
21. Zipunnikov, V., Greven, S., Shou, H., Caffo, B., Reich, D.S., Crainiceanu, C.: Longitudinal high-dimensional principal components analysis with application to diffusion tensor imaging of multiple sclerosis. Ann. Appl. Stat. **8**(4), 2175 (2014)
22. Hastie, T., Tibshirani, R., Friedman, J.H., Friedman, J.H.: The Elements of Statistical Learning: Data Mining, Inference, and Prediction, 2nd edn. Springer, New York (2009). https://doi.org/10.1007/978-0-387-84858-7
23. Dinov, I.D., Heavner, B., Tang, M., Glusman, G., Chard, K., Darcy, M., Toga, A.W.: Predictive big data analytics: a study of Parkinson's disease using large, complex, heterogeneous, incongruent, multi-source and incomplete observations. PLoS ONE **11**(8), e0157077 (2016)
24. Ahmadi Rastegar, D., Ho, N., Halliday, G.M., et al.: Parkinson's progression prediction using machine learning and serum cytokines. NPJ Parkinson's Dis. **5**, 14 (2019)
25. Kraskov, A., Stogbauer, H., Grassberger, P.: Estimating mutual information. Phys. Rev. E **69**, 789 (2004)
26. Leung, K.H., et al.: Using deep-learning to predict outcome of patients with Parkinson's disease. In: 2018 IEEE Nuclear Science Symposium and Medical Imaging Conference Proceedings (NSS/MIC), Sydney, NSW, Australia, pp. 1–4 (2018)

27. Little, M.A., McSharry, P.E., Hunter, E.J., Ramig, L.O.: Suitability of dysphonia measurements for telemonitoring of Parkinson's disease. IEEE Trans. Biomed. Eng. **56**(4), 1015–1022 (2009)
28. Little, M.A., McSharry, P.E., Roberts, S.J., Costello, D.A.E., Moroz, I.M.: Exploiting nonlinear recurrence and fractal scaling properties for voice disorder detection. Biomed. Eng. Online **6**, 23 (2007)
29. Marvanova, M.: Introduction to Parkinson disease (PD) and its complications. Ment. Health Clin. **6**, 229–235 (2016)
30. Munasinghe, K., Karunanayake, P.: Recursive feature elimination for machine learning-based landslide prediction models. In: International Conference on Artificial Intelligence in Information and Communication (ICAIIC 2021), pp. 126–129 (2021)
31. Nilashi, M., et al.: Predicting Parkinson's disease progression: evaluation of ensemble methods in machine learning. J. Healthc. Eng. **2022**, 2793361 (2022)
32. Batina, L., Gierlichs, B., Prouf, E., Rivain, M., Standaert, F., Veyrat-Charvillon, N.: Mutual information analysis: a comprehensive study. J. Cryptol. **24**, 269–291 (2011)
33. Chandrashekar, G., Sahin, F.: A survey on feature selection methods. Comput. Electr. Eng. **40**(1), 16–28 (2014)
34. DeMaagd, G., Philip, A.: Parkinson's disease and its management: Part 1: disease entity, risk factors, pathophysiology, clinical presentation, and diagnosis. Pharmacy Therapeutics **40**(8), 504–532 (2015)
35. Exley, T.W., Moudy, S., Patterson, R., Kim, J., Albert, M.V.: Predicting UPDRS motor symptoms in individuals with Parkinsons disease from force plates using machine learning. J. Biomed. Health Inform. **8**, 150 (2022)
36. Goetz, C.G.: Unified Parkinson's disease rating scale and the movement-disorder society sponsored-unified Parkinson's disease rating scale. In: Kompoliti, K., Metman, L.V. (eds.) Encyclopedia of Movement Disorders, vol. 309, Academic Press (2010)
37. Grover, S., Bhartia, S., Yadav, A., Seeja, K.R.: Learning, predicting severity of Parkinson's disease using deep. Sci. Procedia Comput. **132**, 1788–1794 (2018)
38. Salmanpour, M., Shamsaei, M., et al.: Optimized machine learning methods for prediction of cognitive outcome in Parkinson's disease. Comput. Biol. Med. **111**, 1–8 (2019)
39. Leung, K., Salmanpour, M., et al.: Using deep-learning to predict outcome of patients with Parkinson's disease. In: 2018 IEEE Nuclear Science Symposium Conference Record, Sydney (2018)

Explaining Network Decision Provides Insights on the Causal Interaction Between Brain Regions in a Motor Imagery Task

Davide Borra[1(✉)] and Mirco Ravanelli[2,3]

[1] Department of Electrical, Electronic and Information Engineering "Guglielmo Marconi" (DEI), University of Bologna, Cesena Campus, Cesena, Italy
davide.borra2@unibo.it

[2] Department of Computer Science and Software Engineering, Concordia University, Montreal, Canada

[3] Mila - Quebec AI Institute, Montreal, Canada

Abstract. Neural decoding widely exploits machine learning for classifying electroencephalographic (EEG) signals for brain-computer interface applications. Recent advancements in neural decoding regards the use of brain functional connectivity estimates as input features and the adoption of convolutional neural networks (CNNs) to realize decoders. Moreover, explainable artificial intelligence (XAI) approaches based on CNNs are growing interest in the neuroscience community, for validating the knowledge learned by networks and for using the decoder not only to classify the EEG but also to analyze it in a data-driven way, without a priori assumptions. However, the adoption of connectivity estimates for neural decoding is still in its infancy, as adopts non-directed connectivity measures, limits the analysis of few interactions/frequency ranges, and exploits classic machine learning approaches without exploring CNNs. Moreover, XAI approaches have never been applied to analyze EEG-based functional connectivity. To overcome these limitations, we design and apply a CNN for processing directed connectivity measures estimated via spectral Granger causality. The CNN automatically learns features in the frequency and spatial domains, and it is coupled with an explanation technique (DeepLIFT) for highlighting the most relevant connectivity inflow and outflow associated to each decoded brain state. Our approach is applied to motor imagery decoding, and achieves state-of-the-art performance compared to existing networks. DeepLIFT relevance representations match the directional interactions known occurring when imagining movements, validating the features related to the brain network, as learned by the CNN.

Keywords: Electroencephalogram · Functional connectivity · Convolutional neural networks · Explainable AI · Motor imagery

C. Y. Suen et al. (Eds.): ANNPR 2024, LNAI 15154, pp. 156–167, 2024.
https://doi.org/10.1007/978-3-031-71602-7_14

1 Introduction

Neural decoders largely adopt machine learning algorithms for classifying electroencephalographic (EEG) data, mainly for realising Brain-Computer Interfaces (BCIs) [24]. In a BCI, the user's neural activity is decoded and a feedback is then provided to the user, e.g., for controlling a robotic arm. Traditional decoders exploit a feature extraction stage separated from the classification stage [13]. Here, features are generally handcrafted in the spatial, temporal, or frequency domain, based on a priori assumptions on the neural correlates underlying the recorded signals, e.g., utilizing the common spatial patterns or features based on event-related desynchronization/synchronization at specific EEG sensor locations [14]. Recently, brain functional connectivity features were exploited for neural decoding [9], exploiting the communication between brain regions in addition to or instead of more traditional and simpler features based on regional activations or synchronizations. The use of connectivity measures mostly led to improved performance; however, most of the studies used non-directed connectivity measures, without considering the direction of interactions between brain regions, and exploited connectivity measures in fixed frequency ranges or across the entire spectrum (that is, the total connectivity, discarding the frequency information) [9].

In addition to improving the neural representation driving the decoder in BCIs, decoding algorithms improved in the past years thanks to the design of deep learning approaches specific for EEG signals [15], able to automatically learn the most meaningful neural signatures from raw/lightly pre-processed data. Among deep neural networks, convolutional neural networks (CNNs) are the most used ones [15]. CNNs were mainly proposed and validated using directly EEG time series as input, widely improving the performance over traditional machine learning approaches [1,8,12,16–18,21]. Moreover, the combination of a trained CNN with an explanation technique (e.g., DeepLIFT [20]) provided useful insights about the neural mechanisms underlying a cognitive task [3,6–8], suggesting that neuroscientists can conveniently exploit the explainable AI (XAI) framework as a novel data-driven EEG analysis tool, able to highlight the most relevant signatures of the input neural time series.

However, despite these advantages, CNNs have never been used for decoding EEG-based functional connectivity estimates [9] – as these were widely decoded with classic machine learning approaches – and XAI approaches have never been applied to analyze brain interactions. Thus, in this study we aim at proposing a XAI framework based on a compact CNN and on DeepLIFT devoted at both decoding and analyzing motor imagery from EEG-based directed functional connectivity estimates. Spectral Granger causality is adopted for computing the directed functional connectivity as a function of frequency, which is used as input for the neural decoder. Thus, here the entire information contained in the connectivity estimate is decoded, without limiting the decoding pipeline to specific frequency bands or to the total connectivity (i.e., aggregating across all frequencies). The adopted CNN learns features separately in the frequency and spatial domains, and DeepLIFT representations are exploited to derive the importance of connectivity entering in (inflow) or departing from (outflow) each EEG sensor. Here, we compare the decoding performance of our architecture with state-of-the-art networks, and we illustrate the potentialities of a XAI framework to highlight the most relevant connectivity inflow and outflow characterizing motor imagery, comparing the results with traditional analyses and with connectivity modulations known in the literature.

2 Methods

2.1 Dataset

Data Pre-processing. We exploited the BNCI2014-001 dataset [23], extensively employed in previous studies [15]. Here, 22-channel EEG was recorded from 9 healthy participants across 2 recording sessions. Signals were sampled at 250 Hz, and electrodes placed as in Fig. 1a. The task consisted in the imagination of movements for 4 s, including the imagination of left-hand and right-hand movements. For each participant, each recording session, and each motor imagery condition, 72 trials were recorded (288 trials in total per subject). By considering trials related to right- and left-hand motor imagery, in this study we aimed at solving a binary classification problem ($N_c = 2$ motor brain states). Signals were band-pass filtered between 1 and 40 Hz using a 4th order zero-phase Butterworth filter. Moreover, these were epoched into 4 s-length epochs by extracting time series from 0 s to 4 s with respect to the motor imagery onset. After this procedure, each EEG epoch was $X \in \mathbb{R}^{C \times T}$, where $C = 22$ (electrodes) and $T = 1000$ (time samples), and the corresponding label was $y \in \{$'left hand', 'right hand'$\}$.

Functional Connectivity Estimation. Directional influences between EEG sensors were estimated by using the pairwise spectral Granger causality (GC) [19], with an order $p = 30$ for the bivariate autoregressive model [4]. At each frequency f, the spectral GC is represented by a non-symmetric matrix $\in \mathbb{R}^{C \times C}$, with the off-diagonal (i, j)-th value quantifying the directional influence from

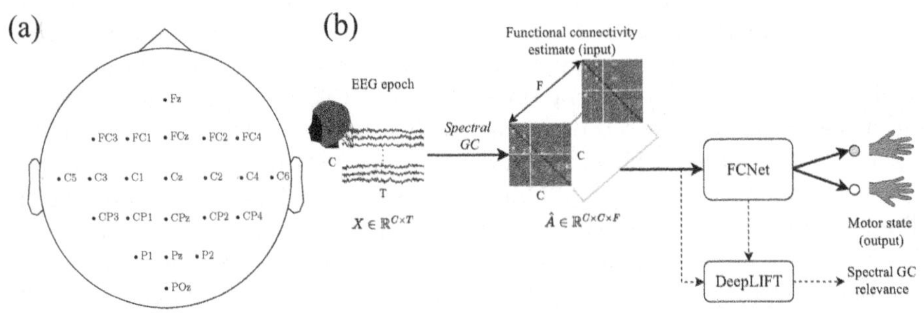

Fig. 1. Panel a – Localization of the EEG sensors on the scalp [23]. Panel b – High-level scheme of the XAI framework used to gain insights about brain causal interactions.

the i-th to the j-th sensor at that frequency ($GC_{i \to j}(f)$). The spectral GC was computed at F frequencies in total. To mitigate volume conduction effects that could potentially generate spurious functional connectivity between sensors, a spherical spline surface Laplacian transformation was applied [2]. For each subject (9) and each EEG epoch (288), the directional influences in the frequency domain were computed. Thus, a 3D matrix $A \in \mathbb{R}^{C \times C \times F}$ was obtained for each EEG epoch X. In our case, $C = 22$, and $F = 81$ (frequencies between 1 and 40 Hz with a frequency resolution of 0.5 Hz). To emphasize how much each connectivity value contributed to the overall connectivity across sensors and reduce inter-trial variability, the connectivity matrix at the k-th frequency bin was normalized such that the sum of all off-diagonal connectivity values was 1, i.e., $\hat{A}[:,:,k] = A[:,:,k] / \sum_{i,j,i \neq j} A[i,j,k]$, $k = 0, ..., 80$ (as done in Borra et al. [4]). The network was fed with $\hat{A}$ as input.

The quality of the obtained functional connectivity was assessed as follows. The matrix $\hat{A}$ was averaged across trials, separately for the right- and left-hand motor conditions. Then, the GC in the alpha- and beta-band was computed ($\hat{A}_{alpha}$, $\hat{A}_{beta}$) – as these frequency ranges are the most relevant during motor imagery [14] – by computing the integral of $\hat{A}$ along the frequency axis in the alpha (from 8 to 12 Hz) and beta (from 12 to 30 Hz) ranges. After this procedure, two connectivity matrices ($\hat{A}_{alpha}$, $\hat{A}_{beta}$) were obtained for each subject (9) and motor condition (2). To analyze the functional connectivity differences related to the motor task, we performed a permutation t test between right- and left-hand motor imagery in the matrix $\hat{A}_{alpha}$ and $\hat{A}_{beta}$, separately for each matrix.

2.2 XAI Framework

The XAI framework is schematized at a high-level in Fig. 1b, and it is described in the following.

Neural Network. We designed and used a compact CNN (named Functional-Connectivity-Net, or 'FCNet'), for decoding motor imagery from brain functional connectivity estimates. FCNet is composed by a *spectral block*, a *dual-branched spatial block*, and a *classification block*, processing the brain functional connectivity matrix $\hat{A} \in \mathbb{R}^{C \times C \times F}$. In the following we present the key characteristics of FCNet blocks, and in Table 1 we provide details of FCNet layers.

1. Spectral block. This block searches features that summarize the causal spectra in the frequency domain. First, a 3D convolutional layer is applied, learning $K = 32$ kernels with size of $W = (1, 1, 9)$, using a unitary stride. Feature maps are pooled (average pooling) in the frequency domain with a pooling size and stride of $W = S = (1, 1, 4)$, and neurons are dropped out during training (dropout probability $p = 0.1$). Then, another 3D convolutional layer continues the learning of spectral features, learning $K = 32$ kernels with size of $W = (1, 1, 9)$. Separable convolutions are employed, utilizing a depth multiplier $D = 1$. Feature maps are pooled (average pooling) in the frequency

Table 1. Network architecture: layers, main hyperparameters, and output shapes.

Block	Layer name	Main hyperparameters	Output shape
	Input	–	$(1, 22, 22, 81)$
Spectral	Conv3D	$K = 32,\ W = (1, 1, 9)$	$(32, 22, 22, 73)$
	BatchNorm3D	–	$(32, 22, 22, 73)$
	ReLU	–	$(32, 22, 22, 73)$
	AvgPool3D	$W = S = (1, 1, 4)$	$(32, 22, 22, 18)$
	Dropout	$p = 0.1$	$(32, 22, 22, 18)$
	SepConv3D	$K = 32,\ D = 1,\ W = (1, 1, 9)$	$(32, 22, 22, 10)$
	BatchNorm3D	–	$(32, 22, 22, 10)$
	ReLU	–	$(32, 22, 22, 10)$
	AvgPool3D	$W = S = (1, 1, 4)$	$(32, 22, 22, 1)$
	Dropout	$p = 0.1$	$(32, 22, 22, 1)$
	Squeeze	–	$(32, 22, 22)$
Spatial-inflow (branch 1)	SepConv2D	$K = 32,\ D = 1,\ W = (22, 1)$	$(32, 1, 22)$
	BatchNorm2D	–	$(32, 1, 22)$
	ReLU	–	$(32, 1, 22)$
	AvgPool2D	$W = S = (1, 22)$	$(32, 1, 1)$
	Flatten	–	$(32,)$
Spatial-outflow (branch 2)	SepConv2D	$K = 32,\ D = 1,\ W = (1, 22)$	$(32, 22, 1)$
	BatchNorm2D	–	$(32, 22, 1)$
	ReLU	–	$(32, 22, 1)$
	AvgPool2D	$W = S = (22, 1)$	$(32, 1, 1)$
	Flatten	–	$(32,)$
Classification	Concatenate	–	$(64,)$
	Fully-connected	$N = N_c = 2$	$(2,)$
	Softmax	–	$(2,)$

domain utilizing a pooling size and stride of $W = S = (1, 1, 4)$, and neurons are dropped out during training (dropout probability $p = 0.1$). This block, with the chosen hyperparameters, provides as output 32 feature maps with shape $(C, C) = (22, 22)$ summarizing the functional connectivity across all frequency bins.

2. Dual-branched spatial block. Here, connectivity inflow- and outflow-related features are learned from the high-level connectivity feature maps extracted by the previous block. A dual-branch structure is exploited, in which each parallel branch operates on the input by separately learning inflow and outflow features. Specifically, the spatial-inflow branch includes a separable 2D convolutional layer (depthwise multiplier of $D = 1$) learning 32 kernels with size $W = (22, 1)$. This way, inflow measures associated to each EEG sensor are learned. These measures are pooled together obtaining a single scalar value for each feature map (average pooling with size and stride of $W = S = (1, 22)$). As concerning the spatial-outflow branch, the same processing is performed,

but in the other spatial dimension. Each parallel branch provides a feature array of 32 elements as output.

3. Classification block. The decoding problem is finalized starting from the feature arrays provided by the dual-branched spatial block. Feature arrays are concatenated together, and then provided as input to a fully-connected layer with $N = N_c = 2$ units. These units are activated via softmax activation function.

In spectral and spatial blocks, batch normalization is included after each convolution, and feature maps are activated via rectified linear activations (ReLU).

Training and Evaluation. The network was trained for 500 epochs, with the cross-entropy as loss function and Adam as optimizer (learning rate of $5 \cdot 1e-4$, mini-batch size of 32). The network was trained separately for each subject, utilizing a within-subject training strategy. Ten-fold cross-validation was adopted, by using in each cross-validation fold 259 training and 29 test examples. Moreover, from the training set we sampled 52 validation examples (20% of the training set) to perform early stopping. Therefore, 207 training examples were effectively used to train the network at each fold. As the used dataset was small, we augmented the training examples as follows. During training, for each training mini-batch we randomly swapped a percentage of the causal spectra across the examples associated with the same motor condition. The percentage used while swapping causal interactions was randomly sampled from a uniform distribution from 20% to 80%. The resulting augmented mini-batch was concatenated to the original non augmented mini-batch, doubling the training examples. The accuracy served as performance metric. It was computed on the test set for each subject and each cross-validation fold, and then averaged across folds.

Decision Explanation. For each trained model (i.e., for each subject and each cross-validation fold), we explained the network decision by using DeepLIFT [20] on the test set examples. We adopted DeepLIFT as a recent benchmark study [22] showed that this explanation technique is superior to other techniques (e.g., saliency maps, GradCAM, guided GradCAM) when processing EEG signals. Here we derived DeepLIFT representations associated to the output neuron of the correct class. Therefore, for each input test example, i.e., functional connectivity estimate, we obtained a relevance representation R with the same shape of $\hat{A}$, that is $R \in \mathbb{R}^{C \times C \times F}$. Each sample $R_{i,j,k}$ quantifies the relevance of the (i, j)-th causal interaction and the k-th frequency bin for discriminating right- and left-hand motor imagery. The relevance representation was averaged across trials (separately for each motor condition), cross-validation folds, and frequency bins. We then averaged the relevance across the first or the last spatial dimension, separately, obtaining two representations quantifying the importance of connectivity inflow and outflow for decoding motor imagery. In these representations, the higher the value at a specific electrode site, the higher the relevance of the interactions entering in (inflow) or departing from (outflow) that electrode site.

Overall, from this procedure, for each subject and each motor imagery condition two relevance arrays $r_{inflow} \in \mathbb{R}^C$ and $r_{outflow} \in \mathbb{R}^C$ were obtained, quantifying the relevance of the connectivity inflow and outflow at each electrode for separating right- vs. left-hand motor imagery. Finally, to equally weight the 9 participants, each array was normalized between –1 and 1.

2.3 State-of-the-Art Neural Networks

We compared the performance achieved with FCNet with five state-of-the-art neural networks. ShallowFBCSPNet and Deep4Net [18] – the first successful CNNs proposed for motor imagery decoding – were included in the benchmark, together with EEGNet [12], a multi-purpose CNN that was applied on a variety of decoding problems such as the decoding of motor imagery, P300, steady-state visual evoked potentials. EEGNet-like architectures consistently outperformed other deep neural networks and traditional machine learning methods during an international scientific competition [21], and were used as reference for CNNs working with invasive recordings [5,10]. Finally, we also included recent CNNs that exploit inception modules for learning temporal features, such as EEGInception [17], and EEGITNet [16]. All these state-of-the-art networks were trained and evaluated with EEG time series as input, using a within-subject strategy employing a 10-fold cross-validation scheme (as done with FCNet). To compare the decoding accuracy scored with FCNet vs. each state-of-the-art decoder, a pairwise Wilcoxon signed-rank test was performed, and the Benjamini-Hochberg procedure was applied to correct p-values for multiple tests (5).

3 Results and Discussion

Figure 2 reports the difference in the functional connectivity matrix between right-hand and left-hand motor imagery conditions, separately for alpha- and beta-band. The largest significant differences in connectivity between motor imagery conditions were observed in beta-band, especially for connections entering into central sites, e.g., at C6 ($FC3 \rightarrow C6$, $C5 \rightarrow C6$), and C5 ($C4 \rightarrow C5$, $C6 \rightarrow C5$, $FC4 \rightarrow C5$). This is an expected results – matching the neural correlates of motor imagery – as the central EEG electrodes play a key role in motor imagery decoding as these are placed approximately over the sensorimotor cortex [14].

The accuracy of participant-specific decoders designed with FCNet spanned from 62.8 (2.6)% to 90.3 (1.6)% (mean and standard error of the mean across folds). Decoders performed well-above the chance level (50%) across subjects, and our network achieved an accuracy across participants of 77.5 (3.1)%. In Fig. 3 we summarize the results obtained while comparing our approach with the state-of-the-art. FCNet significantly outperformed ($p < 0.05$) all the state-of-the-art neural networks, except for ShallowFBCSPNet (achieving an accuracy of 74.3 (5.3)%), that performed similar ($p \geq 0.05$) to FCNet. However, it is worth highlighting that ShallowFBCSPNet is characterized by a network architecture

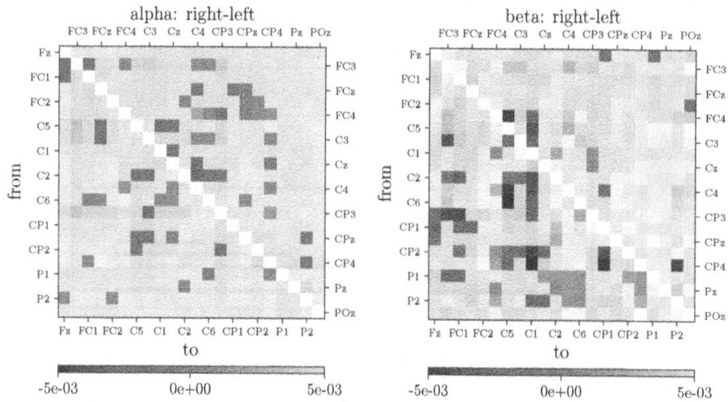

Fig. 2. Functional connectivity. The left and right panels report respectively the difference of functional connectivity between right- and left-hand motor imagery in the alpha- and beta-band (as obtained from $\hat{A}_{alpha}$ and $\hat{A}_{beta}$). The difference is displayed in its mean value across subjects and trials. Only the causal interactions significantly different ($p < 0.05$) between the two conditions are highlighted.

specific for extracting features related to sensorimotor rhythms, and thus, for solving motor imagery decoding. On the other hand, FCNet is designed without any constraint in the learned features, and could be potentially used in the future also for solving other neural decoding problems. These decoding performance results corroborate the idea that functional connectivity estimates are good input representations for addressing neural decoding problems, as they enabled to achieve mostly higher decoding performance than other solutions, based on EEG time series. Future neural decoders could benefit from exploiting the brain functional connectivity for decoding motor states.

Finally, Fig. 4 displays the results obtained from the XAI approach. As we considered a binary classification task (right- vs. left-hand motor imagery), for brevity here we focus on the right-hand motor imagery condition only (a reversed importance was observed for the left imagery condition). From these results, the most relevant absolute connectivity inflow and outflow for discriminating motor imagery was observed at central sites (at C6, C5, C3), with an opposite sign (i.e., positive/negative evidence to network decision) depending on the hemisphere. This contribution of central sites matches the results from the traditional connectivity analysis of Fig. 2. Moreover, the sign in the inflow and outflow relevance of the most relevant electrode sites depended on the hemisphere (right: C6 vs. left: C5, C3). This matches prior findings of Hu et al. [11] on the same dataset, in which a higher connectivity from left-to-right central sites (i.e., increased right-hemisphere inflow) was observed in right-hand motor imagery with traditional EEG analyses, vice versa for left-hand motor imagery.

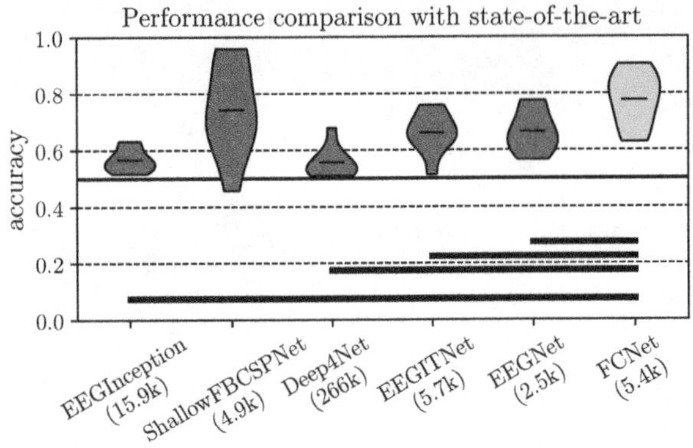

Fig. 3. Decoding accuracy scored with FCNet and state-of-the-art networks. The number of trainable parameters of each network is reported within brackets. Performance distributions are displayed as violin plots, with black bars indicating the mean value. The chance level is reported (0.5, horizontal black line). Significant pairwise comparisons ($p < 0.05$, corrected for multiple tests) are marked with thick horizontal bars.

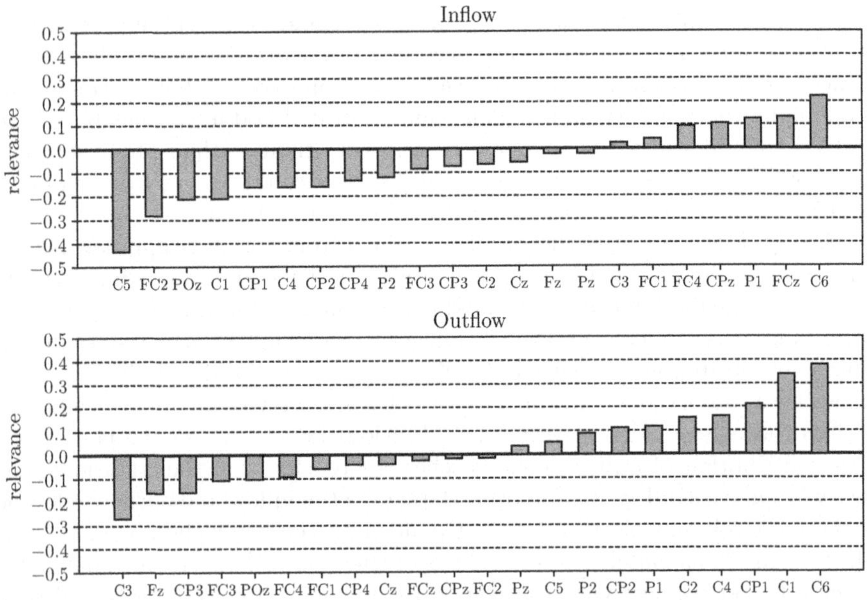

Fig. 4. Connectivity inflow and outflow importance obtained when coupling FCNet with DeepLIFT (right-hand imagery). Relevance arrays r_{inflow} and $r_{outflow}$ are reported respectively in top and bottom panels (bar height: mean across subjects).

4 Conclusions

In this study we designed a CNN for decoding brain functional connectivity in a motor imagery task. The network reached state-of-the-art performance when compared to other five decoders widely used with EEG time series, confirming the effectiveness of connectivity measures as input of neural decoders. By coupling the network with an explanation technique, the XAI approach provided relevance representations that quantified the importance of connectivity inflows and outflows. These representations matched the directional interaction known occurring when imagining right- and left-hand movements and matched also the results obtained with a traditional connectivity analysis, thus validating the feature learning operated by our approach. Even though the proposed methodology exhibited promising results, our results are preliminary as were obtained on a single EEG dataset, and future studies are needed for further validating it on other and larger datasets. Our XAI framework enables to disclose the most relevant interactions between brain regions in a data-driven way, and could be used in prospective by neuroscientists as a new tool for analyzing brain functional connectivity.

Acknowledgments. This research was co-funded by the Italian Complementary National Plan PNC-I.1 "Research initiatives for innovative technologies and pathways in the health and welfare sector" D.D. 931 of 06/06/2022, "DARE - DigitAl lifelong pRevEntion" initiative, code PNC0000002, CUP: B53C22006450001.

Disclosure of Interests. The authors declare no conflict of interest.

References

1. An, J., Chen, X., Wu, D.: Algorithm contest of motor imagery BCI in the world robot contest 2022: a survey. Brain Sci. Adv. **9**(3), 166–181 (2023). http://dx.doi.org/10.26599/BSA.2023.9050011
2. Bastos, A.M., Schoffelen, J.M.: A tutorial review of functional connectivity analysis methods and their interpretational pitfalls. Front. Syst. Neurosci. **9** (2016). http://dx.doi.org/10.3389/fnsys.2015.00175
3. Borra, D., Bossi, F., Rivolta, D., Magosso, E.: Deep learning applied to EEG source-data reveals both ventral and dorsal visual stream involvement in holistic processing of social stimuli. Sci. Rep. **13**(1) (2023). http://dx.doi.org/10.1038/s41598-023-34487-z
4. Borra, D., Fantozzi, S., Bisi, M.C., Magosso, E.: Modulations of cortical power and connectivity in alpha and beta bands during the preparation of reaching movements. Sensors **23**(7), 3530 (2023). http://dx.doi.org/10.3390/s23073530
5. Borra, D., Filippini, M., Ursino, M., Fattori, P., Magosso, E.: Motor decoding from the posterior parietal cortex using deep neural networks. J. Neural Eng. **20**(3), 036016 (2023). http://dx.doi.org/10.1088/1741-2552/acd1b6
6. Borra, D., Filippini, M., Ursino, M., Fattori, P., Magosso, E.: Convolutional neural networks reveal properties of reach-to-grasp encoding in posterior parietal cortex. Comput. Biol. Med. **172**, 108188 (2024). http://dx.doi.org/10.1016/j.compbiomed.2024.108188

7. Borra, D., Magosso, E.: Deep learning-based EEG analysis: investigating P3 ERP components. J. Integr. Neurosci. **20**(4), 791–811 (2021). http://dx.doi.org/10.31083/j.jin2004083
8. Borra, D., Mondini, V., Magosso, E., Müller-Putz, G.R.: Decoding movement kinematics from EEG using an interpretable convolutional neural network. Comput. Biol. Med. **165**, 107323 (2023). http://dx.doi.org/10.1016/j.compbiomed.2023.107323
9. Brusini, L., Stival, F., Setti, F., Menegatti, E., Menegaz, G., Storti, S.F.: A systematic review on motor-imagery brain-connectivity-based computer interfaces. IEEE Trans. Hum.-Mach. Syst. **51**(6), 725–733 (2021). http://dx.doi.org/10.1109/THMS.2021.3115094
10. Filippini, M., Borra, D., Ursino, M., Magosso, E., Fattori, P.: Decoding sensorimotor information from superior parietal lobule of macaque via convolutional neural networks. Neural Netw. **151**, 276–294 (2022). http://dx.doi.org/10.1016/j.neunet.2022.03.044
11. Hu, S., Wang, H., Zhang, J., Kong, W., Cao, Y., Kozma, R.: Comparison analysis: granger causality and new causality and their applications to motor imagery. IEEE Trans. Neural Netw. Learn. Syst. **27**(7), 1429–1444 (2016). http://dx.doi.org/10.1109/TNNLS.2015.2441137
12. Lawhern, V.J., Solon, A.J., Waytowich, N.R., Gordon, S.M., Hung, C.P., Lance, B.J.: EEGNet: a compact convolutional neural network for EEG-based brain–computer interfaces. J. Neural Eng. **15**(5), 056013 (2018). https://doi.org/10.1088/1741-2552/aace8c
13. Lotte, F., et al.: A review of classification algorithms for EEG-based brain-computer interfaces: a 10 year update. J. Neural Eng. **15**(3), 031005 (2018). http://dx.doi.org/10.1088/1741-2552/aab2f2
14. Neuper, C., Wörtz, M., Pfurtscheller, G.: ERD/ERS patterns reflecting sensorimotor activation and deactivation, pp. 211–222. Elsevier (2006). http://dx.doi.org/10.1016/S0079-6123(06)59014-4
15. Roy, Y., Banville, H., Albuquerque, I., Gramfort, A., Falk, T.H., Faubert, J.: Deep learning-based electroencephalography analysis: a systematic review. J. Neural Eng. **16**(5), 051001 (2019). http://dx.doi.org/10.1088/1741-2552/ab260c
16. Salami, A., Andreu-Perez, J., Gillmeister, H.: EEG-ITNet: an explainable inception temporal convolutional network for motor imagery classification. IEEE Access **10**, 36672–36685 (2022). http://dx.doi.org/10.1109/ACCESS.2022.3161489
17. Santamaría-Vázquez, E., Martínez-Cagigal, V., Vaquerizo-Villar, F., Hornero, R.: EEG-inception: a novel deep convolutional neural network for assistive ERP-based brain-computer interfaces. IEEE Trans. Neural Syst. Rehabil. Eng. **28**(12), 2773–2782 (2020). http://dx.doi.org/10.1109/TNSRE.2020.3048106
18. Schirrmeister, R.T., et al.: Deep learning with convolutional neural networks for EEG decoding and visualization. Hum. Brain Mapp. **38**(11), 5391–5420 (2017). https://doi.org/10.1002/hbm.23730
19. Seth, A.K., Barrett, A.B., Barnett, L.: Granger causality analysis in neuroscience and neuroimaging. J. Neurosci. **35**(8), 3293–3297 (2015). http://dx.doi.org/10.1523/JNEUROSCI.4399-14.2015
20. Shrikumar, A., Greenside, P., Kundaje, A.: Learning important features through propagating activation differences (2017). https://doi.org/10.48550/ARXIV.1704.02685, https://arxiv.org/abs/1704.02685
21. Simões, M., et al.: BCIAUT-P300: a multi-session and multi-subject benchmark dataset on autism for p300-based brain-computer-interfaces. Front. Neurosci. **14** (2020). http://dx.doi.org/10.3389/fnins.2020.568104

22. Sujatha Ravindran, A., Contreras-Vidal, J.: An empirical comparison of deep learning explainability approaches for EEG using simulated ground truth. Sci. Rep. **13**(1) (2023). http://dx.doi.org/10.1038/s41598-023-43871-8
23. Tangermann, M., et al.: Review of the BCI competition IV. Front. Neurosci. **6** (2012). https://doi.org/10.3389/fnins.2012.00055
24. Wolpaw, J.R., Birbaumer, N., McFarland, D.J., Pfurtscheller, G., Vaughan, T.M.: Brain-computer interfaces for communication and control. Clin. Neurophysiol. **113**(6), 767–791 (2002). http://dx.doi.org/10.1016/s1388-2457(02)00057-3

Multi-modal Decoding of Reach-to-Grasping from EEG and EMG via Neural Networks

Davide Borra[1(✉)], Matteo Fraternali[1], Mirco Ravanelli[2,3], and Elisa Magosso[1]

[1] Department of Electrical, Electronic and Information Engineering "Guglielmo Marconi" (DEI), University of Bologna, Cesena Campus, Cesena, Italy
davide.borra2@unibo.it

[2] Department of Computer Science and Software Engineering, Concordia University, Montreal, Canada

[3] Mila - Quebec AI Institute, Montreal, Canada

Abstract. Convolutional neural networks (CNNs) have revolutionized motor decoding from electroencephalographic (EEG) signals, showcasing their ability to outperform traditional machine learning, especially for Brain-Computer Interface (BCI) applications. By processing also other recording modalities (e.g., electromyography, EMG) together with EEG signals, motor decoding improved. However, multi-modal algorithms for decoding hand movements are mainly applied to simple movements (e.g., wrist flexion/extension), while their adoption for decoding complex movements (e.g., different grip types) is still under-investigated. In this study, we recorded EEG and EMG signals from 12 participants while they performed a delayed reach-to-grasping task towards one out of four possible objects (a handle, a pin, a card, and a ball), and we addressed multi-modal EEG+EMG decoding with a dual-branch CNN. Each branch of the CNN was based on EEGNet. The performance of the multi-modal approach was compared to mono-modal baselines (based on EEG or EMG only). The multi-modal EEG+EMG pipeline outperformed the EEG-based pipeline during movement initiation, while it outperformed the EMG-based pipeline in motor preparation. Finally, the multi-modal approach was capable of accurately discriminating between grip types widely during the task, especially from movement initiation. Our results further validate multi-modal decoding for potential future BCI applications, aiming at achieving a more natural user experience.

Keywords: EEG · EMG · Multi-modal motor decoding · Reach-to-grasping · Convolutional neural networks · Brain-Computer Interfaces

1 Introduction

Machine learning algorithms represent powerful tools to decode the brain activity from electroencephalographic (EEG) data, both in research and application

C. Y. Suen et al. (Eds.): ANNPR 2024, LNAI 15154, pp. 168–179, 2024.
https://doi.org/10.1007/978-3-031-71602-7_15

areas, especially for designing neural decoders for EEG-based Brain-Computer Interfaces (BCIs) [23]. In an EEG-based BCI, the user's neural activity – as measured by the EEG – is decoded by exploiting a neural decoder, and a feedback is provided back to the user, for example by controlling an external device (e.g., a robotic arm or a neuroprosthesis). Neural decoders based on traditional machine learning approaches exploit a feature extraction stage separated from the classification stage [15]. In these approaches, features are generally handcrafted, based on a priori assumptions about the neural correlates underlying the recorded signals, for example power-based features related to movement (e.g., α and β band event-related desynchronization/synchronization) [17] and time-domain features (e.g., amplitude of motor-related cortical potentials, MRCPs) [9]. Motor decoding from EEG recordings showed promising results, enabling the decoding of movements even during movement preparation [18]. Recent advancements of neural decoding involve the adoption of convolutional neural networks (CNNs) [20]. CNNs proved to outperform traditional machine learning in a variety of scenarios, for example in motor [13,21], P300 [13,22], and steady-state visual evoked potentials [24] decoding. Among CNNs, EEGNet – in its original version [13] and variant designs [1,3,6–8,19,22] – represents the most used one not only with EEG but also with invasive recordings [4,5,10], providing a good trade-off between model compactness, size, training time, and decoding accuracy.

Besides advancements in EEG-based neural decoders, recent research in neural interfacing is moving from a mono-modal neuroelectric interface towards a multi-modal interface [12,14], by exploiting also signals representative of the muscle activity (that is, electromyography (EMG) signals), to improve the neural decoding capabilities and thus to control an external device in a more intuitive and natural way. Multi-modal EEG+EMG decoding of hand movements has been addressed in motor execution paradigms (or attempted movements, in case of patients) [12,14]. However, prior studies only consider simple movements, such as wrist flexion/extension and hand grasp opening/closing in Kim et al. [12], right hand vs. left hand repetitive clutching in Leeb et al. [14]. This is mainly due to the challenge posed by the realization of an ad-hoc recording setup that properly presents stimuli to the user. As an example, in a setup devoted at recording reach-to-grasping movements, objects with different shapes should be presented one at a time during the task, at the same location in front of the user, so to avoid spatial-related features to be encoded in the stimuli presentation (and thus to avoid that spatial-related features reflected onto the neural response can potentially alter the interpretation of results). Therefore, multi-modal EEG+EMG decoding with executed movements was mainly addressed by considering simple movements, while complex movements such as grip prehensions triggered by different object shapes, were not considered in multi-modal decoding setups. This limits the validation of the proposed approaches.

In this study, we aim at contributing to the validation of a multi-modal EEG+EMG decoding pipeline based on a CNN for decoding reach-to-grasping movements. EEG and EMG signals were simultaneously recorded from twelve participants while performing a delayed reach-to-grasping task towards one of

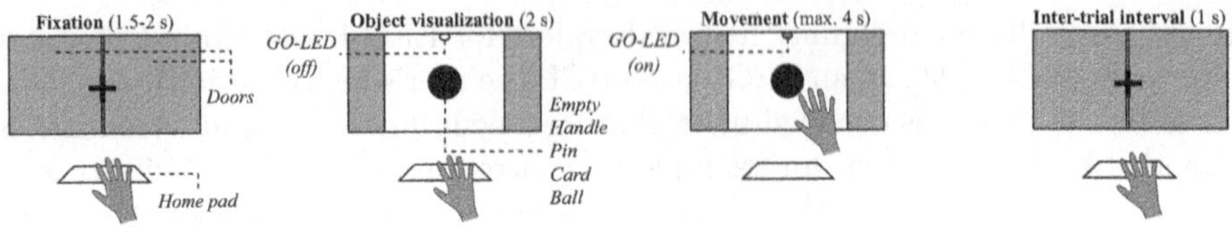

Fig. 1. Reach-to-grasping: trial structure. The four main phases characterizing the trial are depicted, together with their duration.

the following objects: a handle, a ball, a pin, and a card. A dual-branch CNN based on EEGNet was used, by decoding in parallel EEG and EMG recordings. The approach was applied by using a sliding window decoding strategy, in order to simulate a real-time scenario in which the decoder is applied to EEG and EMG recordings continuously over time. Experiments were conducted for assessing the performance of the multi-modal approach, also comparing it with mono-modal approaches (based on EEG or EMG only). We hypothesise that an approach that fuses information derived from both EEG and EMG recordings may be capable of exploiting the advantages of both recording modalities, achieving high decoding performance during movement preparation (by predominantly exploiting modulations in EEG, as no movement is performed in this phase), and during movement (by exploiting modulations in both EEG and EMG).

2 Methods

2.1 Recording Paradigm

Twelve healthy volunteers (8 M and 4 F, aged 23 ± 3.4 years, mean ± standard deviation) participated in the study[1]. Three participants were left-handed, and all participants had normal or corrected-to-normal vision. The recording paradigm consisted of a delayed reach-to-grasping task from a baseline position towards one target object. The baseline position was delimited by a rectangular pad (hereafter termed *home pad*), placed in front of the target object. The latter was presented to the participant by using a custom-made box designed to randomly present one out of 4 possible objects (a *handle*, a *pin*, a *card*, a *ball*) in the same position in front of the participant. Objects were presented to the user when two frontal doors of the box opened, and a LED cued the subject to start the movement (*GO-LED*). See Fig. 1 for a scheme of the box (front view). The home pad and the four target objects embedded touch sensors, thus, they enabled to detect when the participant touched the pad and the objects during the task. The four different object shapes triggered four different prehensions (two power grips: ball, handle; two precision grips: pin, card). In addition to these four conditions, we included a condition without any movement (*no movement* condition), by presenting an empty slot without objects.

[1] The study was approved by the Bioethics Committee of the University of Bologna (protocol code: 61243, date of approval: 15 March 2021).

The recording session consisted of 10 recording blocks, with a short break between blocks, to reduce participant fatigue. In each block, 10 trials were acquired for each condition (50 trials in total); thus, 500 total trials were recorded for each subject. During the task the participant sat on a chair. Each trial (see Fig. 1) started with the participants' dominant hand resting on the home pad while fixating a cross placed in correspondence of the doors of the custom-made box (*fixation phase*, lasting 1.5–2 s, randomly defined). Then, the doors of the box opened, and the participant attended the object and prepared the movement (*object visualization phase*, lasting 2 s), without performing any movement. Finally, the GO-LED turned on, providing the go-signal to the participant, who then performed the forward reaching movement towards the object, grasped the object, and performed the backward reaching movement towards the home pad (*movement phase*, lasting at most 4 s). Once the trial ended, the next trial started after an *inter-trial interval* of 1 s. Objects/empty slot were randomly presented to the user.

During the recording, each participant wore an EEG cap with 60 active electrodes, placed according to the 10/10 system. Moreover, the participant wore an armband on the dominant arm (performing the movement) with 16 monopolar passive electrodes recording surface EMG signals from flexor and extensor muscles of the forearm (8 electrodes above flexors, 8 electrodes above extensors). To this aim, two matrices of electrodes (8 electrodes each) were placed following the setup used in Matran-Fernandez et al. [16]. The 60 EEG and 16 EMG signals were amplified with a g.HIAMP80 Research amplifier (g.tec Medical Engineering GmbH, Schiedlberg, UA, Austria), and sampled at 1200 Hz.

2.2 Signal Pre-processing

Pre-processing steps were applied for preparing the 60 EEG signals for the decoding analysis, similar to the ones applied in our previous study [2] to EEG acquired in another motor task (center-out reaching).

i. Linear detrending of signals belonging to each recording block.
ii. Band-pass filtering between 0.5 and 60 Hz and notch filtering at 50 Hz of signals belonging to each recording block.
iii. Identification of bad EEG channels within each recording block [11].
iv. Concatenation of electrode signals across recording blocks.
v. Removal of channels that were bad (step iii) at least in one recording block.
vi. Removal of artifacts via independent component analysis.
vii. Spherical spline interpolation of the bad channels removed in step v.
viii. Epoching into 5.5 s-length epochs, starting 1.5 s before and ending 4 s after the presentation of the go-signal.
ix. Baseline correction of each trial, by removing the mean from –1.5 s to 0 s.
x. Common average re-referencing.
xi. Downsampling to 150 Hz. This led to EEG epochs $\in \mathbb{R}^{60\times 825}$.

A simpler pre-processing was applied for preparing the 16 EMG signals, including the previous points i, ii (passband between 10 and 500 Hz), viii, x, xi (downsampling to 600 Hz). It is worth remarking that the EMG is characterized

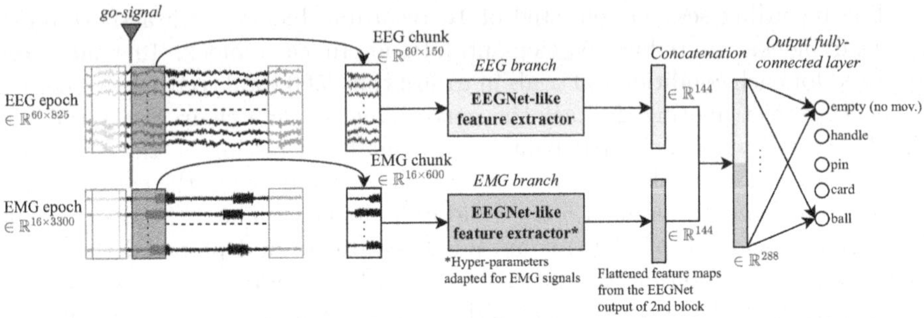

Fig. 2. Scheme of the adopted multi-modal EEG+EMG decoding pipeline. Sliding window decoding was applied. The input layer of the network consisted of EEG and EMG chunks extracted from the EEG and EMG epochs. Chunks were processed independently within two parallel subnetworks, based on EEGNet [13]. EEG and EMG feature maps were then flattened, concatenated together, and provided to the output layer.

by physiological information at higher frequencies than EEG, thus, for EMG we used a wider passband in band-pass filtering and a higher sampling frequency while downsampling. This led to EMG epochs $\in \mathbb{R}^{16\times3300}$.

2.3 Multi-modal EEG+EMG Reach-to-Grasp Decoding

The multi-modal decoding pipeline exploited in this study consisted of a dual-branch CNN that processes EEG and EMG signals in parallel, along an *EEG branch* and an *EMG branch* respectively, and that combines deep feature maps immediately before the output layer. The pipeline was designed to simulate a real-case scenario in which the neural and muscular activity are decoded on small portions of signals (i.e., chunks) during the task. Therefore, here we adopted a sliding window decoding approach, by decoding both neural and muscular activities within 1 s-length chunks sliding over the extracted 5.5-length epochs with a stride of 0.2 s. Therefore, with this procedure the network was fed with EEG chunks $\in \mathbb{R}^{60\times150}$ and EMG chunks $\in \mathbb{R}^{16\times600}$, and the network produced 23 predictions over time. A scheme of the workflow is reported in Fig. 2.

The network architecture was composed by two parallel branches, processing EEG and EMG signals. Both EEG and EMG branches were based on EEG-Net(8,2) (see Lawhern et al. [13] for further details about this network architecture). Specifically, we exploited the convolutional module of EEGNet, i.e., the entire EEGNet network except for the output fully-connected layer. The convolutional module of this network is a shallow network composed by the sequence of 3 convolutional layers: a temporal convolutional layer, a spatial depthwise convolutional layer, and a final temporal separable convolutional layer. Average pooling and dropout are inserted after the last two convolutional layers, and batch normalization is performed after each convolutional layer. The following hyperparameters changed compared to the ones used in Lawhern et al. [13]. The temporal kernel size in the first convolutional layer (temporal convolution) was set at $(1, 16)/(1, 66)$ (EEG branch/EMG branch), the temporal kernel size in the third convolutional layer (separable temporal convolution) was set at $(1,8)/(1,34)$, and the final pooling size and stride – preceding the output layer – were both set at $(1,4)/(1,16)$. Overall, this first part of the network – processing EEG and EMG in parallel – was devoted at extracting neural and muscular representations (EEG and EMG feature maps), useful for motor decoding. Both these representations were flattened as 1D arrays with 144 elements (grey and purple flattened feature maps displayed in Fig. 2), and concatenated together into a single 1D array with 288 elements. Finally, the flattened and concatenated feature maps were provided as input to the output fully-connected layer consisting of N output neurons. Output neurons were activated using the softmax function. In our experiments, we addressed two classification problems: a two-class classification problem to discriminate between no movement and movement condition (collapsing together the four movement conditions, $N = 2$); a five-class classification problem to discriminate among the five different conditions ($N = 5$). The number of trainable parameters was 3794 and 4661, respectively when addressing the first and second decoding problems. Hyperparameters were chosen in preliminary experiments, not reported here for brevity.

Trainable parameters were optimized using the cross-entropy as loss function and Adam as optimizer (learning rate of 1e-4, mini-batch size of 64, maximum number of epochs of 200). Early stopping was performed, and the dropout rate was set to 0.1. Networks were trained by exploiting a within-participant training strategy. Participant-specific decoders, as demonstrated in a previous study [7], outperform cross-participant decoders, primarily due to the high inter-participant variability in EEG signals, and are mainly adopted for designing BCIs. Each participant-specific dataset was split into a training and test set using 5-fold cross-validation. Therefore, in total 5 networks per participant were trained, thus, resulting into 60 trained networks. For each cross-validation fold, 400 epochs (of both EEG and EMG) were included in the training set, and 100 epochs (of both EEG and EMG) were included in the test set. Moreover, 20% of training epochs were held out as validation set for performing early stopping; thus, for each cross-validation fold, 320 epochs were effectively used for designing the training set and 80 epochs were used for the validation set. As previously stated, sliding window decoding was performed. With the adopted sliding window decoding approach, the input for the dual-branch CNN consisted of 23 EEG and EMG chunks of activity sampled from each EEG and EMG epoch. Thus, the number of examples provided as input to the network increased by 23 times compared to the number of epochs. Indeed, sliding window decoding corresponds to performing data augmentation by cropping the original time series. Therefore, for each fold the total number of examples (i.e., EEG and EMG chunks) – effectively presented as input to the network – resulted 7360, 1840, and 2300, respectively for the training, validation, and test set.

3 Experimental Results

In Fig. 3 we report the grand average motor-related cortical potentials (MRCPs) at each electrode and for each condition, to qualitatively assess the recorded EEG

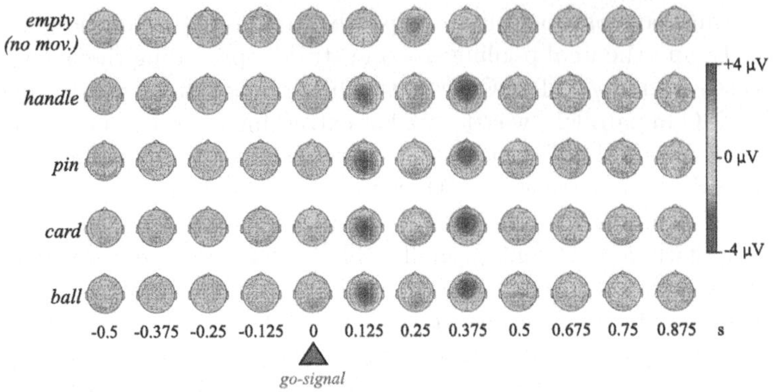

Fig. 3. Motor-related cortical potentials (MRCPs): grand average. This is reported as scalp heatmaps separately for each condition (along rows) between −0.5 s and 0.875 s with respect to the go-signal (0 s) with a step of 0.125 s (time steps along columns). The response is displayed within this interval as it contained most of the motor response.

signals. A clear distinction can be observed between the neural response recorded while preparing and executing reach-to-grasp movements and the no movement condition. As the only difference between movement and no movement conditions regarded the preparation and the execution of a reach-to-grasp movement, the complex of positive and negative deflections observed between 0.125 and 0.375 s with respect to the go-signal was linked to motor-related potentials, and showed a similar dynamics as the ones obtained by Ofner et al. [18] who analyzed MRCPs during upper limb movements.

3.1 Discrimination Between the Presence and Absence of Movement

In the first decoding experiment, we considered the classification between movement vs. no movement conditions. The four movement conditions were merged together as a unique movement condition (2-class classification). Here, we were interested in assessing whether the combination of EEG and EMG is beneficial for detecting motor-related features, and in quantifying the decoding capabilities of these two recording modalities within the main phases of the task (i.e., movement preparation and execution) both considering each modality alone and combined together. To this aim, in addition to the multi-modal EEG+EMG decoding pipeline, we trained and tested also two variant pipelines defined by the EEG branch or by the EMG branch only, thus, realizing mono-modal pipelines based on the neural (EEG) or muscular (EMG) information. For each pipeline (EEG, EMG, EEG+EMG), each participant and each cross-validation fold, we computed the area under the ROC curve (AUROC) on the test set chunks. Then, the performance metric was averaged across folds separately for each participant. We performed paired sample permutation t tests with tmax adjustment for multiple comparisons (10000 iterations). This statistical analysis enabled us to understand if statistical differences emerged across mono-modal (EEG or EMG)

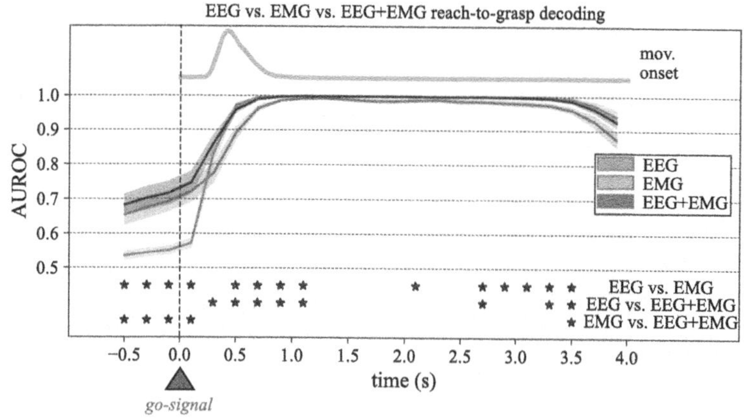

Fig. 4. Decoding performance over time while detecting movement, as scored by mono-modal decoding pipelines (EEG and EMG) and the multi-modal decoding pipeline (EEG+EMG). The bold line represents the mean value, and the shaded area the standard error of the mean across participants. We also report the distribution of the movement onsets (orange bold line, reported on top), as recorded from the home pad. Results from the statistical analysis are reported too (on the bottom), * $p < 0.05$. (Color figure online)

and multi-modal (EEG+EMG) decoding and to disclose in which time period the pipelines differ statistically. Figure 4 reports the AUROC scored by mono-modal pipelines (EEG and EMG), and by the proposed multi-modal pipeline (EEG+EMG). Here, the performance is displayed as a function of time: indeed, the performance was computed for each chunk provided as input to the CNN, resulting in 23 model predictions within the entire epoch time course. The temporal dynamics of the performance showed a common abrupt increase in correspondence of the movement onset (orange curve in Fig. 4), reaching and maintaining a peak during movement execution. During movement preparation (before movement onset), a significant performance improvement was observed in the pipeline exploiting EEG only (blue in Fig. 4) vs. the pipeline exploiting EMG only (red in Fig. 4). On the other hand, the latter was significant more accurate in detecting the movement during the movement execution phase. When using the combination of the neural and muscular activities, the advantages provided by the two single recording modalities add up. Indeed, the EEG+EMG decoding pipeline (black in Fig. 4) outperformed significantly the EMG pipeline during movement preparation, and the EEG pipeline during movement initiation (i.e., around movement onset).

3.2 Discrimination Between Grip Types and Absence of Movement

In the second decoding experiment, we considered the classification of the five different conditions (four movements and no movement condition). As from our

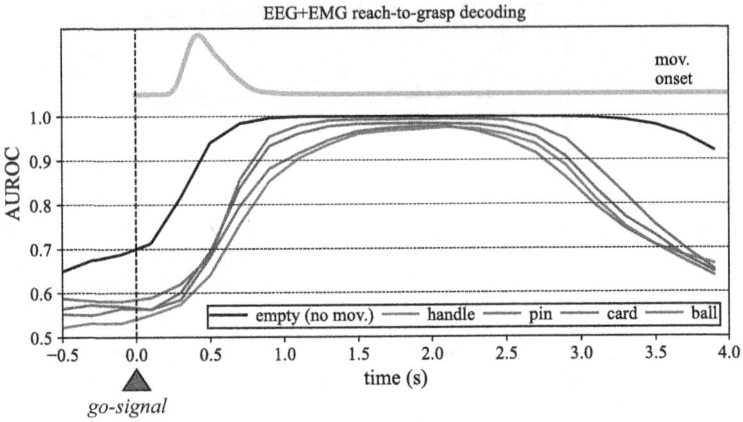

Fig. 5. Decoding performance over time scored while discriminating between the five conditions with the multi-modal decoder (EEG+EMG). To ease the reading we report only the mean value across participants. See also the caption of Fig. 4.

results the best trade-off between performance during movement preparation and during movement execution was obtained using the EEG+EMG decoding pipeline, we addressed this 5-class classification problem using this pipeline. The performance was aggregated as in the previous experiment. Figure 5 reports the test set AUROC scored by the multi-modal pipeline while discriminating each condition. The dynamics of the decoding performance was similar to the one reported in the previous experiment, i.e., the performance for each class exhibited a rapid increase around the movement onset. Moreover, the decoder exhibited the highest performance for discriminating the absence of movement (empty (no mov.) condition in the figure) even during movement preparation, as it resulted the easiest condition to distinguish when compared to movement conditions (see the differences in MRCPs showed in Fig. 3). Moreover, the decoder was able to distinguish between grip types approx. after movement initiation, showing higher performance for discriminating precision grips (triggered by the prehension of pin and card objects) than power grips (triggered by the prehension of handle and ball objects) during movement execution.

4 Conclusions

In this study we recorded EEG and EMG during a delayed reach-to-grasping task towards four different objects, and we designed and tested a multi-modal EEG+EMG decoder based on a dual-branch CNN architecture for discriminating complex hand movements. The multi-modal approach encapsulated the positive aspects of both mono-modal decoding pipelines (EEG or EMG alone), as it was able to decode movement already during the movement preparation phase (exploiting EEG motor-related correlates), and reached the maximum of performance early during movement execution (exploiting both EEG and EMG motor-related correlates). Moreover, our approach was capable of discriminating the grip type approx. after movement initiation. Future studies will extend the validation of the multi-modal approach also to other models (e.g., transformers), providing a performance benchmark for multi-modal motor decoding, and will extend the comparative analysis between mono-modal and multi-modal decoders also in multi-class classification tasks (here presented only in a binary classification task). The results obtained in our experiments – although preliminary – further support the validation of multi-modal decoding frameworks for prospective BCI applications that aim at a more natural and engaging user experience.

Acknowledgments. This work is supported by #NEXTGENERATIONEU (NGEU) and funded by the Ministry of University and Research (MUR), National Recovery and Resilience Plan (NRRP), project MNESYS (PE0000006)—A Multiscale integrated approach to the study of the nervous system in health and disease (DN. 1553 11.10.2022).

References

1. Borra, D., Bossi, F., Rivolta, D., Magosso, E.: Deep learning applied to EEG source-data reveals both ventral and dorsal visual stream involvement in holistic processing of social stimuli. Sci. Rep. **13**(1) (2023). http://dx.doi.org/10.1038/s41598-023-34487-z
2. Borra, D., Fantozzi, S., Bisi, M.C., Magosso, E.: Modulations of cortical power and connectivity in alpha and beta bands during the preparation of reaching movements. Sensors **23**(7), 3530 (2023). http://dx.doi.org/10.3390/s23073530
3. Borra, D., Fantozzi, S., Magosso, E.: EEG motor execution decoding via interpretable Sinc-convolutional neural networks. In: Henriques, J., Neves, N., de Carvalho, P. (eds.) MEDICON 2019. IP, vol. 76, pp. 1113–1122. Springer, Cham (2020). https://doi.org/10.1007/978-3-030-31635-8_135
4. Borra, D., Filippini, M., Ursino, M., Fattori, P., Magosso, E.: Motor decoding from the posterior parietal cortex using deep neural networks. J. Neural Eng. **20**(3), 036016 (2023). http://dx.doi.org/10.1088/1741-2552/acd1b6
5. Borra, D., Filippini, M., Ursino, M., Fattori, P., Magosso, E.: Convolutional neural networks reveal properties of reach-to-grasp encoding in posterior parietal cortex. Comput. Biol. Med. **172**, 108188 (2024). http://dx.doi.org/10.1016/j.compbiomed.2024.108188
6. Borra, D., Magosso, E.: Deep learning-based EEG analysis: investigating P3 ERP components. J. Integr. Neurosci. **20**(4), 791–811 (2021). http://dx.doi.org/10.31083/j.jin2004083
7. Borra, D., Mondini, V., Magosso, E., Müller-Putz, G.R.: Decoding movement kinematics from EEG using an interpretable convolutional neural network. Comput. Biol. Med. **165**, 107323 (2023). http://dx.doi.org/10.1016/j.compbiomed.2023.107323
8. Deng, X., Zhang, B., Yu, N., Liu, K., Sun, K.: Advanced TSGL-EEGNet for motor imagery EEG-based brain-computer interfaces. IEEE Access **9**, 25118–25130 (2021). http://dx.doi.org/10.1109/ACCESS.2021.3056088

9. Dremstrup, K., Gu, Y., Nascimento, O.F.D., Farina, D.: Movement-related cortical potentials and their application in brain-computer interfacing. In: Farina, D., Jensen, W., Akay, M. (eds.) Introduction to Neural Engineering for Motor Rehabilitation (2013). http://dx.doi.org/10.1002/9781118628522.ch13
10. Filippini, M., Borra, D., Ursino, M., Magosso, E., Fattori, P.: Decoding sensorimotor information from superior parietal lobule of macaque via convolutional neural networks. Neural Netw. **151**, 276-294 (2022). http://dx.doi.org/10.1016/j.neunet.2022.03.044
11. Fischler, M.A., Bolles, R.C.: Random sample consensus: a paradigm for model fitting with applications to image analysis and automated cartography. Commun. ACM **24**(6), 381–395 (1981). http://dx.doi.org/10.1145/358669.358692
12. Kim, S., Shin, D.Y., Kim, T., Lee, S., Hyun, J.K., Park, S.M.: Enhanced recognition of amputated wrist and hand movements by deep learning method using multimodal fusion of electromyography and electroencephalography. Sensors **22**(2), 680 (2022). http://dx.doi.org/10.3390/s22020680
13. Lawhern, V.J., Solon, A.J., Waytowich, N.R., Gordon, S.M., Hung, C.P., Lance, B.J.: EEGNet: a compact convolutional neural network for EEG-based brain–computer interfaces. J. Neural Eng. **15**(5), 056013 (2018). https://doi.org/10.1088/1741-2552/aace8c
14. Leeb, R., Sagha, H., Chavarriaga, R., Millán, J.D.R.: A hybrid brain-computer interface based on the fusion of electroencephalographic and electromyographic activities. J. Neural Eng. **8**(2), 025011 (2011). http://dx.doi.org/10.1088/1741-2560/8/2/025011
15. Lotte, F., et al.: A review of classification algorithms for EEG-based brain-computer interfaces: a 10 year update. J. Neural Eng. **15**(3), 031005 (2018). http://dx.doi.org/10.1088/1741-2552/aab2f2
16. Matran-Fernandez, A., Rodríguez Martínez, I.J., Poli, R., Cipriani, C., Citi, L.: Seeds, simultaneous recordings of high-density EMG and finger joint angles during multiple hand movements. Sci. Data **6**(1) (2019). http://dx.doi.org/10.1038/s41597-019-0200-9
17. Neuper, C., Wörtz, M., Pfurtscheller, G.: ERD/ERS patterns reflecting sensorimotor activation and deactivation, pp. 211–222. Elsevier (2006). http://dx.doi.org/10.1016/S0079-6123(06)59014-4
18. Ofner, P., Schwarz, A., Pereira, J., Müller-Putz, G.R.: Upper limb movements can be decoded from the time-domain of low-frequency EEG. PLOS ONE **12**(8), e0182578 (2017). http://dx.doi.org/10.1371/journal.pone.0182578
19. Riyad, M., Khalil, M., Adib, A.: MI-EEGNET: a novel convolutional neural network for motor imagery classification. J. Neurosci. Methods **353**, 109037 (2021). http://dx.doi.org/10.1016/j.jneumeth.2020.109037
20. Roy, Y., Banville, H., Albuquerque, I., Gramfort, A., Falk, T.H., Faubert, J.: Deep learning-based electroencephalography analysis: a systematic review. J. Neural Eng. **16**(5), 051001 (2019). http://dx.doi.org/10.1088/1741-2552/ab260c
21. Schirrmeister, R.T., et al.: Deep learning with convolutional neural networks for EEG decoding and visualization. Hum. Brain Mapp. **38**(11), 5391–5420 (2017). https://doi.org/10.1002/hbm.23730
22. Simões, M., et al.: BCIAUT-P300: a multi-session and multi-subject benchmark dataset on autism for p300-based brain-computer-interfaces. Front. Neurosci. **14** (2020). http://dx.doi.org/10.3389/fnins.2020.568104

23. Wolpaw, J.R., Birbaumer, N., McFarland, D.J., Pfurtscheller, G., Vaughan, T.M.: Brain-computer interfaces for communication and control. Clin. Neurophysiol. **113**(6), 767–791 (2002). http://dx.doi.org/10.1016/s1388-2457(02)00057-3
24. Xu, D., Tang, F., Li, Y., Zhang, Q., Feng, X.: An analysis of deep learning models in SSVEP-based BCI: a survey. Brain Sci. **13**(3), 483 (2023). http://dx.doi.org/10.3390/brainsci13030483

Applications in Computer Vision

VAeViT: Fusing Multi-views for Complete 3D Object Recognition

Ahmed Yasser Eita(✉), Hafsa Ennajari, Akinlolu Oluwabusayo Ojo, and Nizar Bouguila

CIISE, Concordia University, Montreal, Canada
ahmed.eita@mail.concordia.ca, h_ennaja@encs.concordia.ca, {akinloluoluwabusayo.ojo,nizar.bouguila}@concordia.ca

Abstract. The Vision Transformer (ViT) emergence has ushered in a new era in 2D image classification, yet the realm of 3D object recognition presents distinct challenges. To bridge this gap, we introduce VAeViT, a pioneering hybrid model that seamlessly integrates the strengths of Vision Transformers and Variational Autoencoders (VAE). VAeViT leverages VAE's efficiency in feature representation to encode the 3D object views into a lower-dimensional latent space and ViT's superiority for semantic feature extraction and incorporating the positional information to aggregate those low-dimension views into the most representative features for the whole object, providing comprehensive perception of 3D multi-view objects that enables accurate object recognition from any posture. Our experiments on the ModelNet10 and ModelNet40 benchmark datasets show that VAeViT achieves superior accuracy compared to existing models, even when utilizing smaller datasets. This highlights VAeViT's potential to improve 3D object recognition in various computer vision applications significantly. For reproducibility, our code is available at https://github.com/AhmedYEita/VAeViT.

Keywords: Vision Transformer · Variational Autoencoders · 3D object recognition · 3D multi-view

1 Introduction

In the ever-evolving landscape of computer vision, 3D object recognition stands as a pivotal frontier. The capacity to accurately identify and classify 3D objects is integral to numerous fields, including robotics [22], medical imaging [11], and scene monitoring [1] to name a few. This interdisciplinary effort combines progress in deep learning, and sensor technology to understand complex spatial structures in the real world. It positions 3D object recognition as a vital link between digital and physical realms, poised to revolutionize industries, enhance human-machine interaction, and advance artificial intelligence.

Since the inception of 3D object recognition, researchers have been working diligently to develop models in the three prevalent 3D representations:

C. Y. Suen et al. (Eds.): ANNPR 2024, LNAI 15154, pp. 183–194, 2024.
https://doi.org/10.1007/978-3-031-71602-7_16

voxel-based, point cloud, and multi-view. Voxel-based representations [2,15,19], involve partitioning objects into grids to obtain detailed 3D data. The works in [12,16,31] present point cloud representations which constitute scattered points in space. This type is known to be ideal for tasks such as LIDAR-based perception. However, both approaches require expensive resources and generally result in a lower performance than other representation types due to the complexity of their representation. To circumvent these challenges, the multi-view approach has been explored in [3,10,13,17,27]. Multi-view methods excel as they offer interpretability, capture detailed object features from various angles, and are resource-efficient. This motivates our research to primarily adopt the multi-view representations technique.

The successes of CNNs have prompted exploration into 3D object representation and classification tasks, resulting in significant advancements. Recent studies have proposed innovative approaches such as consolidating information from multiple perspectives into unified shape descriptors [21] and exploring view clustering and pooling layers based on dominant sets [24]. However, these approaches often struggle with sensitivity to viewpoint variations, potentially losing crucial visual information.

In the current landscape, Vision Transformers (ViT) [6] have emerged as dominant forces in computer vision, offering exceptional feature capture capabilities without sacrificing essential information. Unlike previous models like RNN and LSTM, ViT addresses short-term memory challenges through self-attention mechanisms inspired by Transformer models [23]. Multi-view Vision Transformer (MVT), the current state-of-the-art model for 3D object recognition [3], adapts ViT for multi-view object representation, but improvements are needed, particularly in reducing dependence on pre-training on large datasets. This paper introduces VAeViT, an enhanced ViT model incorporating a preceding Variational Autoencoder (VAE) [14], leveraging VAE's latent feature representation and ViT's deep feature and positional information capture for comprehensive 3D object representation. Our model leverages VAE's capacity to learn valuable latent feature representation, along with ViT's ability to capture deep features and positional information from multi-views, to comprehensively represent 3D objects. Overall, our contributions are as follows:

- We propose a sequential VAE-ViT model, code-named VAeViT, which offers an innovative approach to learning distinct levels of representations independently. VAeViT has the capacity to capture both low-dimensional and deep features while preserving the positional information.
- The proposed model is specifically designed for 3D multi-view objects, where VAE processes 2D object views into latent features and ViT integrates the deep features of the learned view representation with positional information for a holistic and global 3D object understanding.
- We conduct extensive experiments on two benchmark 3D multi-view datasets, each featuring two variants of varying sizes and numbers of views to evaluate VAeViT alongside state-of-the-art models. The results demonstrate the competitive performance of VAeViT in comparison to existing approaches.

2 Background and Related Work

Numerous CNN-based 3D multi-view object classification methods have been introduced in the past several years, including MVCNN [21] and MVCNN-MultiRes [17], which utilize max-pooling with the conventional 2D CNNs. However, these methods may discard vital visual information due to the maximal activations being retained from only a specific view. Innovative strategies for view feature aggregation have been explored in subsequent works like RCPCNN [24] and GVCNN [8]. Despite advancements, sensitivity to viewpoint variations persists, as seen in works like MHBN [30] and MVLADN [29]. Additionally, Seqviews2seqlabels [10] and 3D2SeqViews [9] introduced Recurrent Neural Networks (RNNs) to model view order. However, they are still sensitive to viewpoint variations. While View-GCN [25] addresses this by employing graph convolution networks, constructing an effective graph poses challenges.

RotationNet [13] addresses rotation variance by estimating pose and object category from multi-view images but is constrained by predefined viewpoints. In contrast, Relation Network [28] enhances patch features by integrating information from all views, though challenges in generalization persist. F-GDA [7] combines Bag of Visual Words (BoVWs) and topic modeling to capture semantic structures in data, leveraging the Generalized Dirichlet distribution for flexibility and generalization. CAR-Net [27] explicitly identifies intra-view and cross-view correspondences via kNN search, with the challenge lying in selecting an appropriate k-value to avoid noisy or irrelevant correspondences. Despite this, CAR-Net exhibits promise in capturing shape features from correspondences through acquired transformations.

Recent studies [4,5,20] have incorporated Transformer encoders into the VAE framework for tasks such as anomaly detection and music generation. This integration leverages ViT's proficiency in capturing sequential information and extracting crucial data features. These resulting hybrid models are, however, computationally expensive, require a huge amount of data to be trained, and are very hard to fine-tune because of the complexity of the model. To address these limitations, we propose VAeViT, a hybrid Transformer-VAE model for 3D multi-view object classification.

3 Model Architecture

In this section, we introduce the architecture of the proposed VAeViT model as illustrated in Fig. 1. As may be seen, the architecture of VAeViT incorporates the structure of VAE, extracts the learning component of interest as an output, and adapts it to fit in the ViT's input via some tweaks in the interconnection layer. The model takes input $\boldsymbol{X} \in R^{V \times N}$, where $\boldsymbol{X}$ is a 2D view. V represents the number of views per object, and N is the number of objects in the dataset. This input is first encoded into a lower-dimensional feature representation by the VAE. Subsequently, the output is combined with spatial information in the positional embedding interconnection layer. Finally, the data undergoes further

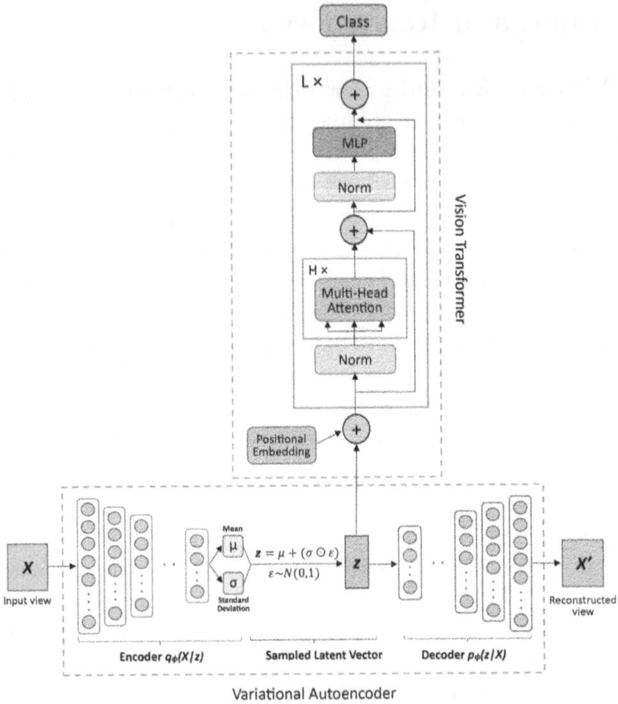

Fig. 1. VAeViT architecture.

processing in the last layer of ViT before object classification is performed by the last layers of ViT. The next three subsections will illustrate in detail the methodology of each of those components and its role in constructing a comprehensive 3D object recognition model.

3.1 Views Latent Feature Embedding

The initial component of the VAeViT model, which receives the input, is a latent feature embedding module that contributes significantly to unsupervised learning, which is mainly based on a VAE architecture [14]. The probabilistic encoder $q_\phi(\boldsymbol{z}|\boldsymbol{X})$ accepts the input views and processes them through all its blocks and outputs two variables, which are the mean μ and the standard deviation σ. The reparameterization trick in Eq. 1 is then deployed to sample the latent vector $\boldsymbol{z}$ since the gradient cannot be pushed through a sampling node during the back-propagation process of the training.

$$\boldsymbol{z} = \mu + (\sigma \odot \varepsilon) \tag{1}$$

where $\varepsilon \sim Normal(0, 1)$. Subsequently, the probabilistic decoder $p_\phi(\boldsymbol{X}|\boldsymbol{z})$ takes the sampled vector $\boldsymbol{z}$, processes it through its blocks, and tries to reconstruct the original object view. In each iteration, the model adapts its updated weights to produce an output $\boldsymbol{X'}$, that closely approximates the original input $\boldsymbol{X}$ to the best of the model's ability. The model keeps on backpropagating aiming to minimize the reconstruction and KL-divergence losses in Eq. 2 until optimum results are reached.

$$L_{VAE}(\theta, \phi; \boldsymbol{X}) = \frac{1}{2}\sum(\mu^2 + \sigma^2 - \log(\sigma^2) - 1) - E_{\boldsymbol{z}\sim q_\phi(\boldsymbol{z}|\boldsymbol{X})}[\log P_\theta(\boldsymbol{X}|\boldsymbol{z})] \quad (2)$$

Since the VAE model is utilized in this framework to represent the 2D object views in a lower dimension, we focus on the sampled vector from each view during the last training epoch. This obtained output represents the embeddings $\boldsymbol{z} \in R^{V\times N}$ that will be combined with the positional embeddings in the next phase to represent a global perception of the 3D object before it is fed to the ViT model.

3.2 Views Positional Embedding

Positional embeddings provide spatial information added to the embedded 2D object views for understanding their order. Unlike the fixed embeddings in [23], in VAeViT, we use learnable embeddings, where we have a set of embeddings matching the number of views per object (12 or 20 in our case). Each number, representing a 2D view's location, is transformed into a vector of the same size as the latent vectors $\boldsymbol{z}$ in an embedding table. Each view embedding and its corresponding position embedding are summed up together to form the complete input $\boldsymbol{g} \in R^{V\times N}$ which carries both spatial and semantic features. Positional embeddings play a crucial role by offering contextual information about the input segment being processed. This aids the model in determining not only the object's class but also its pose from any angle.

3.3 3D Objects Global Feature Learning

Finally, in order to effectively capture global context information and account for long-range dependencies within complex 3D object views, we integrate ViT into our approach. Specifically, ViT takes the input $\boldsymbol{g}$, aggregates all the views for each object together, and feeds the Transformer encoder with an input of the format $[[\boldsymbol{g}] \times V] \times N$. The initial block in the Transformer encoder is layer normalization, a common feature in Transformer architectures to improve training stability and efficiency. It computes mean (μ) and standard deviation (σ) from input vectors, facilitating linear operations for affine transformation parameters. The output vectors are directed to the multi-head attention layer, incorporating the self-attention mechanism for profound semantic representations. This output is then added to the original input $[[\boldsymbol{g}] \times V]$ using a residual connection, aiding gradient flow in training. The output of that goes to another normalization layer for further optimization.

The resulting normalized residual output then enters a Multi-layer perceptron (MLP) where each input token gets processed individually. It involves a linear transformation, where embeddings are multiplied by a learned weight matrix and added to the learned bias vector. A non-linear activation function (GELU in our case) is employed subsequently to allow more complex pattern learning. Incorporating a dropout layer within the ViT's structure is a beneficial practice, proven to enhance the performance by up to 4% in recognition through improved representation propagation across layers. Another residual connection is employed, and the constructed output of class scores is then utilized for classification through a Softmax function, which converts those raw scores into probabilities, indicating the likelihood of each 3D object belonging to a specific class.

4 Experimental Results and Discussion

4.1 Datasets

To evaluate our model, we perform experiments on two benchmark datasets, each with two variations based on object count and captured views. The smaller variants show the model's performance on small datasets. We use a gray-scale style to expedite testing and reduce computation costs.

ModelNet40 [26] consists of 12,311 3D CAD models, categorized into 40 groups. Multiple 2D views are captured for each 3D CAD object from random postures according to the targeted number of views. The two assessed variants are:

- 12 views of 3,200 objects resulting in 38,400 images.
- 20 views of 9,843 objects resulting in 196,860 images.

ModelNet10 [26] is a subset of ModelNet40 that contains 4,899 objects from only 10 categories. The two variants are:

- 12 views of 800 objects resulting in 9,600 images.
- 20 views of 4,899 objects resulting in 97,980 images.

4.2 Ablation Study

In this section, we conduct extensive ablation experiments to learn about the effect of the model's attributes on the overall performance. During all trials, we have considered a batch size of 128 views, each with the shape of $(64 \times 64 \times 1)$. For ViT, the *AdamW* optimizer was utilized with a learning rate of 0.001 and a weight decay of 0.0001. We have also followed a split of 72% for training, 8% for validation, and 20% for testing. In the following three subsections, we will use only the smaller-sized variant of ModelNet10 to streamline computations.

Table 1. Accuracy comparison for different latent dimension d_k

Model variant	256 d_k	384 d_k	512 d_k	768 d_k
VAE	75.36	74.4	77.0	75.36
VAeViT	97.5	88.13	**98.12**	92.5

Table 2. VAeViT epochs accuracy comparison.

Model variant	ModelNet10		ModelNet40	
	12v	20v	12v	20v
VAE	77.0	95.9	67.0	50.36
VAeViT 100 ViT epochs	**98.12**	99.0	90.31	93.71
VAeViT 200 ViT epochs	96.25	**99.67**	94.22	96.1
VAeViT 300 ViT epochs	–	–	**96.88**	**98.02**

Latent Vector Dimension Influence. The learned latent vectors of views serve as fundamental representations encoding the essential features and characteristics of observed objects. Their dimensionality significantly affects performance and the comprehensiveness of representation. Table 1 demonstrates the impact of various vector dimensions on overall accuracy and highlights the complexity of each model. Based on these experimental results, the dimension of 512 is illustrated to be the most optimal one, and it is noticeable that reducing the dimension d_k impacts model quality negatively, while increasing it too much makes the training process more complex and consumes more power.

ViT Epochs Number Influence on Views Number. VAE significantly streamlines the ViT by condensing 2D views into lower-dimensional latent vectors. However, for extensive datasets with numerous categories, VAE may not perform optimally as it requires more time to discern the semantic features for each 3D object class. Since VAE takes a much longer time to train than ViT, it is advisable to allocate more epochs to ViT. See Table 2 for ViT's epoch impact on each dataset variant and the influence of increasing views and dataset size on overall accuracy.

As for the number of views, it can be asserted that the 20-view dataset variants are generally better than the 12-views because they have more information about the object. However, more epochs are required to fully grasp the diverse characteristics of each category in large datasets. Also, from our model's perspective, a good dataset size is larger than any other 3D multi-view object classification model's average dataset size. Simply because it sees a 9,600 3D multi-view image dataset of 12 views as $9600 \div 12 = 800$ objects. Of those 800 objects, only 80% are for training, which means that VAeViT perceives this 9,600 image dataset as 640 of $[[d_k] \times 12]$ vectors.

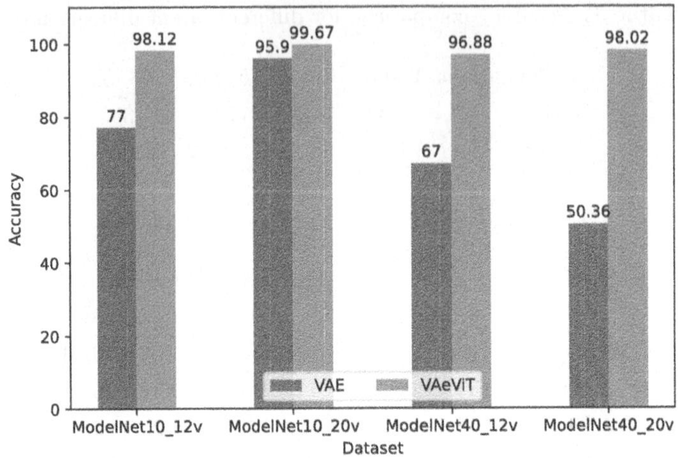

Fig. 2. Comparison of accuracy between VAE-only and VAE-ViT across all dataset variants.

Based on the VAE accuracy in Table 2, the ModelNet40 variant with 20 views exhibits the poorest representation initially. However, both ModelNet40 variants consistently improve with increasing ViT epochs, reaching a remarkably high accuracy of 98.02% for the 20 views variant. Conversely, increasing ViT epochs on ModelNet10 with 12 views leads to overfitting and a subsequent decrease in accuracy. In summary, larger datasets and/or higher numbers of views necessitate more ViT epochs for optimal results, resulting in increased consistency and accuracy.

4.3 Optimum Model

The bar plot in Fig. 2 illustrates the accuracy of all the dataset variants when evaluated with VAE only versus VAeViT's optimum model based on the previous experimentation. A significant enhancement that is up to 47.66% can be observed from the plot if VAeViT is instead employed.

4.4 Comparison with State-of-the-Art Methods

We conducted a comprehensive comparison between VAeViT and state-of-the-art models for each 3D object representation type, as detailed in Sect. 1 and summarized in Table 3. Voxel-based models generally exhibit lower recognition accuracy compared to their multi-view counterparts, as observed in the first section of the table. While VRN Ensemble [2] outperformed some multi-view methods, this exceptional performance can be attributed to its ensemble approach and incorporation of a more advanced base model. The second part shows point-based methods, which demonstrate comparable performance to voxel-based models. Most recent approaches that are based on voxels and

Table 3. 3D object classification accuracy comparison with state-of-the-art models.

Model	Input Modality	ModelNet10	ModelNet40
3DShapeNets [26]	Voxel-based	83.5	77.0
3D-A-Nets [19]		–	90.5
LP-3DCNN [15]		94.4	92.1
VRN Ensemble [2]		97.1	95.5
PointNet++ [18]	Point cloud	–	91.9
DeepCCFV [12]		–	92.5
LDGCNN [31]		–	92.9
RS-CNN [16]		–	93.6
MVCNN [21]	12 views	–	89.9
MVCNN [21]	80 views	–	90.1
MVCNN-MultiRes [17]	20 views	–	91.4
GVCNN [8]	3 views	–	93.1
GVCNN [8]	12 views	–	92.6
RCPCNN [24]	12 views	–	93.8
3D2SeqViews [9]	12 views	94.7	93.4
SeqViews2SeqLabels [10]	12 views	94.8	93.4
MHBN [30]	6 views	95.0	94.7
MHBN [30]	12 views	–	93.4
MVLADN [29]	6 views	94.9	94.6
RotationNet [13]	12 views	94.0	91.0
RotationNet [13]	20 views	98.5	97.4
Relation Network [28]	12 views	95.3	94.3
CAR-Net [27]	12 views	95.8	95.2
CAR-Net [27]	20 views	99.0	97.7
MVT [3]	12 views	95.3	94.4
MVT [3]	20 views	99.3	97.5
View-GCN [25]	20 views	–	97.6
VAeViT (Ours)	**12 views**	**98.12**	**96.88**
VAeViT (Ours)	**20 views**	**99.67**	**98.02**

point clouds tend to achieve similar accuracy, typically falling within the range of 90%–93%. This convergence may be attributed to the inherent complexity or computational demands of these representations.

In evaluating our model against leading approaches for 3D multi-view types, we considered various pioneering methods. Each model demonstrates unique strengths and limitations based on its specific approach and preferred number of views. We notice from models such as GVCNN [8] and MHBN [30] that the higher number of views does not always yield better performance. However, it could be perceived from the baseline models: RotationNet [13], CAR-Net [27], MVT [3], and View-GCN [25] that the 20 view models always perform the highest. Although some of these methods have been pioneering for a few years, VAeViT, with its smaller data requirements as compared to them, has the potential to dominate the field after it has demonstrated a competitive performance of 99.67% and 98.02% on ModelNet10 and ModelNet40 respectively.

5 Conclusion and Future Work

In our research, we introduce a new hybrid Transformer-VAE model designed for sequential 3D multi-view object classification. Our primary aim is to create a methodology for this classification that's highly accurate and robust, without the need for an extensive dataset, unlike the conventional ViT. To achieve this, we propose VAeViT, which harnesses the strengths of both VAE for robustly representing high-level visual features and ViT for capturing semantic image characteristics while incorporating spatial knowledge. In our model, VAE extracts low-level features, encoding each 2D view of a 3D object into lower-dimensional continuous vectors. ViT then utilizes these latent vectors, adding positional information and employing a self-attention mechanism to capture deep relational characteristics. This approach enables the model to acquire a global perception of a 3D object, effectively attending to it from any angle while needing significantly smaller datasets than other methodologies in the field. We thoroughly examined the impact of the model's key attributes on overall performance, comparing the most optimal model with leading methodologies. VAeViT demonstrated its potential by surpassing all state-of-the-art methodologies on two benchmark 3D multi-view datasets. Despite its need for smaller datasets, the model still requires a considerable amount of computational resources for training because of the heaviness of the conventional VAE model, which might be challenging for smaller-scale applications. Therefore, our future work will focus on developing a streamlined version of VAE within the VAeViT framework, ensuring it maintains robust efficacy while being more adaptable to smaller applications.

Acknowledgment. The completion of this research was made possible thanks to the Natural Sciences and Engineering Research Council of Canada (NSERC).

References

1. Bourouis, S., Laalaoui, Y., Bouguila, N.: Bayesian frameworks for traffic scenes monitoring via view-based 3D cars models recognition. Multimedia Tools Appl. **78**, 18813–18833 (2019)
2. Brock, A., Lim, T., Ritchie, J.M., Weston, N.: Generative and discriminative voxel modeling with convolutional neural networks. arXiv preprint arXiv:1608.04236 (2016)
3. Chen, S., Yu, T., Li, P.: MVT: multi-view vision transformer for 3D object recognition. arXiv preprint arXiv:2110.13083 (2021)

4. Chen, T., Li, B., Zeng, J.: Learning traces by yourself: blind image forgery localization via anomaly detection with ViT-VAE. IEEE Sig. Process. Lett. **30**, 150–154 (2023)
5. Choi, B., Jeong, J.: ViV-Ano: anomaly detection and localization combining vision transformer and variational autoencoder in the manufacturing process. Electronics **11**(15), 2306 (2022)
6. Dosovitskiy, A., et al.: An image is worth 16 × 16 words: transformers for image recognition at scale. arXiv preprint arXiv:2010.11929 (2020)
7. Eita, A.Y., Ennajari, H., Bouguila, N.: 3D multi-views object classification based on a fully generalized dirichlet allocation model. In: 2023 IEEE International Conference on Industrial Technology (ICIT), pp. 1–7. IEEE (2023)
8. Feng, Y., Zhang, Z., Zhao, X., Ji, R., Gao, Y.: GVCNN: group-view convolutional neural networks for 3D shape recognition. In: Proceedings of the IEEE Conference on Computer Vision and Pattern Recognition, pp. 264–272 (2018)
9. Han, Z., et al.: 3D2SeqViews: aggregating sequential views for 3D global feature learning by CNN with hierarchical attention aggregation. IEEE Trans. Image Process. **28**(8), 3986–3999 (2019)
10. Han, Z., et al.: SeqViews2SeqLabels: learning 3D global features via aggregating sequential views by RNN with attention. IEEE Trans. Image Process. **28**(2), 658–672 (2018)
11. Hatamizadeh, A., et al.: UNETR: transformers for 3D medical image segmentation. In: Proceedings of the IEEE/CVF Winter Conference on Applications of Computer Vision, pp. 574–584 (2022)
12. Huang, Z., Zhao, Z., Zhou, H., Zhao, X., Gao, Y.: DeepCCFV: camera constraint-free multi-view convolutional neural network for 3D object retrieval. In: Proceedings of the AAAI Conference on Artificial Intelligence, vol. 33, pp. 8505–8512 (2019)
13. Kanezaki, A., Matsushita, Y., Nishida, Y.: RotationNet: joint object categorization and pose estimation using multiviews from unsupervised viewpoints. In: Proceedings of the IEEE Conference on Computer Vision and Pattern Recognition, pp. 5010–5019 (2018)
14. Kingma, D.P., Welling, M.: Auto-encoding variational bayes. arXiv preprint arXiv:1312.6114 (2013)
15. Kumawat, S., Raman, S.: LP-3DCNN: unveiling local phase in 3D convolutional neural networks. In: Proceedings of the IEEE/CVF Conference on Computer Vision and Pattern Recognition, pp. 4903–4912 (2019)
16. Liu, Y., Fan, B., Xiang, S., Pan, C.: Relation-shape convolutional neural network for point cloud analysis. In: Proceedings of the IEEE/CVF Conference on Computer Vision and Pattern Recognition, pp. 8895–8904 (2019)
17. Qi, C.R., Su, H., Nießner, M., Dai, A., Yan, M., Guibas, L.J.: Volumetric and multi-view CNNs for object classification on 3D data. In: Proceedings of the IEEE Conference on Computer Vision and Pattern Recognition, pp. 5648–5656 (2016)
18. Qi, C.R., Yi, L., Su, H., Guibas, L.J.: PointNet++: deep hierarchical feature learning on point sets in a metric space. In: Advances in Neural Information Processing Systems, vol. 30 (2017)
19. Ren, M., Niu, L., Fang, Y.: 3D-A-Nets: 3D deep dense descriptor for volumetric shapes with adversarial networks. arXiv preprint arXiv:1711.10108 (2017)
20. Song, K., Liang, X., Wu, J.: ViT-based VQ-VAE generative network for accompaniment generation. In: Proceedings of the 2021 4th International Conference on Algorithms, Computing and Artificial Intelligence, pp. 1–5 (2021)

21. Su, H., Maji, S., Kalogerakis, E., Learned-Miller, E.: Multi-view convolutional neural networks for 3D shape recognition. In: Proceedings of the IEEE International Conference on Computer Vision, pp. 945–953 (2015)
22. Tang, Y., et al.: Recognition and localization methods for vision-based fruit picking robots: a review. Front. Plant Sci. **11**, 510 (2020)
23. Vaswani, A., et al.: Attention is all you need. In: Advances in Neural Information Processing Systems, vol. 30 (2017)
24. Wang, C., Pelillo, M., Siddiqi, K.: Dominant set clustering and pooling for multi-view 3D object recognition. arXiv preprint arXiv:1906.01592 (2019)
25. Wei, X., Yu, R., Sun, J.: View-GCN: view-based graph convolutional network for 3D shape analysis. In: Proceedings of the IEEE/CVF Conference on Computer Vision and Pattern Recognition, pp. 1850–1859 (2020)
26. Wu, Z., et al.: 3D ShapeNets: a deep representation for volumetric shapes. In: Proceedings of the IEEE Conference on Computer Vision and Pattern Recognition, pp. 1912–1920 (2015)
27. Yong, X., Zheng, C., Ruotao, X., Quan, Y., Ling, H.: Multi-view 3D shape recognition via correspondence-aware deep learning. IEEE Trans. Image Process. **30**, 5299–5312 (2021)
28. Yang, Z., Wang, L.: Learning relationships for multi-view 3D object recognition. In: Proceedings of the IEEE/CVF International Conference on Computer Vision, pp. 7505–7514 (2019)
29. Tan, Yu., Meng, J., Yang, M., Yuan, J.: 3D object representation learning: a set-to-set matching perspective. IEEE Trans. Image Process. **30**, 2168–2179 (2021)
30. Yu, T., Meng, J., Yuan, J.: Multi-view harmonized bilinear network for 3D object recognition. In: Proceedings of the IEEE Conference on Computer Vision and Pattern Recognition, pp. 186–194 (2018)
31. Zhang, K., Hao, M., Wang, J., de Silva, C.W., Fu, C.: Linked dynamic graph CNN: learning on point cloud via linking hierarchical features. arXiv preprint arXiv:1904.10014 (2019)

Leveraging Transformers for Weakly Supervised Object Localization in Unconstrained Videos

Shakeeb Murtaza[1(✉)], Marco Pedersoli[1], Aydin Sarraf[2], and Eric Granger[1]

[1] LIVIA, Department of Systems Engineering, ETS Montreal, Montreal, Canada
shakeeb.murtaza.1@ens.etsmtl.ca
[2] Ericsson, Global AI Accelerator, Montreal, Canada

Abstract. Weakly-Supervised Video Object Localization (WSVOL) involves localizing an object in videos using only video-level labels, also referred to as tags. State-of-the-art WSVOL methods like Temporal CAM (TCAM) rely on class activation mapping (CAM) and typically require a pre-trained CNN classifier. However, their localization accuracy is affected by their tendency to minimize the mutual information between different instances of a class and exploit temporal information during training for downstream tasks, e.g., detection and tracking. In the absence of bounding box annotation, it is challenging to exploit precise information about objects from temporal cues because the model struggles to locate objects over time. To address these issues, a novel method called *transformer based CAM for videos (TrCAM-V)*, is proposed for WSVOL. It consists of a `DeiT` backbone with two heads for classification and localization. The classification head is trained using standard classification loss (`CL`), while the localization head is trained using pseudo-labels that are extracted using a pre-trained `CLIP` model. From these pseudo-labels, the high and low activation values are considered to be foreground and background regions, respectively. Our TrCAM-V method allows training a localization network by sampling pseudo-pixels on the fly from these regions. Additionally, a conditional random field (CRF) loss is employed to align the object boundaries with the foreground map. During inference, the model can process individual frames for real-time localization applications. Extensive experiments on challenging YouTube-Objects unconstrained video datasets show that our TrCAM-V method achieves new state-of-the-art performance in terms of classification and localization accuracy. Code: https://github.com/shakeebmurtaza/TrCAM/.

1 Introduction

The rapid development of video-sharing platforms has led to the availability of extensive video data [34], escalating the need for automated tools that analyze video content. Video object localization plays a crucial role in understanding this

C. Y. Suen et al. (Eds.): ANNPR 2024, LNAI 15154, pp. 195–207, 2024.
https://doi.org/10.1007/978-3-031-71602-7_17

content. It also helps in enhancing the performance of proposed model in different downstream tasks, such as video object detection [8], visual object tracking [22], video summarization [39] and event detection [6].

Most videos are captured in wild and unconstrained environments, exhibiting different properties due to different factors such as differences in camera lenses, multiple viewpoints, decoding distortions, moving objects, and editing effects. Leveraging these videos for training our model to perform different downstream tasks requires bounding boxes or pixel-level labels by human annotators. In contrast to images, labeling videos is a very arduous and expensive task, as each video contains a large number of frames. To reduce the annotation cost, videos are weakly labeled [37] by identifying a global tag/class for each video. These global tags represent the object of interest in the video even though some frames may not contain the object of interest, leading to inconsistent labels for different frames. Moreover, these labels don't represent spatial-temporal information about a particular object across different frames. This results in noisy or corrupted frame-level labels, as labels are attributed to an entire video, even though only a subset of its frames might contain the object of interest.

Using these global noisy labels for object localization can reduce the dependency on bounding box annotations, making this task more challenging. Different techniques have been proposed for weakly supervised video object localization (WSVOL) [17,29,31,40] and weakly supervised video object segmentation [1,9,14,20,36–38,41]. Nevertheless, these techniques necessitate post-processing, rendering them infeasible for object localization in unconstrained videos within real-world scenarios. State-of-the-art WSVOL methods closely adhere to standard protocols [29,34,36,40]. They first generate object proposals using visual cues (e.g., motion) which are then employed to identify the relevant object using different post-processing techniques. Despite their outstanding performance, these methods exhibit several limitations. Typically, these methods involve multiple steps for WSVOL that restrain us from training these models in an end-to-end fashion, thus rendering these models susceptible to suboptimal solutions. They also necessitate a separate model for each class as they are trained to localize a particular object. These limitations deem these models ill-suited for deployment in real-world environments and restrict their scalability to a small number of classes due to various constraints (e.g., resources and inference time). Most of these methods cannot employ class-level labels for extracting initial proposals, so they cannot semantically align different object parts belonging to an object of a particular class/tag. Moreover, these methods often utilize motion cues (e.g., optical flow) to localize an object of interest, which makes them susceptible to the same alignment problem as they often ignore the semantics of a particular object. Also, relying solely on motion cues hinders the network performance as they contain noisy information in unconstrained videos due to various external factors such as camera movements.

In a recent study, a discriminative multi-class deep learning model (DL) for WSVOL has been proposed [1]. This method leverages class activation maps (CAMs) for WSVOL, which have proven effective for weakly supervised object localization (WSOL) in static images [3,24–26]. In WSOL, the model is trained using image-level labels to localize an object of interest corresponding to the underlying class. These CAMs highlight the areas that strongly contribute to predicting a particular class. Nevertheless, these methods are not inherently designed to utilize the temporal information in videos for WSVOL. To deal with this issue, recent studies [1] propose an approach to leverage spatio-temporal information by harvesting CAMs using the LayerCAM [42] method. While this approach harnesses the spatiotemporal relationships within videos, it remains susceptible to the accumulation of inaccurate activations arising due to unconstrained object motion. This prevents the model from forming dependencies between object parts, thereby hindering its ability to localize various parts of the object, as depicted in the results section of their study.

To address these issues, we propose a transformer-based class activation mapping for videos (TrCAM-V) to localize a particular object. Unlike other methods, TrCAM-V requires only video-level annotations for training and is independent of additional assumptions, such as motion and temporal cues. This is inspired by WSOL methods, which employ pseudo-labels to train models for WSOL tasks within static images [3,26]. They utilize pseudo-labels and class-level labels to train the localization and classification head in an end-to-end fashion. Moreover, these approaches scale effectively with a large number of classes, making them applicable to a wide range of applications.

This paper aims to explore the transformer models for WSVOL tasks by levering pseudo-label from a Contrastive Language-Image Pretraining (`CLIP`) model [30]. Our proposed method, called TrCAM-V, consists of a transformer encoder with two heads – one head for classification and the other for localization. The localization head produces a map with the same resolution as the input image, which is trained using pseudo-labels. Furthermore, the pseudo-labels for each frame are extracted from `CLIP` [30] model using GradCAM presented in [19]. This model accepts an image with a text prompt (class label) to produce a pixel-level pseudo-label (activation map). In line with standard protocols [11,42], the strong and weak activations are deemed as FG and BG regions, respectively. Moreover, at each stochastic gradient descent (SGD) step [26], foreground (FG) and background (BG) pseudo-pixels are selected to build pseudo-labels [26] to train localization head. This random sampling enables the network to explore FG/BG regions and fosters the emergence of activation values over different object parts in the localization map. We also employ conditional random field (CRF) loss to align the object boundaries with the boundaries of the localization map by leveraging statistical properties of an image, such as pixel color and proximity among pixels. Moreover, our model does not require exploiting temporal dependencies during either training or inference. This approach is more suitable for real-time applications compared to other state-of-the-art WSVOL methods, as TCAM does not need to process an entire video to localize an object within a particular frame.

The main contributions of this paper are summarized as follows. **(1)** A novel transformer-based CAM (TrCAM-V) method is proposed for WSVOL tasks. It is comprised of two heads – a classification head trained with class labels, and a localization head trained with pseudo pixels, which are obtained from a pre-trained CLIP model. **(2)** Unlike previous WSVOL methods [1], TrCAM-V does not necessitate a pre-trained classifier to harvest activation maps and does not require temporal information during training. **(3)** Following the WSVOL experimental protocol in [1], our empirical results shows that our method achieves state-of-the-art performance the challenging YouTube-Object v1.0 (YTOv1) [29] and v2.2 (YTOv2.2) [16] datasets.

2 Related Work

Despite the success of deep learning methods for WSOL in still images, limited methods have been proposed for WSVOL [1,2,17,29,31,40]. Most of these models typically initialize and refine prominent proposals while incorporating spatiotemporal consistency constraints by exploiting information from visual appearance and motion cues of an object. Some methods have employed proposals as supervision to train localizers [29,40], whereas others have leaned towards segmentation-based approaches [31], further refining the results using GrabCut [32]. Moreover, these methods often only select one or more videos of a particular class to optimize an underlying model.

Similarly, methods have been proposed to localize a prominent object in a video by identifying similar FG regions based on visual appearance and motion cues [17,31]. For instance, [31] generate bounding box proposals within a video and retain relevant ones to build an object appearance detection model by enforcing temporal appearance consistency using maximum a posteriori inference. Similarly, [23] employed region proposals along with appearance and motion consistency to discover FG objects thereby maintaining temporal relationships between consecutive frames. In addition to conventional methods, deep learning (DL) models have been proposed for WSVOL. For instance, [40] proposed SPFTN to jointly learn object segment and localization tasks by employing an optical flow technique [18]. Recently [1,2] proposed an approach for WSVOL that exploits mutual information between different consecutive frames. However, these methods require temporal information during training and also rely on pre-trained classifiers to harvest pseudo-labels and low-level representations that hinder their localization abilities as they tend to minimize the mutual information between instances of the same class. Compared to these methods, TrCAM-V does not require temporal information during training, yet is capable of achieving state-of-the-art performance because it does not rely on discriminative pseudo-labels, harvested from a pre-trained classifier, for the training of the localization head.

3 Proposed Approach

Notation. Given a training set $\mathbb{T} = \{\text{v}_i, \text{y}_i\}$ where each $\text{v}_i = \{x_j \in \mathbb{R}^{H \times W \times 3}\}_{j=1}^{f}$ represent an input video composed of f frames and each $\text{y}_i \in \{1, \ldots, C\}$ represent the global class label i-th video, where $i = 1, 2, \ldots, n$ and C is the number of classes in our dataset. It is assumed that each frame within a video encompasses a particular object corresponding to the underlying class y_i. Furthermore, our model can accept a frame f_i and predict class category y_i along with a localization map $\hat{M}$ encompassing a corresponding object.

3.1 Background on Transformers

Vision Transformer (ViTs) have recently achieved state-of-the art accuracy in different tasks such as classification and WSOL [26] tasks. A ViT is composed of N cascaded encoder blocks that capture long-range relationships between different object parts within a scene. Each block consists of different layers, including a multi-headed attention layer, followed by a fully connected layer. To pass an image through the transformer, a frame f_i is divided into M patches of size $(H/S) \times (W/S)$. Subsequently, positional embeddings and a class (`class`) token are added to the P patches and projected into a fixed embedding dimension of size d. The computed embeddings are then passed through the N cascaded encoder blocks of the transformer. From the output of the last encoder layer, the `class` token is extracted and passed through a fully connected layer to produce classification scores for the relevant classes.

Self-supervised Transformers (SSTs) can learn object representations within an image without relying on explicit supervision. In this context, self-distillation with no labels (DINO) has been proposed to learn meaningful representation from data without supervision [5]. To learn a representation in an unsupervised fashion, the authors employed a student and a teacher network. The objective is to match their probability distributions corresponding to two different views of an input image. The output of these networks is normalized using temperature softmax. Moreover, after each optimization step, the parameters of the student are transferred to the teacher network using an exponential moving average.

Transformer-Based CAM for Videos. The SST [5] can identify objects in a frame without explicit supervision. However, they tend to decompose different object parts each corresponding to different attention maps as they are trained without explicit supervision. These attention maps highlight different object parts based on their attention coefficients learned during network training. For extracting relevant object parts and suppressing BG regions from these maps, we propose TrCAM-V. It decodes the patch embeddings and `class` tokens to produce a localization map that is optimized using pseudo-labels derived from the pre-trained `CLIP` model. Furthermore, these pseudo-labels allow us to directly optimize the loss over generated pseudo-labels instead of just minimizing the mutual information between different instances of a particular class.

More specifically, our model consist of a transformer backbone T with two heads; **(i)** classification head F trained using global video labels via standard cross-entropy, $\min_\theta - \log(\Pr(y|f_i))$. It accept `class` token for an input video frame f_i from transformer backbone and produce classification probabilities for each class $F(T(f_i)) = \Pr(c|f_i)$ where $F(T(f_i)) \in [0,1]_c$. **(ii)** A localization head G that accepts patch embeddings and `class` tokens from the last layer of the transformer to produce a high-resolution localization map M. It consists of two channels M_0 and M_1, which represent the BG and FG maps, respectively. The localization head is trained using pseudo-labels extracted from a pre-trained CLIP model (PCM). To extract the pseudo-label, TrCAM-V first obtains a saliency map from CLIP, after which it determines the FG and BG regions using Otsu's threshold [28]. The densest region among the different regions extracted by the Otsu method is used to sample FG and BG regions. Given that these sampled regions may be noisy and uncertain regarding our concerned object, we refrain from directly using these maps to minimize the loss over the generated map. Instead, we stochastically sample a few FG and BG pixels [12] and utilize them to train our model using partial cross-entropy. This allows the network to explore relevant object parts, resulting in a more robust localization map. This sampling process can also be viewed as a *fill-in-the-gap* approach. It enables our localization head to explore similar-looking object parts while considering the statistical properties of the object to generate a robust localization map. Moreover, in contrast to baseline methods [1,2], we do not take into account the temporal information for producing pseudo-labels.

Initial Pseudo-label Generation. Following [19], we employ GradCAM to harvest maps from CLIP for FG maps while suppressing regions belonging to BG categories. Initially, attention map regions are identified using a text-driven approach, leveraging a sharpness-based prompt as opposed to relying on class categories. To extract weights for generating GradCAM, a softmax function is applied to the computed attention between text and image embeddings. To compute the final activation maps, class-aware attention-based affinity (CAA) maps are extracted and fused with initial CAMs. These harvested maps can be viewed as pseudo-labels that are used to harvest pseudo-pixels.

Selection of FG/BG Pixels. To train the localization head G, we leverage pseudo pixels from saliency maps S obtained from pre-trained CLIP model. We produced a separate pseudo-label for each frame f_i, without considering temporal information into account. In each saliency map, weak and strong activations are likely to represent the BG and FG regions, respectively [1,3,11,26,42]. Relying on this assumption, we can estimate BG $\mathbb{C}^-$ and FG regions $\mathbb{C}^+$ using Otsu threshold [28] as follows, $\mathbb{M}_t^+ = \mathcal{O}^+(\boldsymbol{M})$, $\mathbb{M}_t^- = \mathcal{O}^-(\boldsymbol{M})$. Here, operation $\mathcal{O}^+$ is employed to select top $n\%$ pixels from the most dense connected area in activation map $\boldsymbol{M}$. It selects pixels that have a magnitude larger than the Otsu threshold. Conversely, the operation $\mathcal{O}^-$ is used to select the bottom $-n\%$ pixels from $\boldsymbol{M}$, which are ordered from the lowest to highest activation value. These selected regions are uncertain about object location and may contain incorrect labels. Considering the potential uncertainty, a few BG and FG pixels are sampled probabilistically to construct our pseudo-labels:

$$\boldsymbol{M}' = \mathcal{P}(\mathbb{M}^+) \cup \mathcal{P}(\mathbb{M}^-) , \tag{1}$$

where $\mathcal{P}$ represents the sampling process of a few pixels from selected FG and BG regions. To generate the final pseudo-label, the values of the selected FG and BG pixels are replaced with 1's and 0's, respectively.

Training Loss. Cross-entropy is employed to train our classification head. It relies on `DeiT` features for producing class category $\Pr(c|\boldsymbol{f}_i)$ as follows:

$$\min_{\boldsymbol{\theta}_c} \; -\log(\Pr(y|\boldsymbol{f}_i)) \; . \tag{2}$$

For the localization head, we employ a loss function consisting of three terms:

(i) The pixel alignment loss ($\mathcal{L}_{PAL}$) for learning FG/BG regions[1]. It aims to align the output map $\hat{M}$ with the selected pixels within $\mathcal{P}(\mathbb{M}^+) \cup \mathcal{P}(\mathbb{M}^-)$ via a partial cross-entropy loss denoted as $\mathcal{L}_{CPA}(\mathcal{P}(\mathbb{M}^+) \cup \mathcal{P}(\mathbb{M}^-), \hat{M}^r)$, where r represents the pseudo-pixels: $\mathcal{L}_{PAL} = -(1 - \hat{M})\log(\mathbb{M}^+) - \hat{M}(\log(\mathbb{M}^-))$

(ii) The absolute size loss ($\mathcal{L}_{ASL}$) for localizing all object parts belonging to FG/BG regions, as defined by [3]. It is formulated as a constraint for pushing FG and BG regions away and optimized using the log-barrier method [4].

(iii) The conditional random field (CRF) loss is employed to align the localization map with the boundaries of the object by utilizing colour similarity and proximity of nearby pixels as $\mathcal{L}_{CRF}(\boldsymbol{S}, \boldsymbol{f}_i) = \sum_{r\in\{0,1\}} \boldsymbol{S}^{r\top} \; \boldsymbol{W} \; (\boldsymbol{1} - \boldsymbol{S}^r)$. $\boldsymbol{W}$ is the affinity matrix to capture colour similarity and pixels' proximity between pixels of frame $\boldsymbol{f}_i$ and $\boldsymbol{W}$. The overall loss is defined as:

$$\mathcal{L} = \min_{\theta} \lambda_{PAL}\mathcal{L}_{PAL} + \lambda_{ASL}\mathcal{L}_{ASL} + \lambda_{CRF}\mathcal{L}_{CRF}, \tag{3}$$

where λ_{CLS} and λ_{CPA} are hyperparameters that lie in the interval $[0, 1]$. $2e^{-9}$ and λ_{CRF} is set to $2e^{-9}$ [35].

Inference. During the inference phase, we discard `CLIP` and its associated components utilized during training to harvest pseudo-labels. We only retain `DeiT` backbone T along with the classification and inference head. The backbone network T inputs a frame f_i and predicts a class label y_i and the corresponding localization map $\hat{M}$.

4 Results and Discussion

4.1 Experimental Methodology

Datasets. To validate the performance of our model, we conducted extensive experiments on two challenging datasets comprised of unconstrained videos from YouTube-Object v1.0 (`YTOv1`) [29] and v2.2 (`YTOv2.2`) [16]. For training, only global video labels are available, which correspond to the prominent object in different video frames. Further details of these datasets are given below:

`YTOv1` consists of various videos collected from YouTube. Ten categories were considered to query different videos containing a particular object. Each class consists of a varying number of videos ranging from 9–24 frames with durations spanning from 0.5–3 min. The dataset comprises 155 videos that are segmented

[1] A similar approach for exploiting few pixels to localize a concerned object is also employed in [12].

into short clips, referred to as shorts. After dividing the videos into clips, we ended up with 5507 shorts encompassing 571,089 frames. Within each short, a few frames have bounding box annotations corresponding to the underlying class. The annotated frames can be used as validation and test sets for hyper-parameter and model selection, respectively. Moreover, the authors reserved 128 videos for training and 27 for testing, which collectively encompass a total of 396 labeled bounding boxes. We reserved some frames from the training videos for a validation set. To build the validation set, we followed [3] and randomly chose five videos per class, resulting in a total of 50 videos.

`YTOv2.2` is an extension of `YTOv1` and consist of 722,040 frames. Compared to `YTOv1`, this dataset has a large number of bounding box annotations for different categories. The dataset comprises 155 videos divided into 9 categories. In terms of dataset partitioning, the authors reserved 106 videos for training and 49 videos for testing. Following [1], we built our validation set by randomly selecting three videos per class from the training set. Compared to `YTOv1`, this dataset has a larger number of bounding box annotations; the test set comprises 1,781 frames with bounding box annotations, resulting in a total of 2,667 bounding boxes. The expanded test set and annotations render `YTOv2.2` more challenging.

Evaluation Measures. We employ two metrics for measuring the performance of our model; **(i)** `CorLoc` [10] is used to measure the localization performance. `CorLoc` indicates the proportion of predicted bounding having an intersection over union (IoU) greater than 50%. (ii) Standard classification accuracy (`CL`) is measured to evaluate the classification performance of our model. `CL` is measured over frames that have bounding box annotations.

Implementation Details. We follows the protocols of [1] for all of our experiments. Specifically, we train our model for 100 epochs with a mini-batch size of 32. All images are resized to 256×256, then randomly cropped to a size of 224×224. After cropping, we augment the image by random horizontal flipping, followed by normalization. The weight λ for CRF is set to $2e^{-9}$ as defined in [35]. For log-barrier optimization, we use the same hyperparameter value as suggested in [1]. We optimize our network using the Stochastic Gradient Descent (SGD) algorithm with learning rates ranging between 0.1 and 0.00001 [26]. Moreover, we use the `DeiT` backbone with pre-trained weights from SST [5]. To account for the large number of redundant frames per video, we randomly selected a few frames for each shot for each gradient descent step. This approach enabled us to train our model over a large number of video frames within a reasonable timeframe.

Baseline Models. To validate our TrCAM-V method, its performance is compared with various state-of-the-art methods, e.g., FPPVOS [38], SPFTN [40] POD [17], and [9,14,29,31,36,37]. The performance of our model is also compared with different CAM methods: LayerCAM [15], XGradCAM [13], GradCAM++ [27], GradCAM [33] and CAM [42]. Note that our method for generating pseudo-label and pseudo-pixel selection is generic, and can be integrated to train any model for WSOL.

4.2 Comparison with State-of-Art Methods

The classification and localization performance of our model is presented in Table 1 and 2, respectively. Regarding classification performance, our model exhibits an improvement of 4.9% and 12.7% on the YTOv1 and YTOv2.2 datasets, respectively, when compared with the baseline model. Similarly, our model is able to surpass baseline models in terms of localization performance on both datasets. In contrast to the baseline methods, our method is able to achieve state-of-the-art performance without the need for exploiting temporal information, thereby enabling a consistent pipeline at both training and inference time. In addition to this, we trained our model using pseudo-label harvested from CLIPSeg [21] that helps in improving the performance of our model on YTOv2.2 dataset as shown in Table 1. Figure 1 depicts localization predictions of TrCAM-V compared to baselines and state-of-the-art methods. Results illustrate that our model can generate robust maps that encompass both FG and BG regions, demarcated by sharp boundaries. Activation-based methods focus on discriminative areas that are common among different instances of the same class. Moreover, TCAM generates localization maps with sharp boundaries over various object parts, facilitating the prediction of bounding boxes that fully enclose the object. However,

Table 1. CorLoc performance on YTOv1 [29] and YTOv2.2 [16] test sets.

DS	Method *(venue)*	Aero	Bird	Boat	Car	Cat	Cow	Dog	Horse	Mbike	Train	Avg
YTOv1	POD [17] *(cvpr, 2016)*	64.3	63.2	73.3	68.9	44.4	62.5	71.4	52.3	78.6	23.1	60.2
	Tsai et al. [37] *(eccv, 2016)*	66.1	59.8	63.1	72.5	54.0	64.9	66.2	50.6	39.3	42.5	57.9
	CAM [42] *(cvpr, 2016)*	75.0	55.5	43.2	69.7	33.3	52.4	32.4	74.2	14.8	50.0	50.1
	GradCAM [33] *(iccv, 2017)*	86.9	63.0	51.3	81.8	45.4	62.0	37.8	67.7	18.5	50.0	56.4
	Haller et al. [14] *(iccv, 2017)*	76.3	71.4	65.0	58.9	68.0	55.9	70.6	33.3	69.7	42.4	61.1
	GradCAM++ [7] *(wacv, 2018)*	79.8	85.1	37.8	81.8	75.7	52.4	64.9	64.5	33.3	56.2	63.2
	Smooth-GradCAM++ [27] *(corr, 2019)*	78.6	59.2	56.7	60.6	42.4	61.9	56.7	64.5	40.7	50.0	57.1
	Croitoru et al. [9] (LR-N_{iter1}) *(ijcv, 2019)*	77.0	67.5	77.2	68.4	54.5	68.3	72.0	56.7	44.1	34.9	62.1
	Croitoru et al. [9] (LR-N_{iter2}) *(ijcv, 2019)*	79.7	67.5	68.3	69.6	59.4	75.0	78.7	48.3	48.5	39.5	63.5
	Croitoru et al. [9] (DU-N_{iter2}) *(ijcv, 2019)*	85.1	72.7	76.2	68.4	59.4	76.7	77.3	46.7	48.5	46.5	65.8
	Croitoru et al. [9] (MS-N_{iter2}) *(ijcv, 2019)*	84.7	72.7	78.2	69.6	60.4	80.0	78.7	51.7	50.0	46.5	67.3
	SPFTN (M) [40] *(tpami, 2020)*	66.4	73.8	63.3	83.4	54.5	58.9	61.3	45.4	55.5	30.1	59.3
	SPFTN (P) [40] *(tpami, 2020)*	97.3	27.8	81.1	65.1	56.6	72.5	59.5	81.8	79.4	22.1	64.3
	XGradCAM [13] *(bmvc, 2020)*	79.8	70.4	54.0	87.8	33.3	52.4	37.8	64.5	37.0	50.0	56.7
	LayerCAM [15] *(ieee, 2021)*	85.7	88.9	45.9	78.8	75.5	61.9	64.9	64.5	33.3	56.2	65.6
	FPPVOS [38] *(optik, 2021)*	77.0	72.3	64.7	67.4	79.2	58.3	74.7	45.2	80.4	42.6	65.8
	TCAM [1] *(wacv, 2023)*	90.5	70.4	62.2	75.7	84.8	81.0	81.0	64.5	70.4	50.0	73.0
	TrCAM-V with CLIP (ours)	91.7	77.8	91.9	94.0	84.8	81.0	83.8	77.4	77.8	87.5	84.8
	TrCAM-V with CLIPSeg (ours)	94.0	74.1	94.6	90.9	87.9	81.0	89.2	77.4	74.1	75.0	83.8
YTOv2.2	CAM [42] *(cvpr, 2016)*	52.3	66.4	25.0	66.4	39.7	87.8	34.7	53.6	45.4	43.7	51.5
	GradCAM [33] *(iccv, 2017)*	44.1	68.4	50.0	61.1	51.8	79.3	56.0	47.0	44.8	42.4	54.5
	Haller et al. [14] *(iccv, 2017)*	76.3	68.5	54.5	50.4	59.8	42.4	53.5	30.0	53.5	60.7	54.9
	GradCAM++ [7] *(wacv, 2018)*	74.7	78.1	38.2	69.7	56.7	84.3	61.6	61.9	43.0	44.3	61.2
	Croitoru et al. [9] (LR-N_{iter1}) *(ijcv, 2019)*	75.7	56.0	52.7	57.3	46.9	57.0	48.9	44.0	27.2	56.2	52.2
	Croitoru et al. [9] (LR-N_{iter2}) *(ijcv, 2019)*	78.1	51.8	49.0	60.5	44.8	62.3	52.9	48.9	30.6	54.6	53.4
	Croitoru et al. [9] (DU-N_{iter2})*(ijcv, 2019)*	74.9	50.7	50.7	60.9	45.7	60.1	54.4	42.9	30.6	57.8	52.9
	Croitoru et al. [9] (BU-N_{iter2})*(ijcv, 2019)*	82.2	51.8	51.5	62.0	50.9	64.8	55.5	45.7	35.3	55.9	55.6
	Croitoru et al. [9] (MS-N_{iter2})*(ijcv, 2019)*	81.7	51.5	54.1	62.5	49.7	68.8	55.9	50.4	33.3	57.0	56.5
	Smooth-GradCAM++ [27] *(corr, 2019)*	74.1	83.2	38.2	64.2	49.6	82.1	57.3	52.0	51.1	42.4	59.5
	XGradCAM [13] *(bmvc, 2020)*	68.2	44.5	45.8	64.0	46.8	86.4	44.0	57.0	44.9	45.0	54.6
	LayerCAM [15] *(ieee, 2021)*	80.0	84.5	47.2	73.5	55.3	83.6	71.3	60.8	55.7	48.1	66.0
	TCAM [1] *(wacv, 2023)*	79.4	94.9	75.7	61.7	68.8	87.1	75.0	62.4	72.1	45.0	72.2
	CoLo-CAM *(corr, 2023)*	82.9	92.2	85.4	67.7	80.1	85.7	79.2	67.4	72.7	58.2	77.1
	TrCAM-V with CLIP (ours)	87.6	91.6	90.3	74.1	78.7	79.2	76.2	66.9	60.0	62.0	76.7
	TrCAM-V with CLIPSeg (ours)	84.7	95.5	92.4	79.3	78.7	87.9	84.1	66.9	68.4	62.0	80.0

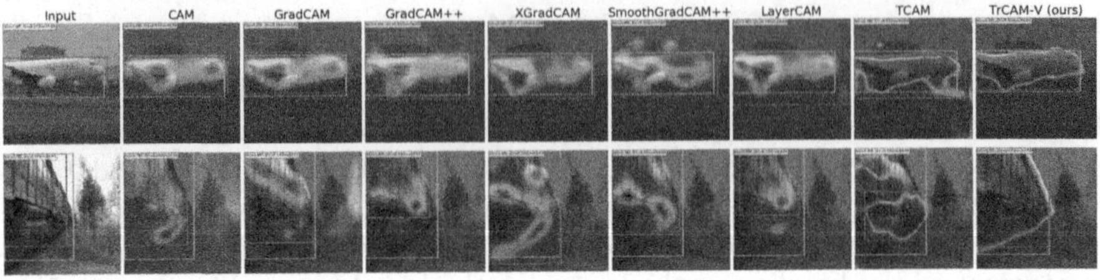

Fig. 1. Visualization of YTOv1 frames. Here, red and green box indicate the predicated and annotated bounding-box. (Color figure online)

activation maps fail to encompass all parts of an object, resulting in inaccurate localization in produced maps.

Ablation Studies. Ablations presented in Table 3 are conducted to show the effectiveness of various loss terms. This study indicates that all auxiliary terms of our loss contribute significantly to the model's performance on both datasets, YTOv1 and YTOv2.2. Harvesting pseudo-labels from the CLIP model yields competitive performance compared to the baseline method. Adding size terms to the loss function helps in connecting different object parts. Additionally, CRF terms help to significantly improve localization performance by aligning the boundaries of the localization map with the boundaries of the concerned object. In contrast to the baseline methods, TrCAM-V can achieve state-of-the-art performance without exploiting temporal information.

Table 2. Classification accuracy on YTOv1 and YTOv2.2 test sets.

Methods	YTOv1	YTOv2.2
CAM (cvpr, 2016)	85.3	73.9
GradCAM (iccv, 2017)	85.3	71.3
GradCAM++ (wacv, 2018)	84.4	72.4
Smooth-GradCAM++ (corr, 2019)	82.6	75.2
XGradCAM (bmvc, 2020)	87.3	71.6
TCAM (wacv, 2023)	84.4	72.1
TrCAM-V (ours)	**92.2**	**87.9**

Table 3. CorLoc accuracy of our proposed method with various losses.

Methods	YT-V1	YTOv2.2
TCAM (wacv, 2023)	73.0	72.2
PAL	74.8	72.3
PAL + ASL	75.5	74.2
PAL + CRF	81.4	76.2
PAL + CRF + ASL	84.8	76.7
Improvement	+11.8	**+4.5**

4.3 Conclusion

In this paper, a novel approach is proposed to train transformers for WSVOL tasks. The transformer consists of two heads for classification and localization, which are trained using video-level class labels and pixel-wise pseudo-labels, respectively. These pseudo-labels are harvested from the standard CLIP model. Additionally, a CRF loss is employed to align the boundaries of the localization map and the concerned object. Our model outperforms state-of-the-art methods in terms of both classification and localization accuracy without requiring temporal information across subsequent frames.

References

1. Belharbi, S., Ben Ayed, I., McCaffrey, L., Granger, E.: TCAM: temporal class activation maps for object localization in weakly-labeled unconstrained videos. In: WACV (2023)
2. Belharbi, S., Murtaza, S., Pedersoli, M., Ayed, I.B., McCaffrey, L., Granger, E.: CoLo-CAM: class activation mapping for object co-localization in weakly-labeled unconstrained videos. CoRR abs/2303.09044 (2023)
3. Belharbi, S., Sarraf, A., Pedersoli, M., Ben Ayed, I., McCaffrey, L., Granger, E.: F-CAM: full resolution class activation maps via guided parametric upscaling. In: WACV (2022)
4. Boyd, S., Vandenberghe, L.: Convex Optimization. Cambridge University Press, Cambridge (2004)
5. Caron, M., et al.: Emerging properties in self-supervised vision transformers. In: ICCV (2021)
6. Chang, X., Yang, Y., Long, G., Zhang, C., Hauptmann, A.G.: Dynamic concept composition for zero-example event detection. In: AAAI (2016)
7. Chattopadhyay, A., Sarkar, A., Howlader, P., Balasubramanian, V.N.: Grad-CAM++: generalized gradient-based visual explanations for deep convolutional networks. In: WACV (2018)
8. Chen, Y., Cao, Y., Hu, H., Wang, L.: Memory enhanced global-local aggregation for video object detection. In: CVPR (2020)
9. Croitoru, I., Bogolin, S.V., Leordeanu, M.: Unsupervised learning of foreground object segmentation. IJCV **127**(9), 1279–1302 (2019)
10. Deselaers, T., Alexe, B., Ferrari, V.: Weakly supervised localization and learning with generic knowledge. IJCV **100**(3), 275–293 (2012)
11. Durand, T., Mordan, T., Thome, N., Cord, M.: WILDCAT: weakly supervised learning of deep convnets for image classification, pointwise localization and segmentation. In: CVPR (2017)
12. Feng, X., Yang, J., Laine, A.F., Angelini, E.D.: Discriminative localization in CNNs for weakly-supervised segmentation of pulmonary nodules. In: Descoteaux, M., Maier-Hein, L., Franz, A., Jannin, P., Collins, D.L., Duchesne, S. (eds.) MICCAI 2017. LNCS, vol. 10435, pp. 568–576. Springer, Cham (2017). https://doi.org/10.1007/978-3-319-66179-7_65
13. Fu, R., Hu, Q., Dong, X., Guo, Y., Gao, Y., Li, B.: Axiom-based Grad-CAM: towards accurate visualization and explanation of CNNs. In: BMVC (2020)
14. Haller, E., Leordeanu, M.: Unsupervised object segmentation in video by efficient selection of highly probable positive features. In: ICCV (2017)
15. Jiang, P., Zhang, C., Hou, Q., Cheng, M., Wei, Y.: LayerCAM: exploring hierarchical class activation maps for localization. TIP **30**, 5875–5888 (2021)
16. Kalogeiton, V., Ferrari, V., Schmid, C.: Analysing domain shift factors between videos and images for object detection. TPAMI **38**(11), 2327–2334 (2016)
17. Koh, Y.J., Jang, W., Kim, C.: POD: discovering primary objects in videos based on evolutionary refinement of object recurrence, background, and primary object models. In: CVPR (2016)
18. Lee, Y., Kim, J., Grauman, K.: Key-segments for video object segmentation. In: ICCV (2011)
19. Lin, Y., et al.: CLIP is also an efficient segmenter: a text-driven approach for weakly supervised semantic segmentation. In: CVPR, pp. 15305–15314 (2023)

20. Liu, X., Tao, D., Song, M., Ruan, Y., Chen, C., Bu, J.: Weakly supervised multi-class video segmentation. In: CVPR (2014)
21. Lüddecke, T., Ecker, A.: Image segmentation using text and image prompts. In: CVPR, pp. 7086–7096 (2022)
22. Luo, W., Xing, J., Milan, A., Zhang, X., Liu, W., Kim, T.: Multiple object tracking: a literature review. Artif. Intell. **293**, 103448 (2021)
23. Manen, S., Guillaumin, M., Gool, L.V.: Prime object proposals with randomized prim's algorithm. In: ICCV (2013)
24. Murtaza, S., Belharbi, S., Pedersoli, M., Sarraf, A., Granger, E.: Constrained sampling for class-agnostic weakly supervised object localization. In: MAIS (2022)
25. Murtaza, S., Belharbi, S., Pedersoli, M., Sarraf, A., Granger, E.: DiPS: discriminative pseudo-label sampling with self-supervised transformers for weakly supervised object localization. IVC **140**, 104838 (2023)
26. Murtaza, S., Belharbi, S., Pedersoli, M., Sarraf, A., Granger, E.: Discriminative sampling of proposals in self-supervised transformers for weakly supervised object localization. In: WACV (2023)
27. Omeiza, D., Speakman, S., Cintas, C., Weldemariam, K.: Smooth Grad-CAM++: an enhanced inference level visualization technique for deep convolutional neural network models. CoRR abs/1908.01224 (2019)
28. Otsu, N.: A threshold selection method from gray-level histograms. IEEE Trans. Syst. Man Cybern. **9**(1), 62–66 (1979)
29. Prest, A., Leistner, C., Civera, J., Schmid, C., Ferrari, V.: Learning object class detectors from weakly annotated video. In: CVPR (2012)
30. Radford, A., et al.: Learning transferable visual models from natural language supervision. In: ICML, pp. 8748–8763. PMLR (2021)
31. Rochan, M., Rahman, S., Bruce, N.D.B., Wang, Y.: Weakly supervised object localization and segmentation in videos. IVC **56**, 1–12 (2016)
32. Rother, C., Kolmogorov, V., Blake, A.: "GrabCut": interactive foreground extraction using iterated graph cuts. ACM Trans. Graph. **23**(3), 309–314 (2004)
33. Selvaraju, R.R., Cogswell, M., Das, A., Vedantam, R., Parikh, D., Batra, D.: Grad-CAM: visual explanations from deep networks via gradient-based localization. In: ICCV (2017)
34. Tang, K., Sukthankar, R., Yagnik, J., Fei-Fei, L.: Discriminative segment annotation in weakly labeled video. In: CVPR (2013)
35. Tang, M., Perazzi, F., Djelouah, A., Ayed, I.B., Schroers, C., Boykov, Y.: On regularized losses for weakly-supervised CNN segmentation. In: Ferrari, V., Hebert, M., Sminchisescu, C., Weiss, Y. (eds.) ECCV 2018. LNCS, vol. 11220, pp. 524–540. Springer, Cham (2018). https://doi.org/10.1007/978-3-030-01270-0_31
36. Tokmakov, P., Alahari, K., Schmid, C.: Weakly-supervised semantic segmentation using motion cues. In: Leibe, B., Matas, J., Sebe, N., Welling, M. (eds.) ECCV 2016. LNCS, vol. 9908, pp. 388–404. Springer, Cham (2016). https://doi.org/10.1007/978-3-319-46493-0_24
37. Tsai, Y.-H., Zhong, G., Yang, M.-H.: Semantic co-segmentation in videos. In: Leibe, B., Matas, J., Sebe, N., Welling, M. (eds.) ECCV 2016. LNCS, vol. 9908, pp. 760–775. Springer, Cham (2016). https://doi.org/10.1007/978-3-319-46493-0_46
38. Umer, S., Dawood, H., Yousaf, M.H., Dawood, H., Ahmad, H.: Efficient foreground object segmentation from video by probability weighted moments. Optik **229**, 166251 (2021)
39. Zhang, D., Han, J., Jiang, L., Ye, S., Chang, X.: Revealing event saliency in unconstrained video collection. TIP **26**(4), 1746–1758 (2017)

40. Zhang, D., Han, J., Yang, L., Xu, D.: SPFTN: a joint learning framework for localizing and segmenting objects in weakly labeled videos. TPAMI **42**(2), 475–489 (2020)
41. Zhang, D., Javed, O., Shah, M.: Video object co-segmentation by regulated maximum weight cliques. In: Fleet, D., Pajdla, T., Schiele, B., Tuytelaars, T. (eds.) ECCV 2014. LNCS, vol. 8695, pp. 551–566. Springer, Cham (2014). https://doi.org/10.1007/978-3-319-10584-0_36
42. Zhou, B., Khosla, A., Lapedriza, A., Oliva, A., Torralba, A.: Learning deep features for discriminative localization. In: CVPR (2016)

Palmprint Classification via Filter Faces and Feature Extraction

Guang Yi Chen[1], Adam Krzyzak[1(✉)], and Ventzeslav Valev[2]

[1] Department of Computer Science and Software Engineering, Concordia University, Montreal, QC H3G 1M8, Canada
krzyzak@cse.concordia.ca

[2] Institute of Mathematics and Informatics, Bulgarian Academy of Sciences, 1113 Sofia, Bulgaria
valev@math.bas.bg

Abstract. Palmprint classification is a popular method for today's biometrics, and it can be combined with iris or fingerprint to identify a person's identification. In this paper, we propose a novel method for palmprint classification by using filter faces and feature extraction. We extract such features as histogram of oriented features (HOG) and local binary patterns (LBP). We select to use both nearest neighbor (NN) classifier and collaborative representation classifier (CRC) for palmprint classification. Experiments demonstrate that filter faces always improve the classification accuracy for palmprint classification no matter what features are used and how much noise is present in palmprint images. Based on our experiments, the HOG features are very robust to noise whereas the LBP features are very sensitive to noise. As a result, it is preferable to use the HOG features for noisy palmprint images and the LBP features for clean palmprint images to improve classification accuracy.

Keywords: Histogram of oriented features (HOG) · Local binary patterns (LBP) · Filter faces (FF) · Collaborative representation classifier (CRC) · Nearest neighbor classifier (NN)

1 Introduction

Palmprint classification is a biometric technique to describe the different characteristics in the palms of a person's hands. Palmprint classification systems use a camera to take a picture of a person's palm and compare it with the palmprint images stored in the database for that person. Palmprint classification has become a very popular method for today's biometric applications.

We briefly review several existing methods for palmprint classification here. Zhang et al. [1] systematically studied on-line palmprint identification with very good results. Chen et al. [2] performed palmprint classification by means of dual-tree complex wavelet

C. Y. Suen et al. (Eds.): ANNPR 2024, LNAI 15154, pp. 208–217, 2024.
https://doi.org/10.1007/978-3-031-71602-7_18

transform. You et al. [3] investigated hierarchical palmprint identification via multiple feature extraction. Zhang and Shu [4] proposed two new characteristics in palmprint verification: Datum point invariance and line feature matching. Liu and Kumar [5] analyzed contactless palmprint identification using deeply learned residual features. Fei et al. [6] studied local apparent latent direction extraction for palmprint recognition. Wu et al. [7] worked on triple-type feature extraction for palmprint recognition. Ahmad et al. [8] analysed non-stationary feature fusion of face and palmprint multimodal biometrics. Jaswal et al. [9] invented multiple feature fusion for unconstrained palmprint authentication. Dai and Zhou [10] investigated multi-feature-based high-resolution palmprint recognition. Zhang et al. [11] worked on palmprint identification by combining hierarchical multi-scale complete local binary patterns and weighted sparse representative classifier.

In this paper, we propose a new method for palmprint classification by using filter faces (FF), feature extraction with histogram of oriented gradients (HOG) and local binary patterns (LBP), and collaborative representation classifier (CRC). In our method, the FF improves classification accuracy significantly no matter what features are used and how much noise is added to the palmprint images. We achieve 100% correct recognition rate when five out of six samples are selected for training and one sample is used for testing with the LBP features and CRC classifier. Our method obtains very good classification rates for other testing cases as well when FF, HOG, and CRC are selected.

The organization of this paper is the following. Section 2 proposes s a new method for palmprint classification. Section 3 performs some experiments to show the effectiveness of our proposed method in this paper. Finally, Sect. 4 concludes the paper and proposes future research directions.

2 Proposed Method

Palmprint classification is a very important task in today's biometric applications. It is essential to combine palmprint image with face, iris, fingerprint, etc. to improve overall classification accuracy. Furthermore, a good palmprint classification method should be robust to noise as much as possible, and it should be able to extract good invariant features from the palmprint images.

In this section, we propose a novel method for invariant palmprint classification. For palmprint classification, the original palmprint sample has the fingers and the background, which is not very good for the classification task. We can extract the central region of the palm sample with a size 128×128 pixels. We calculate the FF [12] from the extracted palmprint region. After that, we use these extracted features to train and test the palmprint database. The features used in this paper include both HOG and LBP. We use either HOG features or LBP features as the features to classify the palmprint to one of the known classes. Figure 1 shows a palm image without preprocessing and the extracted central palmprint image.

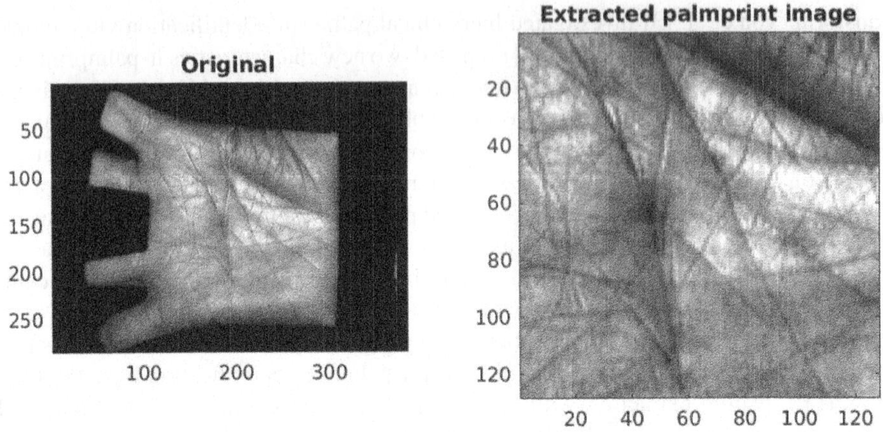

Fig. 1. Original palmprint and its extracted image.

We use the CRC [13] as a classifier for palmprint classification. The authors of [13] find out that it is the collaborative representation (CR) but not the L_1-norm sparsity which makes sparse representative classifier (SRC) powerful for face classification. More importantly, they proposed a very simple but more efficient CRC with regularized least squares. Their experiments demonstrated that CRC achieves very good classification results for classification tasks. We also use the nearest neighbor (NN) classifier in our experiments in comparison with the CRC classifier for palmprint classification.

The HOG [14] is a feature descriptor for pattern recognition. It counts occurrences of gradient orientation in the localized portion of a pattern image. The HOG emphasizes the structure or the shape of an image. It is better than ordinary edge descriptors since it utilizes magnitude as well as angle of the gradient to compute the feature vector. For the regions of the image, it generates histograms using the magnitude and orientations of the gradient. The HOG features are very effective in invariant pattern recognition.

The LBP [15] are a powerful descriptor for pattern recognition. It is a useful feature for texture classification, and it can be combined with the HOG descriptor as well. The LBP features improve the detection performance significantly on many datasets. Because of its discriminative power, the LBP features have become a popular approach in numerous applications including face recognition. One useful property of the LBP features is its robustness to monotonic gray-scale changes such as illumination variations. Another useful property is its computational simplicity to analyze images in complex real-time situations.

We select to use two filter faces (FF_1 and FF_2) in this paper. For any pixel I(x, y) in a palmprint image, the FF [12] is defined as a self-quotient image based on a low-pass filter (LF) and a high-pass filter (HF):

$$FF(I(x,y)) = arctan\frac{HF * I(x,y)}{LF * I(x,y)}$$

where * is the convolution operation. In this paper, we define two HF's (HF_1 and HF_2) and one LF as follows:

$$HF_1 = \begin{bmatrix} 0 & 1 & 1 & 1 & 1 \\ -1 & 0 & 1 & 1 & 1 \\ -1 & -1 & 0 & 1 & 1 \\ -1 & -1 & -1 & 0 & 1 \\ -1 & -1 & -1 & -1 & 0 \end{bmatrix} \qquad HF_2 = \begin{bmatrix} 1 & 1 & 1 & 1 & 0 \\ 1 & 1 & 1 & 0 & -1 \\ 1 & 1 & 0 & -1 & -1 \\ 1 & 0 & -1 & -1 & -1 \\ 0 & -1 & -1 & -1 & -1 \end{bmatrix}$$

$$LF = \begin{bmatrix} 1 & 1 & 1 & 1 & 1 \\ 1 & 1 & 1 & 1 & 1 \\ 1 & 1 & 1 & 1 & 1 \\ 1 & 1 & 1 & 1 & 1 \\ 1 & 1 & 1 & 1 & 1 \end{bmatrix}$$

We explain our processing ideas for both the known palmprint database and the unknown palmprint image here. For every known palmprint image in the database, we extract the central region of the palmprint image and then calculate the filter faces FF_1 and FF_2. We then extract the HOG or LBP features from both filter faces and combine the two feature vectors to save them in the feature database for later query. For every unknown palmprint image, we extract the central region as well, and then compute the filter faces FF_1 and FF_2. We extract the HOG or LBP features from both filter faces, combine the two feature vectors, and then query the feature database to determine the identification of the unknown palmprint image.

The basic steps for building the feature database can be given as follows:

a) For every palmprint sample I(x, y) in the palmprint dataset, extract the central region of the palmprint image with a size of 128 × 128 pixels. Denote it as CEN.
b) Calculate filter face FF_1 from CEN according to HF_1 and LF.
c) Calculate filter face FF_2 from CEN according to HF_2 and LF.
d) Extract HOG or LBP features from both FF_1 and FF_2. Denote them as FEA_1 and FEA_2, respectively.
e) Convert FEA1 and FEA2 into 1D vectors and pack them into one long 1D vector $FEA = [FEA_1, FEA_2]$.
f) Save the feature vector FEA in the feature database for later query.

The basic steps of our proposed method for the unknown palmprint image can be summarized as follows:

a) Given the unknown palmprint sample I(x, y).
b) Extract the central region of the palmprint image with a size of 128 × 128 pixels. Denote it as CEN.
c) Calculate filter face FF_1 from CEN according to HF_1 and LF.

d) Calculate filter face FF_2 from CEN according to HF_2 and LF.
e) Extract HOG or LBP features from both FF_1 and FF_2. Denote them as FEA_1 and FEA_2, respectively.
f) Convert FEA1 and FEA2 into 1D vectors and pack them into one long 1D vector $FEA = [FEA_1, FEA_2]$.
g) Classify the unknown palmprint into one known class with features FEA and the CRC classifier by querying the feature database.
h) Compute correct classification rate for the whole palmprint dataset.

The major contributions of this paper are as follows. We preprocess the palmprint images by FF, which improve the classification accuracy significantly. We extract both the HOG features and the LBP features for palmprint classification. In addition, the CRC classifier contributes the success of our new method significantly when compared to the NN classifier. Our FF always improves classification accuracies no matter what feature is used, what classifier is chosen, and how much noise is present in the palmprint images. Experiments show the advantages of our proposed method in this paper for palmprint classification. According to our experiments, the HOG features are very robust to noise whereas the LBP features are very sensitive to noise. Therefore, it is desirable to use the HOG features for noisy palmprint images and the LBP features for clean palmprint images to improve classification accuracy.

The idea of this paper can be applied to other biometric applications such as fingerprint recognition, face recognition, iris recognition, etc. It is expected that by utilizing the proposed method in this paper better classification results can be obtained for these applications.

3 Experiments

We conduct experiments for palmprint classification using the PolyU Palmprint Database [16], which contains 100 different palms, and each has six samples collected in two sessions. For each palm, we use $k \epsilon [1, 5]$ of the six palmprint samples for training and the rest of samples for testing. The size of the original palms without preprocessing is of 284×384 pixels. We extract the central region of the palm image for palmprint classification with a size of 128×128 pixels.

Table 1, 2, 3, 4 and 5 show the correct classification rates for palmprint classification with 1, 2, 3, 4, 5 samples for training and the rest 5, 4, 3, 2, 1 samples for testing, respectively. The best results are highlighted in bold font. Experiments with both FF and non-FF are performed. We extract such features as HOG and LBP in our experiments. Additive Gaussian white noise (AGWN) are added to the noise-free palmprint images as shown in Fig. 2. We illustrate palmprint image (first column), FF_1 (second column) and FF_2 (third column) for different noise levels in Fig. 3. In this figure, the FF_1 and FF_2 are very robust to noise even when significant amount of noise is added to the palmprint images.

When we use 1, 2, 3 samples from each class for training, FF combined with HOG and CRC achieve the best classification accuracies for all noise levels. When select 4 or 5 samples from each class for training, FF combined with LBP and CRC obtain the best classification accuracies for low noise levels. Nevertheless, FF combined with HOG and CRC achieves the highest classification accuracies for high noise levels. Our experiments demonstrate that FF improves palmprint classification accuracy significantly no matter what features are used and how high the noise levels are. In addition, CRC is always better than NN for palmprint classification in our experiments.

When noise level is low, the LBP features perform as well as the HOG features for palmprint classification. Nevertheless, when noise level is higher, the LBP features perform very poorly. In this case, the HOG features are doing an excellent job for palmprint classification. Furthermore, the FF improves classification accuracy significantly no matter either HOG or LBP features are used. This shows the advantages of our proposed method for palmprint classification.

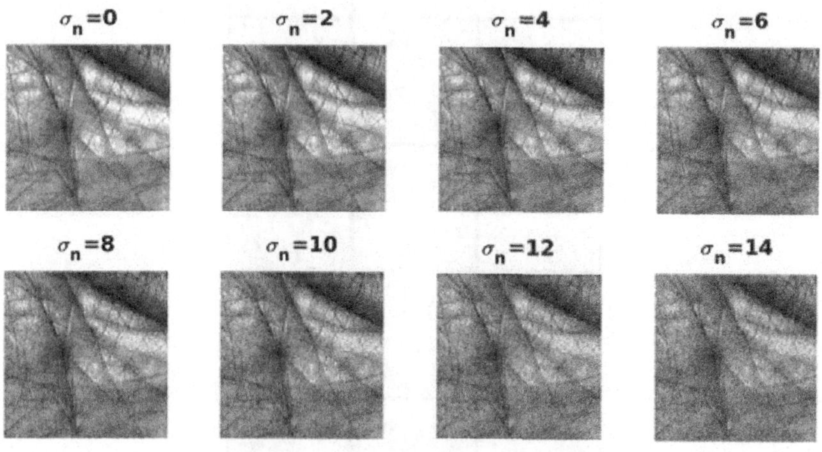

Fig. 2. Palmprint images with different noise levels.

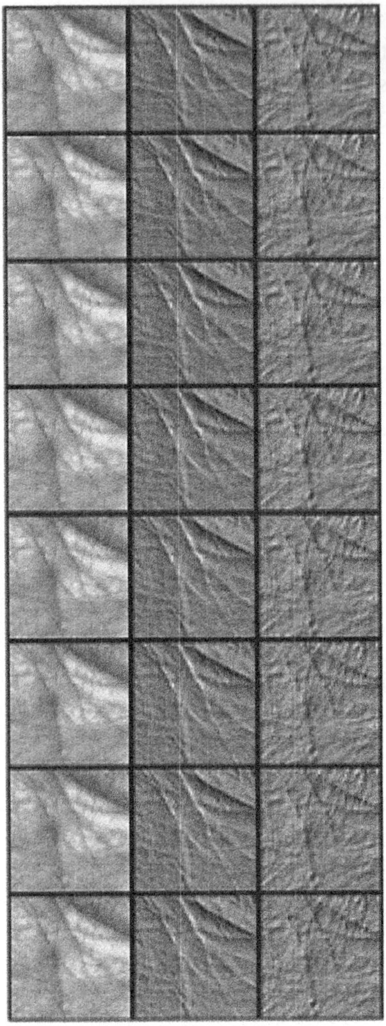

Fig. 3. Palmprint images (column 1), FF_1 (column 2) and FF_2 (column 3) for different noise levels (Rows 1 to 8 correspond to $\sigma_n = 0, 2, 4, 6, 8, 10, 12, 14$). Both filter faces are very robust to noise as can be seen form the second and third columns of this figure.

Table 1. Correct classification rates for one training sample in each class with different features and different noise levels. The best results are highlighted in bold font.

Filter Faces	Features	Classifiers	Noise σ_n							
			0	2	4	6	8	10	12	14
No	HOG	NN	65.15	65.36	63.09	61.44	56.91	51.75	44.33	33.61
		CRC	65.77	66.39	65.98	64.12	62.68	60.82	60.00	56.49
	LBP	NN	53.81	45.15	33.40	18.14	9.07	5.36	4.33	3.51
		CRC	48.04	36.49	35.26	25.77	15.05	9.48	5.98	4.33
Yes	HOG	NN	68.04	67.84	65.98	63.71	61.44	59.79	58.35	55.05
		CRC	**69.45**	**68.87**	**68.45**	**68.25**	**67.84**	**66.80**	**65.36**	**64.33**
	LBP	NN	63.09	58.56	47.84	26.19	11.34	5.36	3.71	2.27
		CRC	68.66	66.80	64.74	52.78	35.67	25.36	16.49	11.34

Table 2. Correct classification rates for two training samples in each class with different features and different noise levels. The best results are highlighted in bold font.

Filter Faces	Features	Classifiers	Noise σ_n							
			0	2	4	6	8	10	12	14
No	HOG	NN	74.23	75.77	73.71	70.62	63.66	57.99	51.55	40.72
		CRC	76.03	75.52	75.52	75.00	73.20	72.16	68.04	63.40
	LBP	NN	50.77	46.65	34.02	12.37	4.64	2.58	1.80	1.55
		CRC	52.84	44.59	39.69	31.19	15.72	8.25	5.41	2.58
Yes	HOG	NN	76.55	75.77	76.03	75.52	74.23	72.68	69.59	65.72
		CRC	**77.84**	**77.58**	**78.09**	**77.84**	**78.09**	**77.84**	**77.06**	**74.74**
	LBP	NN	60.57	57.22	42.27	18.04	8.51	4.90	3.35	1.80
		CRC	70.62	68.81	68.56	56.96	39.95	27.32	16.49	12.11

Table 3. Correct classification rates for three training samples in each class with different features and different noise levels. The best results are highlighted in bold font.

Filter Faces	Features	Classifiers	Noise σ_n							
			0	2	4	6	8	10	12	14
No	HOG	NN	72.85	71.82	70.79	67.01	62.54	53.95	43.64	31.62
		CRC	73.54	72.85	72.16	71.48	67.70	64.95	58.76	54.98
	LBP	NN	44.33	35.40	25.09	6.19	2.41	1.37	1.03	1.03
		CRC	46.74	39.52	36.43	25.77	12.71	6.87	4.81	3.09
Yes	HOG	NN	73.88	73.54	72.51	72.85	71.13	69.42	67.01	61.86
		CRC	**75.95**	**76.63**	**76.98**	**75.95**	**75.26**	**74.57**	**72.16**	**70.45**
	LBP	NN	49.48	46.05	28.18	12.03	5.50	2.75	2.41	1.72
		CRC	65.64	65.29	62.54	53.61	37.11	23.37	16.49	10.65

Table 4. Correct classification rates for four training samples in each class with different features and different noise levels. The best results are highlighted in bold font.

Filter Faces	Features	Classifiers	Noise σ_n							
			0	2	4	6	8	10	12	14
No	HOG	NN	87.63	86.60	85.05	82.99	80.93	74.23	61.34	47.42
		CRC	89.69	89.18	89.18	87.63	86.08	82.47	79.38	75.77
	LBP	NN	78.87	53.61	29.90	6.70	3.09	2.58	1.03	1.03
		CRC	78.35	63.92	63.40	46.91	25.77	18.04	11.34	4.64
Yes	HOG	NN	86.60	87.63	86.60	86.08	85.57	84.02	82.47	76.80
		CRC	91.24	91.75	**90.72**	**90.21**	**89.18**	**89.18**	**87.63**	**86.08**
	LBP	NN	91.24	83.51	52.58	19.07	7.22	3.61	2.58	1.55
		CRC	**93.30**	**91.76**	88.14	73.20	53.09	34.02	22.68	12.89

Table 5. Correct classification rates for five training samples in each class with different features and different noise levels. The best results are highlighted in bold font.

Filter Faces	Features	Classifiers	Noise σ_n							
			0	2	4	6	8	10	12	14
No	HOG	NN	91.75	91.75	89.69	85.57	82.47	80.41	68.04	54.64
		CRC	93.81	92.78	92.78	91.75	87.63	85.57	83.51	81.44
	LBP	NN	87.63	64.95	31.96	6.19	3.09	2.06	1.03	2.06
		CRC	90.72	77.32	72.16	56.70	30.93	17.53	9.28	5.15
Yes	HOG	NN	94.85	93.81	92.78	92.78	91.75	89.69	86.60	83.51
		CRC	95.88	95.88	95.88	**95.88**	**93.81**	**93.81**	**93.81**	**93.81**
	LBP	NN	96.71	93.81	63.92	23.71	6.19	3.09	2.06	2.06
		CRC	**100**	**97.94**	**98.97**	84.54	62.87	43.30	28.87	20.62

4 Conclusions

Palmprint classification is an extremely important task in biometrics. It is essential that a palmprint classification method is robust to noise and at the same time achieves higher classification accuracy. Robust features should be extracted as well so that the palmprint classification system performs well.

In this paper, we have proposed a new method for palmprint classification by means of FF, feature extraction and CRC classifier. Our FF improves classification accuracy significantly no matter what features are used and how much noise is present in palmprint images. In our experiments, the CRC classifier is always better than the NN classifier for palmprint classification. In addition, our method achieves 100% classification accuracy with the combination of FF, LBP, and CRC classifier when no noise is added and five out of six samples in each class are used for training. In addition, FF combined with HOG and CRC achieves the highest classification accuracies for other testing cases. Our experiments confirm that the HOG features are very robust to noise whereas the LBP features are very sensitive to noise. Hence, it is preferable to use the HOG features for noisy palmprint images and the LBP features for clean palmprint images to improve classification accuracy.

Future research will be conducted by using deep convolutional neural networks (DCNN) for palmprint classification. We would also extract modern features with the contourlet transform [17] for palmprint classification to achieve better classification results.

References

1. Zhang, D., Kong, W.K., You, J., Wong, M.: On-line palmprint identification. IEEE Trans. Pattern Anal. Mach. Intell. **25**(9), 1041–1050 (2003)
2. Chen, G.Y., Bui, T.D., Krzyzak, A.: Palmprint classification using dual-tree complex wavelets. In: Proceedings of IEEE International Conference on Image Processing, Atlanta, GA, USA, 8–10 October 2006 (2006)
3. You, J., Li, W., Zhang, D.: Hierarchical palmprint identification via multiple feature extraction. Pattern Recogn. **35**, 847–859 (2002)
4. Zhang, D., Shu, W.: Two novel characteristics in palmprint verification: datum point invariance and line feature matching. Pattern Recogn. **32**, 691–702 (1999)
5. Liu, Y., Kumar, A.: Contactless palmprint identification using deeply learned residual features. IEEE Trans. Biometrics Identity Behav. **2**, 172–181 (2020)
6. Fei, L., Zhang, B., Zhang, W., Teng, S.: Local apparent latent direction extraction for palmprint recognition. Inf. Sci. **473**, 59–72 (2019)
7. Wu, L., Xu, Y., Cui, Z., Zuo, Y., Zhao, S., Fei, L.: Triple-type feature extraction for palmprint recognition. Sensors **21**, 4896 (2021)
8. Ahmad, M.I., Woo, W.L., Dlay, S.: Non-stationary feature fusion of face and palmprint multimodal biometrics. Neurocomputing **177**, 49–61 (2016)
9. Jaswal, G., Kaul, A., Nath, R.: Multiple feature fusion for unconstrained palmprint authentication. Comput. Electr. Eng. **72**, 53–78 (2018)
10. Dai, J., Zhou, J.: Multifeature-based high-resolution palmprint recognition. IEEE Trans. Pattern Anal. Mach. Intell. **33**, 945–957 (2010)
11. Zhang, S., Wang, H., Huang, W.: Palmprint identification combining hierarchical multi-scale complete LBP and weighted SRC. Soft. Comput. **24**, 4041–4053 (2020)
12. Chen, G.Y., Bui, T.D., Krzyzak, A.: Filter-based face recognition under varying illumination. IET Biometrics **7**(6), 628–635 (2018)
13. Zhang, L., Yang, M., Feng, X.: Sparse representation or collaborative representation: which helps face recognition? In: IEEE International Conference on Computer Vision, pp. 471–478 (2011)
14. Dalal, N., Triggs, B.: Histograms of oriented gradients for human detection. In: IEEE Computer Society Conference on Computer Vision and Pattern Recognition, vol. 1, pp. 886–893 (2005)
15. He, D.C., Wang, L.: Texture unit, texture spectrum, and texture analysis. IEEE Trans. Geosci. Remote Sens. **28**, 509–512 (1990)
16. The PolyU Palmprint Database. http://www.comp.polyu.edu.hk/biometrics
17. Do, M.N., Vetterli, M.: The contourlet transform: an efficient directional multiresolution image representation. IEEE Trans. Image Process. **14**(12), 2091–2106 (2005)

Deep Multi-label Classification of Personality with Handwriting Analysis

Marzieh Adeli Shamsabad and Ching Yee Suen(✉)

Centre for Pattern Recognition and Machine Intelligence (CENPARMI), Concordia University, Montreal, QC H3G 1M8, Canada
{m_adelis,suen}@cenparmi.concordia.ca

Abstract. Handwriting analysis has traditionally been used to infer personality traits from the stylistic features of writing. With advances in machine learning, the accuracy and applicability of these analyses have significantly improved. This paper presents a new multi-label classification approach to classify personality traits, such as Extraversion and Conscientiousness, into Low, Average, and High categories. This approach uses Binary Cross-Entropy with Logits Loss and Focal Loss to handle multi-label classification and class imbalance. Image segmentation techniques are also employed to enhance the handling of limited handwriting samples. The paper evaluates the complexity and performance of ResNet-50 and ResNet-101 architectures in recognizing complex handwriting patterns using three optimizers: SGD with momentum, Adam, and AdaBelief. The results demonstrate the efficacy of our proposed method, improving overall accuracy from 67.09% to 90.16% for ResNet-50 and from 69.43% to 90.07% for ResNet-101, with an overall AUC of 0.96. These improvements emphasize the model's capability for practical automated handwriting analysis.

Keywords: Handwriting · Multi-label classification · Class Imbalance · Binary Cross-Entropy with Logits Loss · Focal Loss

1 Introduction

Handwriting is a unique form of personal expression, often considered as distinctive as a fingerprint. Graphologists have traditionally analyzed various handwriting features—such as slant, margin width, size, specific letter shapes, and strokes—to predict personality traits. These features can reveal numerous aspects of an individual's personality, including emotional state, self-esteem, and creativity [15]. Historically, these analyses have been manual, time-consuming, and prone to human error.

To address these challenges, handwriting analysis has increasingly adopted data-driven approaches over the past decades to automate and enhance the accuracy of personality assessments. Initial efforts by Ahmed and Mathkour [2] employed computational methods to reproduce the expert graphological analysis.

C. Y. Suen et al. (Eds.): ANNPR 2024, LNAI 15154, pp. 218–230, 2024.
https://doi.org/10.1007/978-3-031-71602-7_19

This field expanded through machine learning advancements, with Saraswal et al. [19] using the K-Nearest Neighbors (KNN) algorithm to automate feature extraction, significantly simplifying the analysis process. Subsequent studies, such as those by Abd Yusof et al. [1], adopted agglomerative hierarchical clustering to connect handwriting characteristics with personality traits and mental health indicators. Similarly, Chitlangia and Malathi utilized the Histogram of Oriented Gradients (HOG) combined with Support Vector Machines (SVM) to categorize handwriting into various personality profiles [3].

Further studies in handwriting analysis employed more complex algorithms and feature extraction techniques. Gahmousse et al. [6] developed a method using textural features with the Oriented Basic Image Features Column (oBIFC) and RankSVM to predict personality traits from the HWxPI Dataset. Recent innovations have also included the application of the Synthetic Minority Oversampling Technique (SMOTE) with SVM and Convolutional Neural Networks (CNN) classifiers by Afnan Garoot et al. [7], significantly enhancing accuracy and addressing data imbalances. Additionally, using different architectures of CNNs for deeper handwriting analysis has shown promising results. Elngar et al. [5] and Durga et al. [4] improved prediction accuracies through advanced CNN architectures. This reflects more robust preprocessing and model selection strategies, as highlighted by Yan Xu et al. [25] with the convNeXtTiny model following extensive preprocessing efforts.

Despite these advancements, identifying different levels of personality traits through handwriting remains under-explored. This task requires multi-label classification, which often faces the challenges of imbalanced datasets where some traits are more common, and a limited number of handwriting samples, both of which affect model performance.

In this study, a new deep learning approach is proposed to address the challenges. To address the issue of limited handwriting samples among some traits, the images are segmented to generate sufficient training samples, enhancing model performance. Binary Cross Entropy with Logits Loss (BCEWithLogitsLoss) is used for multi-label classification [10] and combined with Focal Loss to address class imbalance [17]. The effectiveness and performance of two pre-trained CNN models, ResNet-50 and ResNet-101, are compared in terms of training time and computation efficiency in recognizing complex handwriting patterns. Finally, the models are evaluated using three different optimizers.

The paper is structured as follows: Sect. 2 describes the dataset, Sect. 3 covers the methodology, Sect. 4 presents the results, and Sect. 5 concludes the paper.

2 Dataset Description

The dataset, sourced from CENPARMI, consists of 1,110 handwriting samples collected using the standardized protocols provided to participants during a structured survey [7] and approved by the Human Research Ethics Committee at Concordia University. These samples are written in various languages and labeled by a professional graphologist for two personality traits from the

Big Five Factor Model (BFFM): Conscientiousness and Extraversion, which are categorized into three levels: low, average, and high.

The BFFM is a widely recognized approach for describing and assessing personality traits [8]. While it includes five main traits, this study focuses on Conscientiousness and Extraversion. Extraversion includes sociability and emotional expressiveness, whereas Conscientiousness involves high levels of thoughtfulness and good impulse control. We aim to correlate BFFM test scores with handwriting analysis results, predicting trait levels from handwriting samples. This involves a multi-label classification problem, where an instance may have multiple labels. The research specifically emphasizes the French language (Table 1), as it is the most represented in the dataset with 869 handwriting samples.

Table 1. Single-Label distribution of two factors in dataset

Total Number of Handwriting Samples = 1110									
French	English	Persian	Korean	Chinese	Arabic	Spanish	Urdu	Bengali	Russian
869	184	22	15	8	5	4	1	1	1

French with 869 Samples					
Extraversion			Conscientiousness		
Low	Average	High	Low	Average	High
132	316	421	40	325	504

3 Methodology

3.1 Data Preprocessing

Initially, the dataset required data cleaning, including removing duplicate samples, labeling modification, separating different languages based on distinct patterns, handling missing values, and standardizing formats to ensure its integrity and consistency. As a large dataset is important for neural networks to learn diverse patterns and generalize well to unseen data to enhance model performance [23], image segmentation is employed to increase the number of samples. From the initial dataset of 869 samples, 40 images from each category (Low, Average, High) are selected based on the least represented class, a total of 120 images. A recent study showed that data processing and segment size affect personality trait prediction accuracy [25]. Therefore, these images are segmented into 6×4 sizes (see Fig. 1) to maintain the original ratio, accurately capture features, and reduce the need for resizing which could degrade the quality of pictures [11].

This process increases the dataset to 2880 sub-samples. Ineffective samples, mainly edges or blank spaces, are identified and removed as they contain no information the model can learn from, resulting in 1928 valuable handwriting images available for this research (Table 2).

Fig. 1. Segmentation of handwriting image into 6 × 4 dimensions

Table 2. Single-label distribution of two factors in segmented images

French with 1928 Samples					
Extraversion			Conscientiousness		
Low	Average	High	Low	Average	High
578	688	662	651	634	643

As handwriting samples are digitized, various preprocessing techniques are explored to enhance feature extraction. Otsu's binarization method combined with bilateral filtering is found to be the most effective (see Fig. 2). Otsu's method simplifies images to black and white based on an optimal threshold [25], and bilateral filtering reduces noise while preserving key structural details like the edges of letters and lines [24]. These processed images are then converted into a three-channel format to be compatible with neural network inputs.

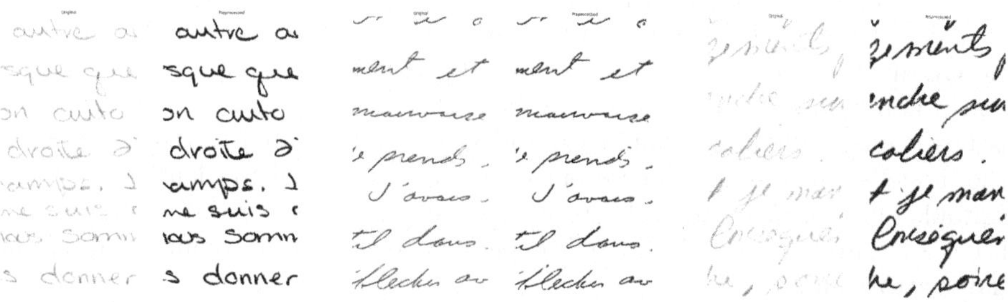

Fig. 2. Handwriting image enhancement via Otsu's Binarization and Bilateral Filtering

Despite the increase in the number of samples through segmentation, further data augmentation is essential due to the limited number of handwriting samples available [21]. Augmentation techniques are carefully selected to introduce realistic variations in handwriting that the model might encounter in real-world scenarios. These techniques include random rotations and flips to simulate natural variations in writing angles, affine transformations to imitate different handwriting scales and distortions, perspective changes and blurring to adjust the viewing angle and simulate scanning artifacts, and random erasing to introduce small areas of missing data. Together, these augmentations help create a robust dataset that trains the neural network to be adaptive and accurate in analyzing diverse handwriting styles, thus enhancing its ability to predict personality traits effectively.

3.2 Models of CNN Architectures

Convolutional Neural Networks (CNNs) are known for their effectiveness in image-processing tasks, such as handwriting recognition, because of their ability to detect spatial hierarchies, from simple edges to complex patterns [13]. In this study, ResNet-50 and ResNet-101 are employed to explore the impact of network depth on feature extraction. These models are selected for their robust feature-detection capabilities, which are important for detailed image analysis in handwriting. Both models are pre-trained on the large-scale ImageNet dataset and utilize residual blocks to efficiently train deeper networks, addressing the challenge of vanishing gradients.

ResNet-50, with 50 layers, is optimized for environments with limited computational resources. Efficiency, speed, and accuracy are balanced, making it suitable for quicker training cycles without significantly compromising performance. On the other hand, ResNet-101, with 101 layers, offers a deeper architecture that captures finer details and more complex patterns, which is important for high-precision tasks such as handwriting analysis. While more computational resources are required, the investment is justified by its ability to extract more detailed features [9].

Optimization Method. Selecting the right optimizer is another important factor in training deep neural networks, particularly with imbalanced handwriting datasets. To evaluate their impact, three optimizers are used: SGD (Stochastic Gradient Descent) with Momentum [16], AdaBelief (Adaptive Belief) [26], and Adam (Adaptive Moment Estimation) [12]. The selection of each optimizer is based on its distinct advantages, and they are evaluated for their effect on the model's performance. These optimizers are employed to identify the most suitable ones for achieving a balance between convergence speed, stability, and accuracy. This enhances the model's capability to classify personality traits from handwriting data.

3.3 Classification Layer

In multi-label classification, each sample can belong to multiple classes, which can increase the challenge of class imbalance. To address label independence and imbalance, BCEWithLogitsLoss is used for the binary classification of each label. Additionally, Focal Loss, introduced by Tsung-Yi et al. [14], is used to address the issue of imbalanced data and enhance the training on difficult-to-classify examples. The original fully connected output layer is replaced with a linear layer that generates predictions for six distinct labels. These loss functions enhance the model's robustness by treating each label independently and addressing class imbalance, making the model more effective and stable in multi-label classification.

BCEWithLogitsLoss. This loss function combines Sigmoid activation and binary cross-entropy loss, making it well-suited for multi-label classification tasks. The Sigmoid function converts raw logits (raw outputs from the network before activation) into probabilities, allowing independent predictions for each label. Binary cross-entropy loss then evaluates the accuracy of these probabilities against actual labels [22]. This function enables simultaneous prediction of multiple labels per instance, essential for tasks where instances can belong to multiple categories, such as identifying various personality traits from one handwriting sample. This study explores two personality traits, each categorized into three levels (low, average, high), resulting in six classes. Therefore, a person can have different values for Extraversion and Conscientiousness simultaneously. This combined approach ensures accurate, flexible predictions, effectively handling the complexity of multi-label classification.

Focal Loss. To complement BCEWithLogitsLoss, Focal Loss is implemented in the training phase to address class imbalance. Focal Loss is defined by two parameters: alpha(α) and gamma(γ): α is a weighting factor for the class, and γ is a focusing parameter that adjusts the rate at which easy examples contribute to the loss. By increasing γ, the loss contribution of well-classified examples is reduced, shifting the training focus to more challenging cases [18].

3.4 Performance Metrics

Five fundamental metrics evaluate our models' performance: Accuracy, Area Under the Curve (AUC), F1-score, Precision, and Recall. These metrics are defined based on the values in the Confusion matrix, which displays the performance of a classification model by showing the true positives, true negatives, false positives, and false negatives for each class i:

- True Positives (TP_i): Instances that are Correctly identified as class i.
- True Negatives (TN_i): Instances that are Correctly identified as not class i.
- False Positives (FP_i): Instances that are Incorrectly identified as class i.
- False Negatives (FN_i): Instances of class i Incorrectly identified as another class.

Precision measures the proportion of true positive predictions among all positive predictions, indicating the model's ability to avoid false positives. Recall quantifies how many of the actual positive instances the model identifies, highlighting its ability to avoid false negatives. The F1 score combines both Precision and Recall for a balanced evaluation. The AUC is the area under the receiver operating characteristic (ROC) curve and represents a two-dimensional plot of the true positive rate (recall on the y-axis) against the false positive rate (x-axis). An AUC close to 1 signifies a perfect model.

Accuracy is the ratio of correct predictions to total predictions. While accuracy is useful, it can be misrepresented by class imbalances, making Precision, Recall, F1-score, and AUC important for a comprehensive evaluation. Higher values of these metrics indicate better model performance [18].

4 Experimental Results

The dataset is divided into a 60:20:20 ratio for training, validation, and testing respectively. The initial dataset, comprising 869 samples (Table 1), present an average imbalance ratio of 12.61 among personality traits. Validation results show that without focal loss, ResNet-50 achieves an accuracy of 30.94%, and ResNet-101 achieves 39.57% using the Adam optimizer. With focal loss, these accuracies improve to 76.79% for ResNet-50 and 75.42% for ResNet-101, marking an approximately 40% increase in validation accuracy for both architectures.

For the segmented dataset with 1928 samples (Table 2) and an average imbalance ratio of 1.19, the validation accuracy without focal loss is 74.43%, which increases to 91.05% for ResNet-50 (Table 3) and from 74.17% to 92.29% with focal loss for ResNet-101 with the Adam optimizer (Table 4). This improvement highlights the importance of a larger dataset, enabling neural networks to learn diverse patterns and generalize better to unseen data. The effectiveness of our approach to multi-label classification using BCEWithLogitsLoss combined with focal loss to address the challenge of class imbalance is evaluated by comparing the performance of ResNet-50 and ResNet-101 architectures with and without focal loss on segmented images. This demonstrates the effect of image segmentation and preprocessing techniques in improving model performance, facilitating the automation of handwriting analysis even with small datasets.

Table 3. Training and Validation Performance of ResNet-50

Optimizer	Train				Validation			
	BCE with Logits	Accuracy	Focal Loss	Accuracy	BCE with Logits	Accuracy	Focal Loss	Accuracy
SGD	0.183	73.97%	0.084	85.47%	0.256	62.78%	0.100	82.69%
AdaBelief	0.122	84.18%	0.031	95.49%	0.212	71.52%	0.076	90.72%
ADAM	0.114	87.10%	0.026	**96.34%**	0.197	74.43%	0.066	**91.05%**

Table 4. Training and Validation Performance of ResNet-101

Optimizer	Train				Validation			
	BCE with Logits	Accuracy	Focal Loss	Accuracy	BCE with Logits	Accuracy	Focal Loss	Accuracy
SGD	0.141	81.18%	0.072	87.67%	0.235	70.87%	0.082	86.19%
AdaBelief	0.111	86.21%	0.023	96.34%	0.224	71.84%	0.058	91.96%
ADAM	0.095	89.05%	0.023	**96.88%**	0.206	74.17%	0.051	**92.29%**

Additionally, the impact of different optimizers—Adam, SGD with momentum, and AdaBelief—is assessed on the models' performance to evaluate the best fit for handwriting analysis. The evaluation uses accuracy, precision, recall, F1-score, and AUC.

4.1 Impact of Focal Loss

Through experimentation with different values and the evaluation of their performance on the validation dataset, the optimal values for focal loss parameters are determined to be $(\alpha) = 1$ and $(\gamma) = 2$. These parameters help the model prioritize harder examples, improving robustness and reducing bias towards more frequent labels. As shown in Tables 3 and 4 and discussed about the initial dataset, significant improvements in validation accuracy are observed with focal loss compared to using only BCEWithLogitsLoss. The training curves in Fig. 3 show that focal loss leads to faster and more stable accuracy improvements and loss reductions. These curves demonstrate the model's learning process and support the improvements noted in the validation metrics.

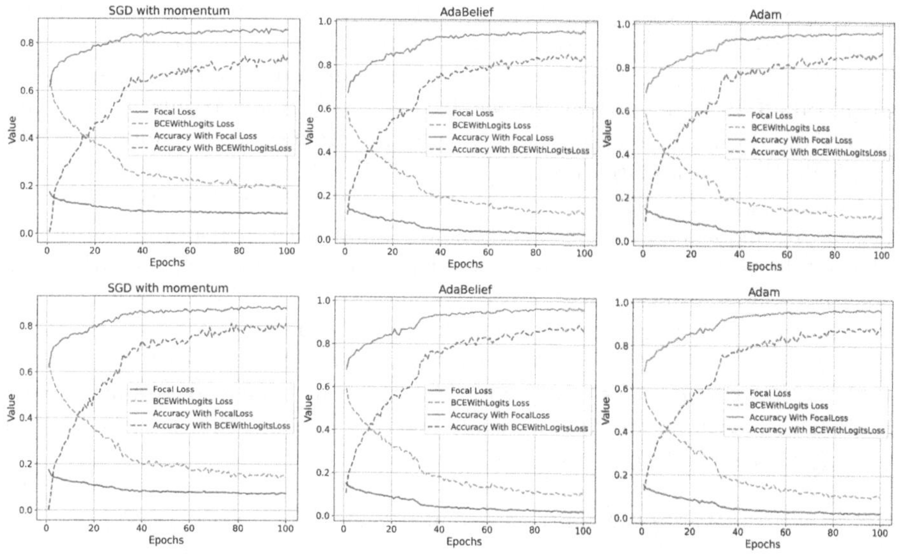

Fig. 3. Training curves comparing the performance with and without focal loss: Top - ResNet-50, Bottom - ResNet-101.

In addition, based on the observations in other metrics shown in Table 5, the performance metrics across both ResNet-50 and ResNet-101 architectures for all optimizers (SGD, AdaBelief, and Adam) are increased. Specifically, the average percentage increase in precision is 3.32%, in recall is 5.13%, and in F1-score is 3.72%.

These results highlight the effectiveness of focal loss in enhancing the overall performance of the models to classify personality traits from handwriting data.

4.2 Performance Comparison of CNN Architectures

The performance comparison between the ResNet-50 and ResNet-101 models highlights the differences in their effectiveness for multi-label classification tasks.

Table 5. Focal loss performance on ResNet-50 and ResNet-101 in the test phase

Optimizer	Loss Function	Precision		Recall		F1-score	
		ResNet50	ResNet101	ResNet50	ResNet101	ResNet50	ResNet101
SGD	Without Focal	0.7796	0.8278	0.7675	0.7992	0.7728	0.8166
	With Focal	0.8367	0.8349	0.7901	0.8138	0.8127	0.8200
AdaBelief	Without Focal	0.8702	0.8741	0.8341	0.8367	0.8578	0.8550
	With Focal	0.8903	0.8851	0.8837	0.8792	0.8869	0.8820
ADAM	Without Focal	0.8637	0.8644	0.8212	0.8264	0.8419	0.8450
	With Focal	**0.8923**	**0.8892**	**0.8844**	**0.8867**	**0.8881**	**0.8879**

As shown in Table 6, ResNet-50 achieved a higher accuracy of 90.16% compared to ResNet-101's 90.07% when optimized with Adam. Both models displayed strong classification capabilities with AUC values of 0.96, suggesting that a deeper network does not always result in better performance. In terms of training time, ResNet-50 was more efficient, taking 1 h and 40 min to train for 100 epochs on a GPU-T4, whereas ResNet-101 required 2 h and 10 min for the same number of epochs. This reflects the increased computational complexity of the deeper ResNet-101 architecture.

Evaluating SGD with momentum, Adam, and AdaBelief on ResNet-50 and ResNet-101 models using focal loss shows Adam with a 0.0001 learning rate achieved the highest precision, recall, and F1-score, closely followed by AdaBelief. Despite using a MultiStepLR scheduler and a 0.005 learning rate with 0.9 momentum, SGD with momentum performed the lowest. Accuracy and loss curves (Fig. 4 and Fig. 5) show that Adam and AdaBelief provided higher, more stable accuracy and lower loss values. Adam is the preferred optimizer, with AdaBelief as a strong alternative, while SGD with momentum is ineffective for multi-label classification tasks.

Table 6. Performance of models in the test phase

Optimizer	ResNet50			ResNet101		
	Accuracy	Loss	AUC	Accuracy	Loss	AUC
SGD	80.22%	0.021	0.87	84.33%	0.022	0.91
AdaBelief	90.03%	0.010	0.96	89.59%	0.016	0.96
ADAM	**90.16%**	0.011	0.96	**90.07%**	0.011	0.96

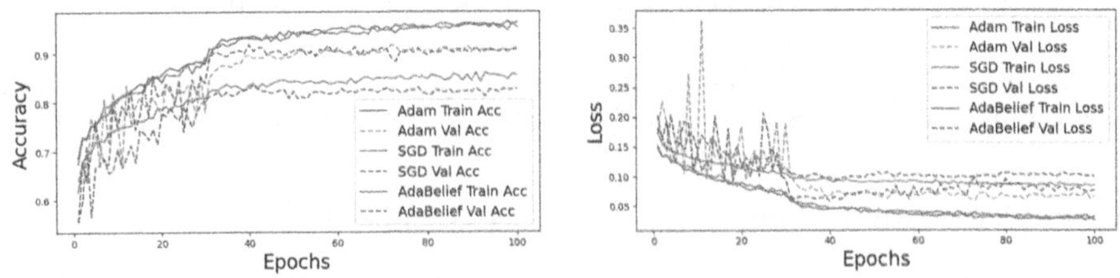

Fig. 4. Performance of Optimizers: Training and Validation Accuracy vs. Loss for ResNet-50

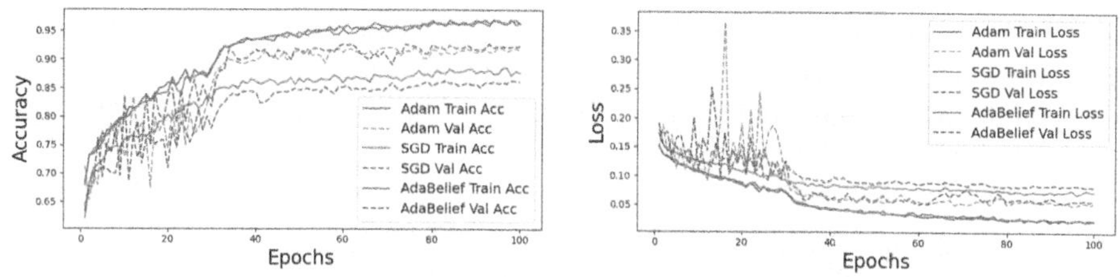

Fig. 5. Performance of Optimizers: Training and Validation Accuracy vs. Loss of ResNet-101

The confusion matrix (Fig. 6) of the models with Adam optimizer (as the best performance among other optimizers), provides further clarification and indicates that while ResNet-101 has better recall, indicating it captures more positive cases, ResNet-50 demonstrates slightly better precision and F1-scores.

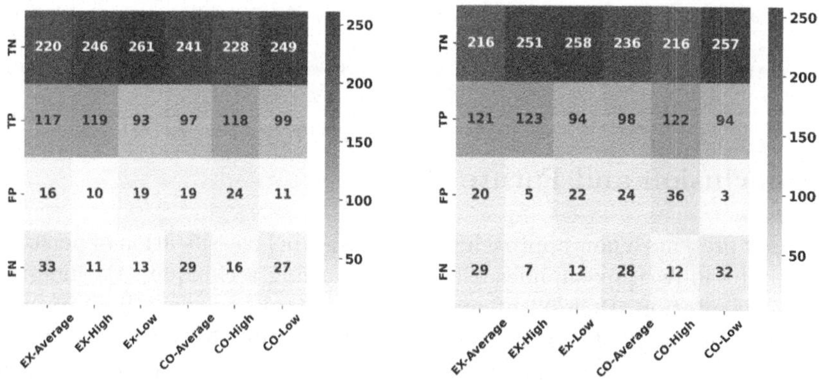

Fig. 6. Performance in the test phase with Confusion Matrix: Left - ResNet-50, Right - ResNet-101 (EX: Extraversion and CO: Conscientiousness)

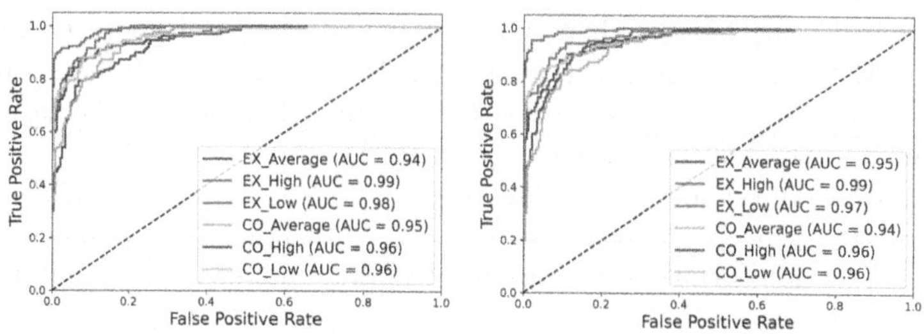

Fig. 7. AUC Multi-label classes of ADAM Optimizer: Left - ResNet-50, Right - ResNet-101

The ROC curves for the best-performing models (Adam optimizer), shown in Fig. 7, indicate high AUC values ranging from 0.94 to 0.99 for both architectures, performing exceptionally well across various classes. The EX-High (high-level Extraversion) class achieved the highest AUC of 0.99 for both architectures. These high AUC values confirm the robustness of both models in handling multi-label classification tasks.

The results demonstrate the potential of the proposed technique for identifying personality traits and comparing the effectiveness of the solution in this study with the recent state-of-the-art systems summarized in Table 7.

Table 7. Performance Comparison: Proposed Model with Focal Loss vs. recent State-of-the-Art Methods

Paper	Features	Classification	Dataset	Accuracy
Chitlangia et al. (2019) [3]	HOG technique	SVM	50 Private samples	80.00%
Gahmousse et al. (2020) [6]	oBIFC	Rank-SVM	418 Private samples	58.60%
Durga et al. (2022) [4]	Binary word vector	CNN	200 Private samples	80.88%
Saraswal et al. (2022) [19]	Graphological	KNN	EMNIST	89.00%
Sati et al. (2023) [20]	Graphological	ANN + MLP	Private	82.64%
Yan Xu et al. (2024) [25]	Graphological	convNeXtTiny	CENPARMI 1828 samples	86.84%
Proposed Method	**Graphological**	**CNN+Focal Loss**	**CENPARMI 1928 samples**	**90.16%**

5 Conclusion and Future Work

This study presents a new approach to the multi-label classification of personality traits from handwriting samples using deep learning techniques. By integrating image segmentation, BCEWithLogitsLoss, and focal loss, data limitations and class imbalance were effectively addressed. Evaluating ResNet-50 and ResNet-101 architectures with the Adam, SGD with momentum, and AdaBelief optimizers showed that Focal loss significantly enhanced model performance, improving validation accuracy by an average of 18.58% across different setups. ResNet-50 achieved an accuracy of 90.16% and faster training times compared to ResNet-101 with an accuracy of 90.07%. However, ResNet-101 had better true positive rates, which can be beneficial in scenarios where deeper feature extraction is important. Both models demonstrated high classification capabilities with AUC values of 0.96. The Adam and AdaBelief optimizers provided the best overall performance in the task of handwriting analysis.

Future work will include expanding the dataset to ensure better balance and generalizability. Since data processing and certain transformations can impact metrics and model performance, potentially introducing bias, a comprehensive bias analysis on the dataset and evaluation of the influence of data processing tasks on bias and performance will be conducted. Additionally, advanced data augmentation and transfer learning strategies will be integrated to enhance model performance. Other deep learning architectures like ResNet18, 34, or 152, as well as optimization techniques, will be explored to further model improvement. Lastly, the work will be extended to include handwriting samples in different languages and to analyze a broader range of personality traits.

References

1. Abd Yusof, N.F., Zulkarnain, N.Z., Syed Ahmad, S.S., Othman, Z., Hashim, A.: Extracting graphological features for identifying personality traits using agglomerative hierarchical clustering algorithm, pp. 1–5 (2022). https://doi.org/10.1109/IICAIET55139.2022.9936858
2. Ahmed, P., Mathkour, H.: On the development of an automated graphology system, pp. 897–901 (2008)
3. Chitlangia, A., Malathi, G.: Handwriting analysis based on histogram of oriented gradient for predicting personality traits using SVM. Procedia Comput. Sci. **165**, 384–390 (2019)
4. Durga, L., Deepu, R.: A self adaptive cognitive deep learning framework for classifying graphology features to big five personality traits. Int. J. Adv. Technol. Eng. Explor. **9**, 1151–1167 (2022). https://doi.org/10.19101/IJATEE.2021.875577
5. Elngar, A., Jain, N., Sharma, D., Negi, H., Trehan, A., Srivastava, A.: A deep learning based analysis of the big five personality traits from handwriting samples using image processing. J. Inf. Technol. Manag. **12**, 3–35 (2020). https://doi.org/10.22059/jitm.2020.78884
6. Gahmousse, A., Gattal, A., Djeddi, C., Siddiqi, I.: Handwriting based personality identification using textural features. In: 2020 International Conference on Data Analytics for Business and Industry: Way Towards a Sustainable Economy (ICDABI), pp. 1–6 (2020). https://doi.org/10.1109/ICDABI51230.2020.9325664
7. Garoot, A., Suen, C.Y.: Measuring the Big Five factors from handwriting using ensemble learning model AvgMLSC. In: Carmona-Duarte, C., Diaz, M., Ferrer, M.A., Morales, A. (eds.) IGS 2022. LNCS, vol. 13424, pp. 159–173. Springer, Cham (2022). https://doi.org/10.1007/978-3-031-19745-1_12
8. Gavrilescu, M., Vizireanu, N.: Predicting the big five personality traits from handwriting. EURASIP J. Image Video Process. **2018**(1), 57 (2018)
9. He, K., Zhang, X., Ren, S., Sun, J.: Deep residual learning for image recognition. In: 2016 IEEE Conference on Computer Vision and Pattern Recognition (CVPR), pp. 770–778 (2016). https://doi.org/10.1109/CVPR.2016.90
10. Heaton, J.: Ian goodfellow, yoshua bengio, and aaron courville: deep learning. Genet. Program Evolvable Mach. **19**(1), 305–307 (2018)
11. Huang, Y.L.: Wavelet-based image interpolation using multilayer perceptrons. Neural Comput. Appl. **14**, 1–10 (2005). https://doi.org/10.1007/s00521-004-0433-0
12. Kingma, D., Ba, J.: Adam: a method for stochastic optimization. In: International Conference on Learning Representations (2014)
13. Krizhevsky, A., Sutskever, I., Hinton, G.: ImageNet classification with deep convolutional neural networks. In: Advances in Neural Information Processing Systems, vol. 25 (2012). https://doi.org/10.1145/3065386
14. Lin, T.Y., Goyal, P., Girshick, R., He, K., Dollár, P.: Focal loss for dense object detection. In: 2017 IEEE International Conference on Computer Vision (ICCV), pp. 2999–3007 (2017). https://doi.org/10.1109/ICCV.2017.324
15. Mekhaznia, T., Djeddi, C., Sarkar, S.: Personality traits identification through handwriting analysis. In: Djeddi, C., Kessentini, Y., Siddiqi, I., Jmaiel, M. (eds.) MedPRAI 2020. CCIS, vol. 1322, pp. 155–169. Springer, Cham (2021). https://doi.org/10.1007/978-3-030-71804-6_12
16. Qian, N.: On the momentum term in gradient descent learning algorithms. Neural Netw. Off. J. Int. Neural Netw. Soc. **12**, 145–151 (1999). https://doi.org/10.1016/S0893-6080(98)00116-6

17. Qin, R., Wang, L., Zeng, L., Chen, J., Yan, B.: Weighted focal loss: an effective loss function to overcome unbalance problem of chest X-ray14. IOP Conf. Ser. Mater. Sci. Eng. **428**, 012022 (2018). https://doi.org/10.1088/1757-899X/428/1/012022
18. Romdhane, T.F., Alhichri, H., Ouni, R., Atri, M.: Electrocardiogram heartbeat classification based on a deep convolutional neural network and focal loss. Comput. Biol. Med. **123**, 103866 (2020)
19. Saraswal, A., Saxena, U.: Analysis and recognition of handwriting patterns for personality trait prediction using unsupervised machine learning approach (2022). https://doi.org/10.3233/ATDE220778
20. Sati, K., Kumar, D.: Identifying the personality traits based on handwriting using ANN and MLP concept, pp. 524–528 (2023). https://doi.org/10.1109/ICSCSS57650.2023.10169354
21. Shorten, C., Khoshgoftaar, T.: A survey on image data augmentation for deep learning. J. Big Data **6**, 60 (2019). https://doi.org/10.1186/s40537-019-0197-0
22. Sukumar, S., Harish, A., Shahina, A., Sanjana, B., Khan, A.N.: Deep learning based pulmonary embolism detection using convolutional feature maps of CT pulmonary angiography images. Procedia Comput. Sci. **233**, 317–326 (2024)
23. Sun, C., Shrivastava, A., Singh, S., Gupta, A.: Revisiting unreasonable effectiveness of data in deep learning era. In: 2017 IEEE International Conference on Computer Vision (ICCV), pp. 843–852 (2017). https://doi.org/10.1109/ICCV.2017.97
24. Tomasi, C., Manduchi, R.: Bilateral filtering for gray and color images. In: Sixth International Conference on Computer Vision (IEEE Cat. No. 98CH36271), pp. 839–846 (1998). https://doi.org/10.1109/ICCV.1998.710815
25. Xu, Y., Tang, Y., Suen, C.Y.: Two key factors in handwriting analysis for personality prediction. In: Su, R., Liu, H. (eds.) Fifth International Conference on Image, Video Processing, and Artificial Intelligence (IVPAI 2023), vol. 13074, p. 130740G. International Society for Optics and Photonics, SPIE (2024). https://doi.org/10.1117/12.3023796
26. Zhuang, J., et al.: Adabelief optimizer: adapting stepsizes by the belief in observed gradients (2020)

License Plate Detection and Character Recognition Using Deep Learning and Font Evaluation

Zahra Ebrahimi Vargoorani(✉) and Ching Yee Suen

Concordia University, Montreal, QC, Canada
{zahra.ebrahimivargoorani,chingyee.suen}@concordia.ca

Abstract. License plate detection (LPD) is essential for traffic management, vehicle tracking, and law enforcement but faces challenges like variable lighting and diverse font types, impacting accuracy. Traditionally reliant on image processing and machine learning, the field is now shifting towards deep learning for its robust performance in various conditions. Current methods, however, often require tailoring to specific regional datasets. This paper proposes a dual deep learning strategy using a Faster R-CNN for detection and a CNN-RNN model with Connectionist Temporal Classification (CTC) loss and a MobileNet V3 backbone for recognition. This approach aims to improve model performance using datasets from Ontario, Quebec, California, and New York State, achieving a recall rate of 92% on the Centre for Pattern Recognition and Machine Intelligence (CENPARMI) dataset and 90% on the UFPR-ALPR dataset. It includes a detailed error analysis to identify the causes of false positives. Additionally, the research examines the role of font features in license plate (LP) recognition, analyzing fonts like Driver Gothic, Dreadnought, California Clarendon, and Zurich Extra Condensed with the OpenALPR system. It discovers significant performance discrepancies influenced by font characteristics, offering insights for future LPD system enhancements.

Keywords: Deep Learning · License Plate · Font Evaluation

1 Introduction

Automatic License Plate Recognition (ALPR) systems face challenges due to environmental factors like light conditions, rain and dust, high speeds, varied angles, and poor image quality [1]. The font style on license plates also significantly affects recognition [2]. ALPR is crucial for traffic control, security enhancement, and law enforcement. Despite various techniques, including image processing and machine learning, there is still potential to improve accuracy in ALPR systems.

C. Y. Suen et al. (Eds.): ANNPR 2024, LNAI 15154, pp. 231–242, 2024.
https://doi.org/10.1007/978-3-031-71602-7_20

Faster R-CNN [3] is a notable computer vision model that improves object detection by integrating a Region Proposal Network (RPN) with a detection network. This model increases detection speed and accuracy through shared computation across the entire image. It works in two stages: first, the RPN generates region proposals [4], which are then classified and refined by the detection network. Known for its end-to-end training, high accuracy, and fast processing, Faster R-CNN is versatile in surveillance and autonomous navigation applications, effectively handling various object sizes and scenarios.

Deep neural networks, particularly Convolutional Neural Networks (CNNs) and Recurrent Neural Networks (RNNs), have been harnessed in image and natural language processing, achieving significant improvements in object detection via deep learning. The integration Of these networks with Connectionist Temporal Classification (CTC) Loss [5] has notably enhanced the accuracy and efficiency of these systems. CNNs, with their ability to handle spatial data processing, and RNNs, which excel in sequence prediction, are particularly suitable for complex scenarios. CTC Loss aligns input sequences with outputs, thereby improving traditional methods and expanding applications in automation and intelligence, such as autonomous driving and surveillance.

Typography plays a crucial role in text presentation and recognition systems. Specific typefaces designed for unique applications, like mathematical notation or music scores, enhance text presentation and comprehension. The font choice impacts the accuracy and efficiency of computer-based text recognition systems, significantly influencing their reliability and operational success.

The main contributions of this work are:

1. Introducing a new dataset(CENPARMI) of license plate images collected from Quebec and Ontario provinces in Canada, California, and the State of New York in the United States of America.
2. Proposing a new two-stage solution for the license plate detection phase using a Faster-RCNN model and CNN+RNN Model with Connectionist Temporal Classification(CTC) Loss networks for recognizing letters.
3. Perform an analytical evaluation of five fonts commonly used on license plates and explore how typeface design can impact computer-based studies in this context.

1.1 Datasets and Fonts

UFPR ALPR Dataset: The UFPR-ALPR [6] dataset, developed by the Federal University of Paraná, includes 4,500 diverse images of license plates from Paraná, Brazil, captured using GoPro Hero4 Silver, Huawei P9 Lite, and iPhone 7 Plus cameras. The dataset is designed for machine learning and divided into 40% training, 20% validation, and 40% testing splits. It features various real-world conditions, such as different distances, sunlight, and shadows, to improve the robustness of automatic license plate recognition systems. Although generally clear, the mandatory font used on Brazilian license plates can still be confusing in specific contexts. Figure 1 highlights potential confusion with the Mandatory font used on license plates, illustrating how '8' might be mistaken for 'B,' and '4' could be confused with 'A' and 'O' with number '0'. These examples emphasize the need to carefully evaluate the fonts on license plates to ensure clarity and avoid misinterpretation in both digital and real-world settings.

Fig. 1. Confusing glyphs of Mandatory font in license plate context

CENPARMI Dataset: The Centre for Pattern Recognition and Machine Intelligence (CENPARMI) has developed a dataset containing 1,600 license plates from diverse regions, including California, New York, Ontario, and Quebec, captured using a mobile phone camera. This dataset addresses challenging lighting conditions such as harsh sunlight, shadows, and dimly lit environments, significantly affecting license plate visibility. It includes plates photographed from 1 to 30 m and features about 20% of the plates shot at angles, adding complexity due to their non-standard orientations. This diverse collection is a valuable tool for advancing license plate recognition technologies.

Moreover, the dataset is not just a theoretical exercise, but a practical resource that encompasses a range of typefaces used on license plates, including Driver Gothic for Quebec and Dreadnought for Ontario, Clarendon for California, and Zurich Extra Condensed for New York. It also spans multiple vehicle types-motorcycles, cars, and public transport vehicles-and illustrates variations in image brightness, vehicle distance, plate sizes, and background colors, all contributing to the complexities faced by recognition systems. This diversity ensures the dataset's applicability in real-world scenarios, making it a critical resource for developing robust recognition technologies.

The paper includes a review of related literature, introduces a proposed model, discusses the results, and concludes with future directions in license plate recognition technology (Fig. 2).

Fig. 2. Sample images from CENPARMI Dataset

2 Related Work

In reference [7], researchers used a Convolutional Neural Network (CNN) to extract relevant features from license plates, identifying distinct visual patterns through convolutional layers. These features were then processed by a Recurrent Neural Network (RNN) with 36 hidden units to sequence the detected characters and accurately reconstruct the license plate information. By combining the spatial pattern recognition capabilities of CNNs with the sequential learning abilities of RNNs, they achieved a 76% accuracy rate for recognizing all characters on a license plate and a 95.1% accuracy rate per individual character. This approach highlights the complementary strengths of CNNs and RNNs: CNNs capture intricate visual details, while RNNs process sequences of characters in the correct order.

In reference [8], the Single Shot Detection (SSD) algorithm, coupled with Residual Network (ResNet) instead of the typical VGG Network, was used for license plate detection. ResNet helped prevent gradient vanishing, enhancing performance and achieving an 85.5% accuracy rate, better than VGG's 83.6%. This study emphasizes the importance of refining algorithms to improve effectiveness in computer vision tasks.

Reference [9] described modifications to the FAST-YOLO network for detecting vehicles and their license plates simultaneously, adapting YOLO for character detection and integrating heuristic strategies. This resulted in a 63.18% accuracy rate in detecting and recognizing license plates, surpassing the 55.47% accuracy of the Sighthound method. This improvement underscores the benefits of customizing neural network architectures and using heuristic techniques to enhance precision in computer vision tasks.

The research in [10] highlighted the impact of font characteristics on word recognition and reading comprehension in digital publishing. It introduced a new shape descriptor for recognizing ancient characters with high accuracy, accommodating variations like translation, rotation, and scaling. This study underlined the relationship between typography and readability, showing how design choices enhance digital reading experiences.

The study in [11] outlined the development of Sitka, a new serif typeface designed to enhance font legibility on digital displays. This typeface was part of a broader initiative aimed at facilitating legibility studies and iterative design processes, emphasizing trade-offs in design, especially regarding x-height, which affects the legibility of letters with ascenders and descenders. This research advances typeface design, highlighting challenges in optimizing legibility for narrow letters.

3 Proposed Methodology

While this work leverages established approaches in the field of license plate detection and recognition, it enhances them through strategic combinations and novel applications to achieve improved performance. The methodological framework consists of the following key components:

3.1 License Plate Detection

The license plate detection is performed using a Faster R-CNN [3] (Region-based Convolutional Neural Network) architecture, which is widely recognized for its robustness and high detection accuracy, even in challenging conditions such as varying lighting, occlusions, and diverse plate designs. Our approach utilizes a ResNet-50 backbone pre-trained on ImageNet to balance accuracy and computational efficiency while extracting high-level features from input images. The Region Proposal Network (RPN) generates region proposals likely to contain license plates, refining and filtering these proposals to reduce false positives. These proposals undergo ROI (Region of Interest) pooling and are passed through fully connected layers for precise bounding box regression and classification.

3.2 License Plate Recognition

For the OCR task, we developed a CNN-RNN model integrated with Connectionist Temporal Classification(CTC) loss [5]. This combination is particularly effective for sequence-based recognition tasks such as reading the characters on a license plate. The CNN feature extractor, inspired by MobileNetV3, is lightweight yet powerful, designed to run efficiently on edge devices while extracting spatial features from the detected license plate images. The RNN sequence modeler, using a bidirectional LSTM (Long Short-Term Memory) network, processes the sequential features extracted by the CNN, capturing contextual information across the character sequence. The Connectionist Temporal Classification(CTC) loss allows the model to be trained end-to-end for sequence prediction without requiring pre-segmented training data, crucial for handling varying lengths and unaligned character sequences typical of license plates.

3.3 Model Training and Optimization

The recognition model underwent 100 iterations or epochs during the training phase to ensure optimal convergence and proficiency. We chose to use 100 iterations for training the model based on empirical observations. This number of iterations allowed for optimal convergence and proficiency without overfitting the model, balancing training time and performance. The learning rate was set to 0.01, determining the step size at which the stochastic gradient descent optimizer adjusted the model's parameters during the back-propagation process. This choice of learning rate was informed by empirical observations and experimentation, balancing the need for stable convergence and fine-grained parameter updates. To train the model's ROI (Region of Interest) heads, we utilized a batch size of 64. This batch size indicates the number of ROI samples per image used in the training process. By carefully selecting this batch size, we aimed to balance computational efficiency and the model's ability to generalize to varying regions of interest within license plate images.

In our study, OpenALPR serves as a baseline for comparison. This widely recognized system, developed by OpenALPR Technology Inc. and launched in May 2014, can detect license plates from countries such as Canada, Brazil, the USA, Japan, and China. By comparing our proposed model's performance against OpenALPR [12], we aim to highlight the improvements and advantages of our approach. While OpenALPR provides a general solution, our model demonstrates enhanced performance, particularly under challenging conditions and with diverse font styles, as evidenced by higher recognition and lower character error rates.

Three sets of results will be obtained. The initial set will present the Average Precision (AP) values for the detection task, utilizing Faster-RCNN [3]. The second set will detail the mAP(mean Average Precision) values for the recognition tasks implemented with Connectionist Temporal Classification(CTC) networks [5]. The final set will include qualitative results, showcasing the confusion matrices for each dataset, which will aid in determining the readability of one font over another. We recommend enhancing automatic license plate recognition systems and font features based on these findings.

3.4 Data Augmentation and Synthetic Data Generation

To enhance the robustness of the recognition model, various data augmentation techniques were employed. These included rotation, perspective transform, color channel transform, and adding noise. Rotation was applied in the range of -15 to 15° to simulate different viewing angles. Perspective transforms were used to mimic various distortions. Color channel transformations involved adjustments to hue (hsv_h: 0.015), saturation (hsv_s: 0.7), and value (hsv_v: 0.4) to account for different lighting conditions. Additionally, translations (translate: 0.1) and scaling (scale: 0.5) were applied to simulate different image perspectives, and a mosaic effect (mosaic: 1.0) was added to increase variability. These techniques significantly increased the dataset's variability, allowing the model to learn from a broader range of examples. The data augmentation and synthetic data generation contributed to improved model performance, particularly in handling challenging conditions such as occlusions, different lighting, and varied plate designs.

3.5 Font Type Evaluation Aspects

Our research evaluates font types based on several key characteristics from reference [2]: similar apexes, defined as the junction of a stem; positioning of crossbars, which are horizontal strokes connecting parts of a letter, with varied positions enhancing legibility; similar top and bottom counters, the open spaces within a letter, with identical bottom counters increasing the likelihood of confusion; similar bowls, the rounded parts of letters; identical spurs, small projections at the end of a curved stroke; identical horizontal strokes at the bottom or top of letters; unclear tails, short downward strokes that can be mistaken for other elements when not clear; and similar diagonal strokes, where identical placement and design can lead to confusion. These criteria were chosen because they directly impact the readability and recognizability of different fonts. By analyzing these aspects, we aim to determine which font features contribute to better recognition performance and why certain fonts may lead to higher error rates.

4 Results and Discussion

For the training process of detection and recognition models, we leveraged the power of the NVIDIA Tesla T4 GPU. The NVIDIA Tesla T4 has 16GB of GPU memory, offering substantial computational resources and parallel processing capabilities. This GPU enabled us to accelerate the training and optimization of our models, effectively handling large-scale datasets.

4.1 Evaluation Metrics

The performance of our models on the test set was evaluated using various metrics. The average precision (AP) at different IOU thresholds, areas, and maximum detections provide insights into the model's object detection capabilities. For the recognition model, we used recognition rate, character error rate (CER) and recall ratio as evaluation metrics. Additionally, confusion matrices were computed to analyze the types of errors made by the recognition model.

4.2 Detection Results

The detection model maintained a high precision at a higher IoU threshold of 0.75. When considering different object sizes, the model exhibited varying performance. For small objects, the average precision needed to be computed (-1.000), indicating a lack of reliable detection. However, for medium-sized objects, the model achieved a reliable average precision. The model excelled in detecting large objects, achieving an Impressive average precision. These results demonstrate the effectiveness of our Faster R-CNN [3] model in accurately detecting license plates in various scenarios, highlighting its potential for real-world license plate recognition applications(see Fig. 3). The results of the detection models for four datasets are shown in Table 1.

Fig. 3. Detection sample results from left to right for Quebec and Ontario provinces and States of New York and California

Table 1. AP50:95, and AP75 scores for various datasets

Dataset	AP50:95	AP75
Quebec	79.54 %	94.16 %
Ontario	80.37 %	94.95 %
New York	78 %	92.14 %
California	83.87 %	95.72 %

4.3 Recognition Results

The recognition model was evaluated on two datasets: the UFPR-ALPR [6] dataset and our CENPARMI dataset. The results indicate a significant improvement in performance when using the CENPARMI dataset, attributed to the extensive data augmentation and synthetic data generation techniques employed during training. The results are summarized in the Table 2.

The higher recognition rate and lower CER on the CENPARMI dataset suggest that the model generalizes well to the varied conditions and plate designs encountered in real-world scenarios, and the effectiveness of our advanced training techniques. This improvement is particularly noteworthy given the diversity of the CENPARMI dataset, which includes images from different states with varying plate designs and used fonts, lighting conditions, and environmental factors.

One of the key innovations in our approach is the use of a lightweight backbone inspired by MobileNetV3 for the recognition model. This design choice ensures that the model can run efficiently on edge devices with limited computational resources. The lightweight architecture reduces the model's complexity and computational cost without compromising accuracy, making it suitable for real-time applications on devices such as smartphones, tablets, and embedded systems.

Table 2. Performance metrics for different models and datasets

Model	Dataset	Recognition Rate	Character Error Rate (CER)	Recall
our model	UFPR-ALPR	84.2 %	5.3 %	92.1%
	CENPARMI	89.8 %	4.5 %	90.0 %
open alpr	UFPR-ALPR	71.7 %	15.5 %	55.8%
	CENPARMI	65.5 %	19.6 %	80.2 %

4.4 State-Wise Recognition Performance

The recognition model's performance was further analyzed on datasets from four different datasets in Canada and United States of America. The results for each dataset are detailed in the Table 3.

Table 3. Performance metrics for different states

State	Recognition Rate	Character Error Rate (CER)	Recall
Quebec	91.6%	3.8 %	91.5 %
Ontario	91.6%	3.8 %	91.5 %
New York	80.1%	8.1 %	80.6 %
California	96%	2.3 %	96.4 %

The model demonstrated consistent performance across different dataset, with slight variations attributed to differences in plate designs, font styles, and environmental conditions. California State achieved the highest accuracy and the lowest CER, likely due to clearer and more consistent plate designs. State of New York also performed exceptionally well, indicating that the model can handle diverse plate designs with high accuracy.

4.5 Font Analysis Results

In this section, we present five distinct sets of outcomes from evaluating fonts based on the recognition results of individual letters(see Fig. 4). From our confusion matrices and detailed analysis of the task of recognizing license plates, we can draw several conclusions:

CENPARMI Dataset: Our CENPARMI dataset includes 1600 license plates from diverse California, New York, Ontario, and Quebec environments. We conducted license plate detection and recognition in each province separately because each dataset has a unique font.

1. **Quebec Dataset:** The characters ‘Q’ and ‘0’ often need clarification due to their similar shapes, with the ‘Q’s tail sometimes indistinct. The resemblance between ‘6’ and ‘G’ stems from their similar curves and inner spaces. Confusion between ‘Q’ and ‘D’ can occur if the ‘Q’s tail blends with its bowl, resembling a ‘D,’ especially if both have a similar spur. ‘W’ and ‘M’ are also confusing due to their similar diagonal strokes and internal structure. Additionally, ‘6’ and ‘4’ can be mistaken for each other due to similarities in their upper structures and angles.
2. **Ontario Dataset:** The characters ‘Q’ and ‘0’ might be mistaken for each other due to their rounded bowls, especially if the ‘Q’ has a subtle tail. Similarly, ‘5’ and ‘2’ could be confused because of their upper curves and strokes, while ‘6’ and ‘G’ might look alike due to their shape and internal space. Characters like ‘1’ and ‘I’, as well as ‘4’ and ‘A’, can be indistinguishable when depicted as simple vertical lines. ‘V’ and ‘Y’ can also appear similar if the ‘Y’ has a short or blended tail.
3. **California Dataset:** The characters ‘O’ and ‘0’ can easily be confused due to their similar rounded shapes. Similarly, ‘F’ and ‘E’ might be mistaken for each

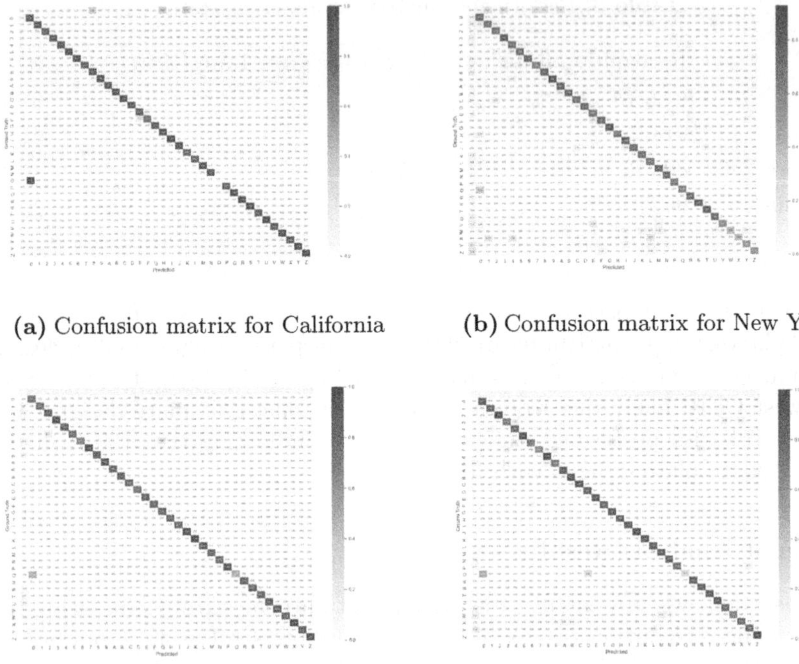

(a) Confusion matrix for California **(b)** Confusion matrix for New York

(c) Confusion matrix for Ontario **(d)** Confusion matrix for Quebec

Fig. 4. Confusion matrix of our proposed model on the CENPARMI and UFPR-ALPR dataset

other because of their horizontal strokes, making it hard to notice the missing middle stroke in 'F'. 'L' and 'T' can also appear similar due to their vertical and horizontal line components, especially in bold weights. Additionally, '7' and 'Z' might be confused when '7' has a pronounced horizontal stroke and 'Z' is styled with straight, minimalistic angles.

4. **New York Dataset:** 'O' and '0' can look very similar due to their tightly rounded shapes. The letters 'J,' 'I,' and 'T' might be mistaken for one another due to similar vertical strokes, especially when 'J' has a subtle curve and 'T' has a short horizontal stroke. 'V' and 'U' can also be confused if 'V' has a sharp vertex resembling a 'U' without the middle crossbar. 'X' and 'L' may appear similar if 'X's diagonal strokes blend together, looking like intersecting 'L's. In a condensed font, 'M' and 'W' might look like mirror images with unclear middle peaks and valleys.

UFPR-ALPR Dataset: The confusion matrix highlights common character misidentifications due to their similar appearances(see Fig. 5). 'O' and '0' are often confused as both have round, unmarked shapes. 'I', '1', and 'T' can be mistaken for each other when depicted as straight lines, especially if 'T' has a short horizontal stroke. 'M' and 'W' look alike due to their mirrored structures and bold lines. 'B' and '8', as well as 'E' and '8', are confusing when 'B' and 'E' resemble the double loops of '8'. 'C' and 'G' might be confused if 'G's tail is subtle, making it look like 'C'.

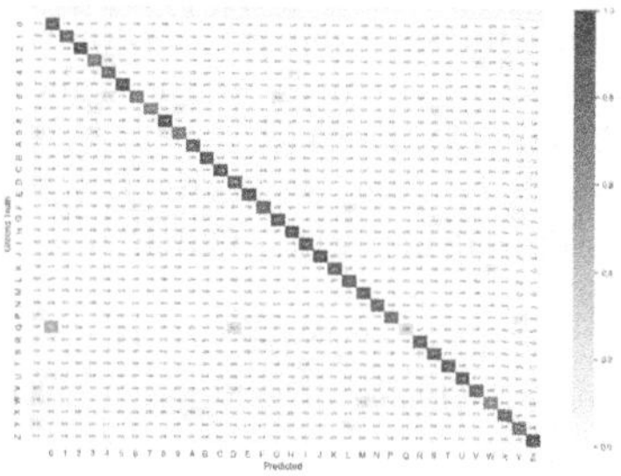

Fig. 5. Confusion matrix of our proposed model on the UFPR-ALPR dataset

To better understand the results, we conducted a detailed statistical analysis to evaluate the performance of our model. We compared our approach with the OpenALPR baseline. The results highlight our method's strengths and demonstrate its superiority in terms of recognition rate and character error rate. We conducted a detailed error analysis to identify the causes of false positives and negatives. Common errors included misclassification due to similar font characteristics and issues with plate visibility under extreme lighting conditions. Our experimental results demonstrate the robustness and accuracy of the proposed method across diverse conditions. While our approach builds on established techniques, the strategic combination and novel application of these methods significantly advance the field of license plate recognition, sparking new interest and possibilities. The statistical and comparative analyses further validate the effectiveness of our approach, providing strong evidence for its superiority over existing methods.

5 Conclusion and Future Work

In summary, this research concentrates on detecting license plates and recognizing their characters through a dual-phase approach that utilizes a Region-based Convolutional Neural Network (R-CNN) for detection and a Connectionist Temporal Classification(CTC) network for character recognition. It also examines how font selection affects ALPR system performance. The paper introduces a novel dataset and underscores the importance of precise LPD systems. Future work could include enlarging the dataset, employing additional deep learning methods, expanding the system to encompass license plate recognition and generation, and selecting a more legible font for license plates. This research contributes to advancing intelligent transportation systems, supporting improved traffic management and security.

References

1. Al-qudah, R., Suen, C.Y.: Enhancing YOLO deep networks for the detection of license plates in complex scenes. In: Proceedings of the Second International Conference on Data Science, E-Learning and Information Systems. Presented at the Dubai, United Arab Emirates (2019). https://doi.org/10.1145/3368691.3368712
2. Al-qudah, R., Suen, C.Y.: Impact of font on computer recognition of license plates on automobiles. In: Proceedings of the 3rd International Conference on Vision, Image and Signal Processing. Presented at the Vancouver, BC, Canada (2020). https://doi.org/10.1145/3387168.3389112
3. Ren, S., He, K., Girshick, R.B., Sun, J.: Faster R-CNN: towards real-time object detection with region proposal networks. CoRR, abs/1506.01497 (2015). http://arxiv.org/abs/1506.01497
4. Ahmed, M., Pagani, A., Liwicki, M., Stricker, D., Afzal, M.: Survey and performance analysis of deep learning based object detection in challenging environments. Sensors **21**(15), 5116 (2021)
5. Shi, B., Bai, X., Yao, C.: An end-to-end trainable neural network for image-based sequence recognition and its application to scene text recognition. arXiv arXiv:1507.05717 (2015)
6. Laroca, R., et al.: A robust realtime automatic license plate recognition based on the YOLO detector. In: International Joint Conference on Neural Networks (IJCNN), Rio de Janeiro, Brazil (2018). https://doi.org/10.1109/IJCNN.2018.8489629
7. Cheang, T.K., Chong, Y.S., Tay, Y.H.: Segmentation-free vehicle license plate recognition using convnet-RNN. arXiv preprint arXiv:1701.06439 (2017)
8. Wu, P., Lin, Y.: Research on license plate detection algorithm based on SSD. In: Proceedings of the 2nd International Conference on Advances in Image Processing, pp. 19–23 (2018)
9. Silva, S.M., Jung, C.R.: Real-time Brazilian license plate detection and recognition using deep convolutional neural networks. In: 2017 30th SIBGRAPI Conference on Graphics, Patterns and Images (SIBGRAPI), pp. 55–62. IEEE (2017)
10. Suen, C.Y., Dumont, N., Dyson, M., Tai, Y., Lu, X.: Evaluation of fonts for digital publishing and display. In: 2011 International Conference on Document Analysis and Recognition, Beijing, pp. 1424–1436 (2011). https://doi.org/10.1109/ICDAR.2011.307
11. Larson, K., Carter, M.: Chapter 3: Sitka: a collaboration between type design and science. In: Dyson M., Suen, C. (eds.) Series on Language Processing, Pattern Recognition, and Intelligent Systems: Digital Fonts and Reading, pp. 37–53.World Scientific Publishing Co Pte Ltd. (2016). https://doi.org/10.1142/9789814759540_0003
12. OpenALPR. https://platerecognizer.com/. Accessed 04 Oct 2018

Applications in NLP, Speech, and Music

Experiments in Modeling Disagreement

Narjes Tahaei(✉) and Sabine Bergler

CLaC Lab Concordia University, Montreal, QC, Canada
{n_tahaei,bergler}@encs.concordia.ca

Abstract. This research examines how annotators' features influence the labeling of sexist content in social media datasets, with a specific focus on the EXIST dataset, which includes direct sexist messages, reports, and descriptions of sexist experiences and stereotypes. By comparing the use of gold labels derived by majority vote with individual annotator labels, we found that incorporating annotator labels into the input tokens enhances model performance in predicting gold labels. Our study further investigates the impact of additionally integrating annotators' demographic information into BERT models to enhance performance in subjective natural language processing tasks. We find that integrating such demographic data into the input leads to improved model performance.

Keywords: demographics · sexism · disagreement

1 Introduction

Data labeled by multiple annotators forms training data for most supervised AI systems. Annotators often disagree, particularly in subjective natural language processing tasks where determining a definitive ground truth is challenging or impossible. Topics like *sexism, sarcasm*, or *hate speech* are such areas leading to diverging opinions among annotators [1,6,16].

Traditional datasets manage annotator disagreements by either appointing an expert to assign a single *gold* label or taking a majority vote of the annotators. These datasets ignore the varied backgrounds and perspectives of the annotators and dismiss minority opinions. Consequently, they are often criticized for being (partially) subjective and not fully reliable for downstream applications [18,21].

The EXIST shared task [16], which focuses on identifying and categorizing sexism on Twitter, addresses this issue by providing the actual label assigned by each annotator. The training set comprises 6,920 Spanish and English tweets, each annotated by six individuals. There are 725 distinct annotators from 45 different countries. Additionally, EXIST annotations include the annotators' demographic information, enabling a nuanced understanding of labels within this dataset.

We propose a straightforward approach for detecting sexism and determining the author's intention in the EXIST dataset: when adding demographics

C. Y. Suen et al. (Eds.): ANNPR 2024, LNAI 15154, pp. 245–255, 2024.
https://doi.org/10.1007/978-3-031-71602-7_21

and annotator labels to the input for a pre-trained BERT model, performance improves in both sexism detection and identifying the author's intent. We compare performance when using the gold label (derived by majority vote) versus the six individual annotator labels available in the EXIST dataset. Our preliminary research analyzes the impact of both annotator labels and the gold label, providing new insights into the interplay between annotator diversity and the performance on the EXIST shared task dataset.

2 Related Work

Recent years have seen an increase in the availability of sexism-related datasets. [24] introduced a dataset of tweets labeled as *racist*, *sexist*, or *neutral*. Similarly, MeTwo [17] provides a Spanish Twitter corpus with three labels, *sexist, non-sexist*, or *uncertain*.

Some datasets focus on classifying variants of sexism such as misogyny or hatred towards women [3,7,14]. [4] introduces a dataset of French tweets annotated to identify either reports of sexist experiences or direct sexist messages. The tweet *My boss asked me: "Who's going to cook for your husband when you're away?"* is classified as a report, not a sexist tweet, which potentially triggers different responses from the recipient compared to a directly sexist message. The EXIST dataset considers a wide range of tweets as sexist, from explicit to reports of sexist experiences or even benevolent expressions that involve implicit sexist behaviors.

Building on the assumption that annotators' judgments are influenced by their background, specifically in subjective tasks such as *sexism* and *hate speech*, a growing body of research focuses on developing training methods to learn from disagreement in crowd annotations [9,10,13,21].

[8] found significant variations in perceptions of the harmfulness of sexually explicit language across eight countries. [2] developed an Arabic Twitter dataset on sexism and misogyny, demonstrating that annotators' religious beliefs affect their labeling. [19] reported strong correlations between annotator identity, beliefs, and toxicity ratings. [15] integrated gold labels with information from crowd annotations to account for item difficulty or annotator ability. We adopt this approach, incorporating annotators' demographics into the training process for sexism identification using gold (hard) labels.

[23] predicts disagreement levels using task text and annotators' demographics. By analyzing annotators' voting histories and fine-tuning language models, it finds that demographic information like gender, ethnicity, and education helps predict annotator disagreements. [22] uses the EXIST2023 dataset to predict sexism and the authors' intentions. The study incorporates annotators' age and gender as additional model inputs. It demonstrates that these demographics do not improve performance for the first task but are helpful for the second task. The study does not report on other demographic features, nor does it compare results when annotators' labels are included in the training process.

The EXIST dataset includes a range of demographic information not typically found in previous datasets. We explore the impact of all these demographics on the labeling process within the EXIST dataset.

3 Data Description

For this study, we chose the EXIST 2023 dataset, which has been used for the EXIST shared task in 2023 and 2024 [16]. EXIST stands for sEXism Identification in Social neTwork and focuses on identifying explicit and implicit sexism in social media posts. The training dataset includes 6920 tweets, 3660 in Spanish, and 3260 in English. Each tweet is annotated by six annotators drawn from a set of 725 annotators from 45 countries, resulting in $6 \times 6,920 = 41,520$ annotated instances.

Task 1 is a binary classification to determine whether a given tweet refers to sexist expressions or behaviors in an encompassing sense. The sample can be sexist itself, describe a sexist situation where discrimination against women occurs, or criticize a sexist behavior. The training set for Task 1 has 18,753 'sexist' and 22,767 'non-sexist' labels.

Task 2 explicitly distinguishes between a *directly* sexist tweet, one that *reports* a sexist situation, and one that condemns or *judges* a sexist situation, essentially classifying the author's intention. For some samples, a majority vote cannot be established. Some samples are labeled 'non-sexist'. Task 2 is thus a 5-way classification task, which we model as a three-way classification following [16]: we disregard samples where no majority can be established and we classify only tweets that our system for Task 1 labeled 'sexist'.

For Task 2, the 'direct' label is the most frequent with 9,004 instances, followed by the 'judgemental' label with 5,015 instances, and the 'reported' label with 4,652 instances.

Direct The tweet itself is sexist.
Example: *A woman needs love, to fill the fridge, if a man can give this to her in return for her services (housework, cooking, etc.), I don't see what else she needs.*

Reported The tweet reports a sexist situation experienced by a woman or by women.
Example: *Today, one of my year 1 class pupils could not believe he'd lost a race against a girl.*

Judgemental The tweet describes sexist situations or behaviors to condemn them.
Example: *As usual, the woman was the one quitting her job for the family's welfare ...*

The EXIST 2024 dataset includes demographic information for each annotator, namely their *id, gender, age, level of study, country*[1]*,ethnicity, and language*. The demographic features and their distributions in the training set are described in Table 1.

[1] 45 countries are represented in the dataset but are omitted in Table 1.

Table 1. Demographic features and their distribution

Feature	Range	#instances
Gender	M	20760
	F	20760
Age	18-22	13840
	23-45	13840
	>46	13840
Levels of study	less than a high school diploma	684
	high school degree or equivalent	12483
	bachelor's degree	20794
	master's degree	6635
	doctorate	639
Ethnicity	Black or African American	2348
	Hispano or Latino	11742
	White or Caucasian	26221
	Multiracial	468
	Asian	342
	Asian Indian	0
	Middle Eastern	171
Language	English	19560
	Spanish	21960

4 Methodology

4.1 Preprocessing

We exchange URLs and @user mentions in the tweets with new special tokens [url] and [user] in the BERT tokenizer. Following [12,20], we exchange each emoji symbol with its own new emoji word embedding by using the emoji package[2] for the list of all emojis and then adding them to the BERT tokenizer. We refer to the preprocessed sample text as st_i.

The input vector to BERT is a concatenation of st_i with selected meta-information provided by the EXIST data to assess their performance implications. I.e., to identify the language of the tweet, new special tokens [english] or [spanish] are prepended to each st_i for all our experiments. For our different experiments, we append different demographic features in various encodings, described in the following.

[2] https://pypi.org/project/emoji/.

4.2 Encodings

For each st_i (where $1 \leq i \leq 6920$) there are six separate target label annotations l_{ijk} ($1 \leq j \leq 6$) by a selection from the set of annotators a_k, $1 \leq k \leq 725$. Each annotator has an associated set f_k of demographic features (see Table 1). Correspondingly, we replicate each sample st_i six times, corresponding to each labeling l_{ijk}, forming a matrix of six rows for each st_i. Each row assumes the same language token (omitted in Table 2), followed by the same st_i, followed by the demographic features selected from f_k^E for the experiment E, followed by the same gold label ($gold_i$) determined by majority voting. Examples can be found in Sect. 4.3.

We report here on three of our experiments, which vary in the selection of demographic features f_k^E:

E_b Input to BERT: [<language>] st_i
Baseline representation where the input to BERT consists solely of the sample text st_i. No demographic features are included.

E_a Input to BERT: [<language>] st_i f_k^E
Input to BERT includes selected demographic features f_k^E of the annotator for experiment E.

E_{a+l} Input to BERT: [<language>] st_i f_k^E l_{ijk}
The label l_{ijk} assigned by annotator a_k is appended in addition to demographic features f_k^E.

4.3 Feature Representation

We encoded features f_k^E three different ways: I. the original strings found in the data (see Table 1), II. numerical encodings, and III. a modified textual encoding which we found to be more likely to correspond to word embeddings of related meaning for the age and gender features. For III, we replaced F' with female' and M' with male' for the gender features, and we replaced 18-22' with GenZ', 23-45' with Millennials', and ¿46' with GenX' for the age features. The numerical features ('0' for 'F', '1' for 'M', etc.) did not perform well.

Feature encoding III performs better than the other two across our experiments and was used for the experiments reported in Table 2.

4.4 Architecture

Our architecture is based on BERT. As previously described, we vary the input to BERT for the reported experiments. We use BERT's CLS token as input to a classifier consisting of two feed-forward networks. We use the Cross-entropy loss to calculate the loss and the Adam optimizer for training the model.

4.5 Experimental Setup

We used multilingual BERT (mBERT) [5], a model pre-trained on the top 104 languages in Wikipedia. We contrasted this with the cardiffnlp/twitter-roberta-base-sentiment-latest model [11], a RoBERTa-base model trained on approximately 58 million tweets and fine-tuned for sentiment analysis. The fine-tuned RoBERTa-base model performed best for Task 1 and mBERT for Task 2. For comparison, we also experimented with XLM-RoBERTa; however, since the performance did not significantly improve, we use RoBERTa-base for Task 1 and mBERT for Task 2 in Table 2.

We use the HuggingFace implementations of both models, trained with a batch size of 1 for 5 epochs. The results are reported for both tasks based on the average performance over 5 epochs. The learning rate of **gold-gold** and **annotation-annotation** is $5e^{-6}$. The learning rate of **annotation-gold** is $5e^{-7}$.

4.6 Label Predictors

Because EXIST provides six labels from six annotators plus a way to determine the hard gold label (majority vote), we experimented with using different combinations during the training and test phases. We present three experiments:

gold-gold The training and testing phases use only gold labels.
During training, the model is trained to predict $\widehat{gold}_i$. The input to BERT can be any of E_b, E_a, E_{a+l}.
The loss function is computed for each sample s_i using cross-entropy loss between the predicted gold label $\widehat{gold}_i$ and $gold_i$:

$$\text{Loss}_{st_i} = \text{Loss}(\widehat{gold}_i, gold_i),$$

allowing the model to predict the majority gold label for each tweet.
During testing, the model takes the sample text st_i, in the form of one of the encodings, and outputs a prediction for $gold_i$.

annotation-gold Similar to **gold-gold**, both the training and testing phases use gold labels $gold_i$ as target.
The input to BERT remains any of E_b, E_a, E_{a+l}.
Predictions are now individual annotations, i.e. $\hat{l}_{ijk}$.
This differs from the previous model in the calculation of the loss function. The output for each of the six replicated instances is collected. The loss for each instance l_{ijk} is computed using cross-entropy loss between the predicted label $\hat{l}_{ijk}$ and $gold_i$. These losses are then aggregated by averaging to form a single loss value for the sample.

$$\text{Aggregated Loss}_{\text{st}_i} = \frac{1}{6}\sum_{j=1}^{6} \text{Loss}(\hat{l}_{ijk}, gold_i)$$

This aggregated loss is used to update the model parameters.

For the test data, the model predicts a label for each of the six instances l_{ijk}. These predictions are aggregated using majority voting to produce a single predicted label $\widehat{gold}_i$ for the sample and is then compared to $gold_i$ for evaluation. The aggregation method ensures that the model's prediction for each tweet considers the variability across annotators while still being evaluated against the majority-voted gold label.

annotation-annotation When no gold label $gold_i$ exists, each annotator label is predicted independently.
This model is not compatible with E_{a+l} encodings since it is assumed that there is not any aggregated gold label.
It predicts the label $\hat{l}_{ijk}$ assigned by the annotator a_k. The loss function for this configuration is computed using cross-entropy loss between the predicted label $\hat{l}_{ijk}$ for each instance and the actual label l_{ijk}.

$$\text{Loss}_{ijk} = \text{Loss}(\hat{l}_{ijk}, l_{ijk})$$

This forces the model to learn from the individual perspectives of each annotator rather than converging on a single consensus label.
During the testing phase, the model outputs a prediction $\hat{l}_{ijk}$. The performance is evaluated by comparing the predicted labels to the actual labels l_{ji}. This approach provides insights into how well the model can replicate a spread of judgments of individual annotators.

An initial baseline system predicted both sexism and the three intention categories *direct, reported, judgmental* within a single model. However, we obtain better results when we reduce the number of non-sexist tweets for Task 2 by first running our Task 1 system in order to identify the sexist tweets and subsequently categorizing only those with our different approaches.

5 Evaluation

Each instance has seven labels: one given by each annotator and the gold label.

For Task 1, a binary classification task, the hard label is determined by a majority vote of the annotators, following the shared task evaluation setting. The class annotated by more than three annotators is selected. If no class receives a majority for sample s_i, meaning no class exceeds the threshold, s_i is removed from the training process. We evaluated the binary task based on the F1 measure.

Task 2 is a three-class classification task for sexist tweets, that is annotators can label tweets as *direct*, *reported*, or *judgmental* or *non-sexist*. The gold label $gold_i$ is assigned to sample s_i if two or more annotators agree. In case of a tie, s_i is removed from the training process. We evaluate Task 2 using the average F1 measure (Macro F1).

6 Results

Table 2 shows averaged results for five runs. For encodings E_{a+l}, the column 'Labels' indicates for which task the annotator label was included.

Table 2. Results on development set. Task 1 evaluated with $F1$. Task 2 evaluated with $MacroF1$

Model	Encoding	Features used	Labels	Task 1	Task 2
gold-gold	E_b	–	–	0.80 ± 0.009	0.51 ± 0.01
	E_a	age,gender	–	0.82 ± 0.01	0.55 ± 0.05
	E_a	age,gender,study,ethnicity	-	0.82 ± 0.01	0.55 ± 0.02
	E_{a+l}	age,gender,study,ethnicity	T1	$\mathbf{0.84} \pm 0.009$	$\mathbf{0.58} \pm 0.03$
	E_{a+l}	age,gender,study,ethnicity	T1,T2	0.83 ± 0.009	$\mathbf{0.65} \pm 0.05$
annotation-gold	E_b	–	–	0.80 ± 0.01	0.53 ± 0.06
	E_a	age,gender	–	0.82 ± 0.01	0.52 ± 0.01
	E_a	age,gender,study,ethnicity	–	0.82 ± 0.009	0.52 ± 0.01
annotation-annotation	E_{base}	–	–	0.74 ± 0.008	0.45 ± 0.03
	E_a	age,gender	–	0.74 ± 0.01	$0.46 \pm 0.0.1$
	E_a	age,gender,study,ethnicity	–	0.74 ± 0.01	0.43 ± 0.03

Our comparisons here are against our own baseline, a BERT model with no additional features (E_b) used because test data is not made public. The best performance at EXIST 2023 obtained 0.81 for Task 1 and 0.57 for Task 2 on the test set but has not published results on the development set. We conclude that our results here are comparable.

We first observe that the **annotation-annotation** models perform nearly 10% below the models scored against hard gold labels and we will not further analyze the results here.

The baseline encodings perform identically for both, **gold-gold** and **annotation-gold** models for Task 1. There is also no difference for the different feature encodings shown. For Task 2, however, we notice an inversion: **annotation-gold** performs better than **gold-gold** for the baseline systems (+0.02) but worse when demographic features are encoded (-0.03). Further analysis is needed to find out whether this is a consistent pattern, whether it is due to the characteristics of the development set, the encodings, or the increased complexity for the three-way task.

For the **gold-gold** model we noticed (not shown in Table 2) that including the feature *country* lowered performance, likely due to sparcity issues for the 45 countries.

We observe, however, that appending the Task 1 annotator's label to the BERT input leads to a significant improvement on both tasks. We attribute the greater increase of 0.03 when providing Task 1 labels for Task 2 to our two-phase classification: by providing the actual gold label, Task 2 is less vulnerable to our Task 1 system's predictions during training. We see an even greater gain for Task 2 when Task 2 labels are additionally appended to the BERT input (+0.07). This is of course a form of information leaking. We will analyze further whether the finetuned BERT word embeddings are rich enough to benefit from knowing about the task at the outset without leading to shortcut learning and overfitting in general.

Computation is more expensive for **gold-gold** with six independent forward passes and six independent backpropagation steps per tweet. On the other hand, **annotation-gold** combines all annotator feedback for each tweet, leading to six independent forward passes but only a single backpropagation step per tweet based on aggregated loss. This puts the small improvements for the second task into perspective, but suggests that increased computation provides the model with more opportunities to capture individual tweet features.

Both **annotation-annotation** and **annotation-gold** models show greater difficulty in dealing with demographic features compared to **gold-gold** because the model is being trained on six very similar inputs for different annotator labels during the training phase. This presents a challenge, particularly for the second task, where greater disagreement among annotators is observed. For instance, in cases where the six annotator labels for the second task are *reported, judgemental, reported, no, reported, direct*, the model receives conflicting signals from the same input, introducing noise to training data.

In Table 3, we compare the performance of **gold-gold** when non-sexist tweets are either included or omitted. The four-class task inherently contains more noise due to the inclusion of the non-sexist category. Adding additional information such as annotator labels and demographics introduces even more noise, complicating the model's ability to accurately categorize the content. Conversely, in the three-class task almost half of the data is identified as non-sexist and is filtered out, thus the model benefits from the more focused dataset (rows 1-3). Again, Table 3 shows that incorporating demographic and annotator label information provides more relevant context and enhances the model's performance.

Table 3. Micro F_1 for Task 2 when non-sexist tweets are included or omitted

Model	Classification	Encoding	Features	Labels	Task 2
gold-gold	only sexist	E_b	–	–	0.5
		E_b	–	T1,T2	0.63
		E_{a+l}	age,gender, study,ethnicity	T1,T2	0.65
	non-sexism included	E_b	–	–	0.52
		E_b	–	T1,T2	0.48
		E_{a+l}	age,gender, study,ethnicity	T1,T2	0.47

7 Conclusion

This paper demonstrates that appending annotators' demographic features to BERT input can enhance performance on the EXIST dataset.

Our findings highlight the potential benefits of leveraging meta-data to enrich the training on a dataset with significant annotator disagreement. However, performance drops when too many or the wrong features are included. This is but a first exploration of this potential and more studies on a greater variety of datasets are needed.

References

1. Alm, C.O.: Subjective natural language problems: motivations, applications, characterizations, and implications. In: HLT 2011, Proceedings of the 49th Annual Meeting of the Association for Computational Linguistics & Human Language Technologies: Short Papers - Volume 2, p. 107–112 (2011)
2. Almanea, D., Poesio, M.: ArMIS - the Arabic misogyny and sexism corpus with annotator subjective disagreements. In: Proceedings of the Thirteenth Language Resources and Evaluation Conference, pp. 2282–2291 (2022)
3. Anzovino, M., Fersini, E., Rosso, P.: Automatic identification and classification of misogynistic language on twitter. In: Silberztein, M., Atigui, F., Kornyshova, E., Métais, E., Meziane, F. (eds.) NLDB 2018. LNCS, vol. 10859, pp. 57–64. Springer, Cham (2018). https://doi.org/10.1007/978-3-319-91947-8_6
4. Chiril, P., Moriceau, V., Benamara, F., Mari, A., Origgi, G., Coulomb-Gully, M.: He said "who's gonna take care of your children when you are at ACL?": reported sexist acts are not sexist. In: Proceedings of the 58th Annual Meeting of the Association for Computational Linguistics, pp. 4055–4066 (2020)
5. Devlin, J., Chang, M.W., Lee, K., Toutanova, K.: Bert: pre-training of deep bidirectional transformers for language understanding. In: North American Chapter of the Association for Computational Linguistics (2019)
6. Gordon, M.L., et al.: Jury learning: integrating dissenting voices into machine learning models. In: Proceedings of the 2022 CHI Conference on Human Factors in Computing Systems. CHI 2022, Association for Computing Machinery (2022)
7. Guest, E., Vidgen, B., Mittos, A., Sastry, N., Tyson, G., Margetts, H.: An expert annotated dataset for the detection of online misogyny. In: Proceedings of the 16th Conference of the European Chapter of the Association for Computational Linguistics: Main Volume, pp. 1336–1350 (2021)
8. Jiang, J.A., Scheuerman, M.K., Fiesler, C., Brubaker, J.R.: Understanding international perceptions of the severity of harmful content online. PLOS ONE **16**(8), 1–22 (2021)
9. Kairam, S., Heer, J.: Parting crowds: characterizing divergent interpretations in crowdsourced annotation tasks. In: Proceedings of the 19th ACM Conference on Computer-Supported Cooperative Work & Social Computing, pp. 1637–1648. CSCW 2016, Association for Computing Machinery (2016)
10. Leonardelli, E., et al.: SemEval-2023 task 11: learning with disagreements (LeWiDi). In: Proceedings of the 17th International Workshop on Semantic Evaluation (SemEval-2023), pp. 2304–2318 (2023)
11. Loureiro, D., Barbieri, F., Neves, L., Espinosa Anke, L., Camacho-collados, J.: TimeLMs: diachronic language models from twitter. In: Proceedings of the 60th Annual Meeting of the Association for Computational Linguistics: System Demonstrations, pp. 251–260 (2022)
12. Mostafavi, M., Porter, M.D.: How emoji and word embedding helps to unveil emotional transitions during online messaging. In: 2021 IEEE International Systems Conference (SysCon), pp. 1–8 (2021)

13. Mostafazadeh Davani, A., Díaz, M., Prabhakaran, V.: Dealing with disagreements: looking beyond the majority vote in subjective annotations. Trans. Assoc. Comput. Linguist. **10**, 92–110 (2022)
14. Pamungkas, E.W., Basile, V., Patti, V.: Misogyny detection in twitter: a multilingual and cross-domain study. Inf. Process. Manage. **57**(6), 102360 (2020)
15. Plank, B., Hovy, D., Søgaard, A.: Learning part-of-speech taggers with inter-annotator agreement loss. In: Proceedings of the 14th Conference of the European Chapter of the Association for Computational Linguistics, pp. 742–751 (2014)
16. Plaza, L., et al.: Overview of EXIST 2023 - learning with disagreement for sexism identification and characterization. In: Arampatzis, A., et al. (eds.) Experimental IR Meets Multilinguality. Multimodality, and Interaction, pp. 316–342. Springer Nature Switzerland, Cham (2023). https://doi.org/10.1007/978-3-031-42448-9_23
17. Rodríguez-Sánchez, F., Carrillo-de Albornoz, J., Plaza, L.: Automatic classification of sexism in social networks: an empirical study on twitter data. IEEE Access **8**, 219563–219576 (2020)
18. Rottger, P., Vidgen, B., Hovy, D., Pierrehumbert, J.: Two contrasting data annotation paradigms for subjective NLP tasks. In: Carpuat, M., de Marneffe, M.C., Meza Ruiz, I.V. (eds.) Proceedings of the 2022 Conference of the North American Chapter of the Association for Computational Linguistics & Human Language Technologies, pp. 175–190 (Jul 2022)
19. Sap, M., Swayamdipta, S., Vianna, L., Zhou, X., Choi, Y., Smith, N.A.: Annotators with attitudes: how annotator beliefs and identities bias toxic language detection. In: Proceedings of the 2022 Conference of the North American Chapter of the Association for Computational Linguistics & Human Language Technologies, pp. 5884–5906 (2022)
20. Tahaei, N., Verma, H., Bagherzadeh, P., Farahnak, F., Sheikh, N., Bergler, S.: Identifying author profiles containing irony or spreading stereotypes with SBERT and emojis. In: CLEF 2022 Labs and Workshops, Notebook Papers (2022)
21. Uma, A., et al.: SemEval-2021 task 12: learning with disagreements. In: Proceedings of the 15th International Workshop on Semantic Evaluation (SemEval-2021), pp. 338–347 (2021)
22. Vallecillo-Rodríguez, M.E., del Arco, F., Ureña-López, L.A., Martín-Valdivia, M.T., Montejo-Ráez, A.: Integrating annotator information in transformer fine-tuning for sexism detection. In: Working Notes of the Conference and Labs of the Evaluation Forum (CLEF 2023), Thessaloniki, Greece, September 18th to 21st, 2023. CEUR Workshop Proceedings, vol. 3497, pp. 1024–1036 (2023)
23. Wan, R., Kim, J., Kang, D.: Everyone's voice matters: quantifying annotation disagreement using demographic information. In: The Thirty-Seventh AAAI Conference on Artificial Intelligence (AAAI-23) (2023)
24. Waseem, Z., Hovy, D.: Hateful symbols or hateful people? Predictive features for hate speech detection on Twitter. In: Proceedings of the NAACL Student Research Workshop, pp. 88–93 (2016)

Deep Multiresolution Wavelet Transform for Speech Emotion Assessment of High-Risk Suicide Callers

Alaa Nfissi[1,2,4(✉)], Wassim Bouachir[1,4], Nizar Bouguila[2], and Brian Mishara[3,4]

[1] Data Science Laboratory, University of Québec (TÉLUQ), Montréal, Quebec, Canada
[2] Concordia Institute for Information Systems Engineering, Concordia University, Montréal, Quebec, Canada
[3] Psychology Department, University of Québec at Montréal, Montréal, Quebec, Canada
[4] Centre for Research and Intervention on Suicide, Ethical Issues and End-of-Life Practices, Montréal, Quebec, Canada
alaa.nfissi@mail.concordia.ca

Abstract. This paper introduces an end-to-end (E2E) deep learning framework for speech emotion recognition (SER), aimed at assisting suicide helpline counselors during their interventions. Despite progress, SER still faces challenges such as system complexity, feature distinctiveness, and noise interference. Our approach uses a learnable architecture for fast discrete wavelet transform (FDWT) multi-resolution analysis, for directly extracting features from raw speech waveforms. It combines a 1D dilated convolutional neural network (1D dilated CNN) with spatial attention (SA) and bidirectional gated recurrent units (Bi-GRU) with temporal attention (TA) to capture spatial and temporal characteristics. The framework handles variable-length speech without segmentation, simplifying preprocessing. It introduces a learnable architecture for wavelet bases and coefficient denoising with a learnable asymmetric hard thresholding (LAHT) activation function, enhancing noise resilience and feature distinctiveness. We validate our model on the NSPL-CRISE dataset, which includes recordings from individuals with psychological challenges and potentially suicidal ideations. Our results show significant performance improvements over state-of-the-art SER methods, demonstrating the effectiveness of our multi-resolution deep learning framework in identifying emotional states related to suicide risk. The source code of our model is shared on: https://github.com/alaaNfissi/DeepWaveSpeechEmotion-SuicideRiskAssessment.

Keywords: Speech Emotion Recognition (SER) · Suicide Ideation Detection · Deep Learning Framework · Multi-Resolution Analysis · Fast Discrete Wavelet Transform (FDWT) · Bidirectional Gated Recurrent Units (Bi-GRU)

C. Y. Suen et al. (Eds.): ANNPR 2024, LNAI 15154, pp. 256–268, 2024.
https://doi.org/10.1007/978-3-031-71602-7_22

1 Introduction

The advancement of Speech Emotion Recognition (SER) technology marks a significant leap in human-computer interaction, paving the way for more efficient and effective tools. SER's ability to analyze and interpret the emotional content of speech has profound implications across several fields, notably enhancing decision-making processes in emergency call centers by assessing caller stress levels and identifying signs of psychological distress. This also holds promise in healthcare, where it could play a role in detecting psychological conditions and preventing suicidal behaviors through early identification of suicidal ideation [1]. Furthermore, the integration of SER into emotional AI chatbots offers the potential for personalized online therapy, showcasing the technology's versatility.

Amidst the backdrop of suicide as a leading cause of death, the need for innovative prevention strategies is more pressing than ever [1]. The complexity of suicide risk factors necessitates a multifaceted approach to prevention, highlighting the importance of objective, non-invasive monitoring methods such as SER [12]. By analyzing speech patterns indicative of suicidal ideation, SER offers a promising tool for early detection and intervention, potentially saving lives by enabling timely support for those at risk [5].

Speech signals, characterized by their complexity and high-frequency nature, convey rich information about the speaker, including emotional state, gender, and language. The challenge of SER is compounded by individual variability in speech patterns and the need for capturing the full spectrum of emotional expression, requiring analysis of extended speech segments [22]. Traditional approaches have relied on the extraction of hand-crafted features for input into machine learning models, a process that can be both time-consuming and susceptible to noise interference [8]. Wavelet transforms have emerged as a powerful tool for feature representation, offering advantages such as frequency-time localization and customizable wavelet functions [10]. Despite these benefits, their adoption has been limited compared to methods like the Fourier transform, partly due to the challenges associated with selecting appropriate wavelet functions [13]. The advent of CNNs has introduced a new era of SER research, demonstrating that temporal CNNs can function as effective digital filters, learning convolution kernels similar to those used in Fourier and wavelet transforms [29].

2 Related Work

In SER research, a blend of feature extraction methods and classification models enhances recognition accuracy. This includes using handcrafted features with traditional machine learning (ML) and deep learning (DL) classifiers, as well as adopting E2E frameworks that autonomously extract features from raw speech waveforms. In [11], a signal segmentation approach using the depth-first search (DFS) algorithm was introduced to set segment durations and overlaps. Local features such as pitch, MFCC, LSP, intensity, and ZCR were extracted, smoothed, and normalized; then, global features were computed using the OpenSMILE toolkit [9]. For classification, the authors used a linear kernel SVM with sequential minimal optimization (SMO). In [28], the authors used a 3-channel log-mel spectrogram, encompassing static, delta, and delta-delta components similar to an RGB image.

Other studies have utilized DL to enhance emotion classification by extracting higher-level features from raw data. While DL models like 1D CNNs are adept at learning directly from raw speech, they often require large datasets and can be inherently unstable. In [20], a deep-stride CNN is specifically designed to classify emotions, converting 1D audio signals into 2D spectrograms via the Short-Time Fourier Transform (STFT) and incorporating adaptive threshold-based noise removal in preprocessing. In [17], the scattering transform was detailed, which serves as a convolutional network using predefined kernels to achieve temporal stability in signal representation. Research in [4] computes scattering coefficients in the log-frequency domain to achieve frequency transposition invariance, enhancing speaker-independent representation. This is adaptable to both 1D and 2D data, with [3] introducing a joint time-frequency scattering that merges multiscale frequency energy distributions into a time-invariant format.

Wavelet analysis, applied in various domains such as denoising, detection, compression, and classification, leverages a multi-resolution framework to capture complex signal interactions [24]. This technique is adept at analyzing speech signals for emotion recognition. For example, [27] uses wavelet-based features along with MFCC and Teager-Energy-Operator (TEO) for classifying emotions. Furthermore, [31] employs tuned Q-factor wavelet transform (TQWT) and wavelet packet transform (WPT) to assess emotions in stroke patients. Traditionally, audio signals have been segmented into fixed-length segments to meet the uniform input size requirement of classification models, which is conducive for extracting prosodic and spectral features but may ignore emotion-rich areas in the audio. Research in [18] found that variable-length, voice-based segments perform better in emotion recognition than fixed-length segments. Additionally, wavelets are employed for dynamic signal analysis such as speech, where discrete wavelet transform (DWT) provides adjustable window sizes and selection flexibility for the analysis function, offering advantages over discrete Fourier transform (DFT) and STFT which are less adaptable to signal dynamics [25]. Numerous studies have utilized wavelets for SER, e.g., [23] integrates wavelet coefficients with LPCC and MFCC for emotion classification. [2] combines wavelet with prosodic and spectral features for Arabic SER, using an autoencoder for reducing dimensionality. [26] compares wavelet packet coefficients with MFCC, highlighting the superiority of wavelet features via a sequential floating-forward search for feature selection. Authors in [14] enhance classifier accuracy by incorporating wavelet-derived features, demonstrating their robustness in noisy environments.

3 Proposed Method

3.1 Motivation

Despite advancements in SER, issues such as system complexity, indistinct features, and vulnerability to noise continue to hinder its precision and real-world usability. Current systems often fail to adequately manage the complexity of emotional speech, with traditional methods relying on handcrafted features and fixed-length frame segmentation being unable to capture the full range of emotional cues throughout the speech. This highlights the necessity for advanced feature extraction techniques that can detect subtle emotional variations in continuous voice segments. Additionally, existing SERs' susceptibility to noise poses significant challenges, particularly in noisy environments, necessitating improvements in noise robustness for dependable performance.

3.2 Method

Fast Discrete Wavelet Transform (FDWT): The wavelet transform is a linear time-frequency transform that addresses the limitations highlighted by the Uncertainty Principle, which states that a function and its DFT cannot both be tightly localized, mathematically expressed as $\|f\|_0 \cdot \|Df\|_0 \geq n$, where $\|f\|_0$ and $\|Df\|_0$ measure the sparsity of f and its DFT, respectively, illustrating a fundamental trade-off between time and frequency localization. Drawing on the Heisenberg Uncertainty Principle [6] from quantum mechanics, which states $\Delta x \cdot \Delta p \geq \frac{\hbar}{2}$, where Δx and Δp denote the standard deviations of position and momentum distributions, respectively, while $\hbar$ stands for the reduced Planck's constant. Wavelets, offering a solution to these challenges, enable localized signal analysis in both time and frequency through a multi-scale signal decomposition essential for analyzing complex signals such as speech. The wavelet transformation utilizes the dilation and translation of a mother wavelet $\psi(t)$, as described in Eq. 1, where a and b denote the scale and translation parameters, respectively.

$$\psi_{a,b}(t) = \frac{1}{\sqrt{|a|}}\psi\left(\frac{t-b}{a}\right) \quad (1) \qquad , \quad \psi_j[n] = \frac{1}{2^j}\psi\left(\frac{n}{2^j}\right) \quad (2)$$

Specifically, we leverage the FDWT in our methodology, which constructs an orthonormal family of wavelets through specific scaling and translation operations by a power of 2, as detailed in [16]. The principal mother wavelet, ψ, is characterized by a zero mean and unit norm, undergoing variations through the process of dilation and scaling as described by Eq. (2), where $n, j \in Z$.

The DWT for a discrete signal x is defined in Eq. (3), where wavelets function as a band-pass filter bank, facilitating signal decomposition. Equation (4) describes a low-pass scaling function, ϕ, which collectively represents all wavelets beyond a certain scale j, and its Fourier transform, $\hat{\phi}$.

$$Wx\left[n, 2^j\right] = \sum_{m=0}^{N-1} x[m]\psi_j^*[m-n] \quad (3) \qquad , \quad |\hat{\phi}(\omega)|^2 = \int_1^{+\infty} \frac{|\hat{\psi}(s\omega)|^2}{s} ds \quad (4)$$

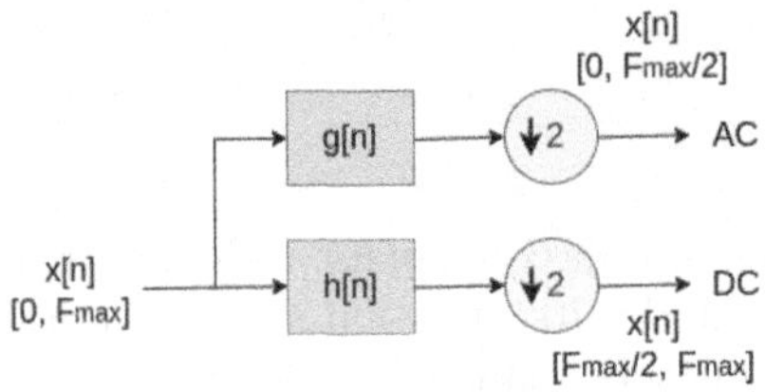

Fig. 1. FDWT traditional decomposition block

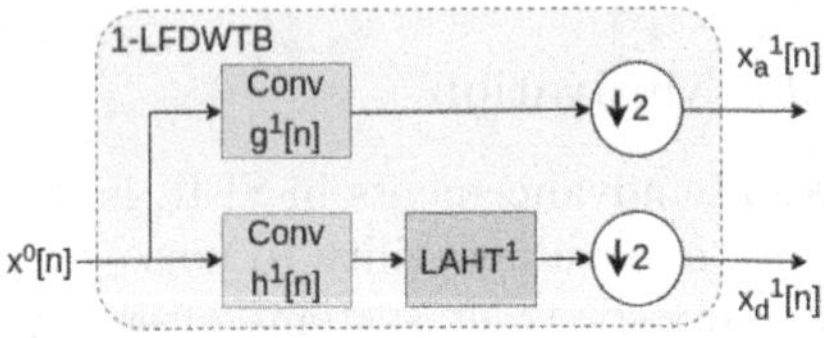

Fig. 2. SigWavNet decomposition block (1-LFDWTB)

This setup allows the FDWT to decompose a signal x into coarser approximations a (low-pass) and detail coefficients d (high-pass), with reversibility assured by the orthonormality in $\mathbf{L}^2(R)$. Each decomposition level l is a function of the preceding level, sub-sampled by a factor of 2, employing the base non-dilated wavelet. FDWT uses a series of wavelets, each a scaled and translated version of a mother wavelet by powers of 2, creating an orthonormal family. This approach allows for the extraction of detailed features from speech signals essential for capturing emotional nuances. The FDWT uses a cascading algorithm, illustrated in Fig. 1, for the efficient computation of the wavelet transform. As defined in Eqs. (5) and (6), h denotes the scaling function and g the wavelet, linking the wavelet coefficients with the filters h and g and enabling a recursive algorithm for performing the wavelet transform. This foundational framework underpins our method for analyzing speech signals to recognize emotions.

$$h[n] = \left\langle \frac{1}{\sqrt{2}}\phi\left(\frac{t}{2}\right), \phi(t-n) \right\rangle \quad (5) \quad , \quad g[n] = \left\langle \frac{1}{\sqrt{2}}\psi\left(\frac{t}{2}\right), \phi(t-n) \right\rangle \quad (6)$$

Conjugate Quadrature Filter (CQF): Our approach utilizes the CQF bank feature of the FDWT, which ensures symmetric filter responses around the cutoff frequency, thus reducing aliasing effects. This characteristic arises from designing the wavelet function g as the alternating flip of the scaling function h, expressed in Eq. (7), where $h[-n]$ denotes the reversed sequence of the nth coefficient of h:

$$g[n] = (-1)^n \cdot h[-n] \tag{7}$$

Our method relies on the wavelet filter bank decomposition to compute approximation and detail coefficients, a and d, through recursive scaling and wavelet filtering as formulated in Eqs. (8) and (9):

$$a_{j+1}[p] = \sum_{n=-\infty}^{+\infty} h[n-2p]a_j[n] \quad (8) \quad , \quad d_{j+1}[p] = \sum_{n=-\infty}^{+\infty} g[n-2p]a_j[n] \quad (9)$$

Our algorithm begins with the initial signal x as a_0, systematically dividing it into high- and low-frequency components through convolution with scaling

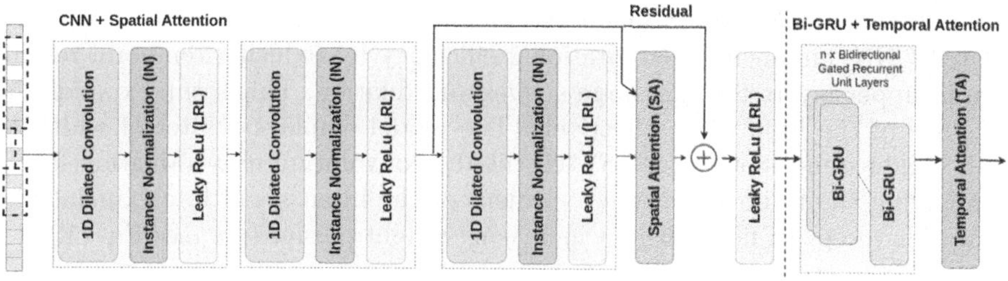

Fig. 3. Architecture of a 1D Dilated CNN with Spatial Attention, Bi-GRU, and Temporal Attention

(h) and wavelet (g) filters. The low-frequency outputs become the inputs for the next level of decomposition, each iteration halving the data size to enhance computational efficiency. This process employs just two filters, avoiding a full filter bank, thus effectively segregating the signal into distinct frequency bands.

Our DL framework uses three main components, led by the L-Level FDWT Block (L-LFDWTB), highlighted in Fig. 2. We incorporate an FDWT cascade within our model to learn the necessary filters and optimal denoising thresholds at each level. By employing orthogonal wavelet filter banks with CQF properties, our method structures the learning to emulate FDWT operations effectively. The architecture includes L-level wavelet cascade composed of pairs of learnable convolutional layers, $Conv_g$ and $Conv_h$, that downsample data, mirroring the FDWT's scaling and wavelet filters. This design results in a compact architecture with far fewer parameters than typical DL models, which often involve learning extensive parameters.

Learnable Asymmetric Hard-Thresholding ($LAHT$): We've integrated a novel LAHT layer/function that is continuous and differentiable, automating the hard-thresholding process in wavelet denoising. This layer processes coefficients to prepare them for further decomposition, learning optimal thresholding parameters automatically, thus eliminating manual intervention and enhancing denoising. The LAHT function is defined by combining two contrasting sigmoid functions, aimed at adjusting activation dynamics flexibly (Eqs. (10) and (11)).

$$\forall x \in \mathbb{R}, \qquad S(x) = \frac{1}{1+e^{-x}} \tag{10}$$

$$LAHT(x) = x \cdot \left[S\left(\alpha \cdot (x + bias^{-})\right) + S\left(\beta \cdot (x - bias^{+})\right)\right] \tag{11}$$

where α and β represent sharpness factors for negative and positive values, respectively, with opposite signs, while bias^{+} and bias^{-} are positive, learnable biases setting asymmetric thresholds. This allows for traditional FDWT emulation by adjusting biases to zero and keeping α and β as real numbers.

1D Dilated CNN and Spatial Attention: Our system incorporates a 1D-dilated CNN, enhancing computational efficiency by reducing temporal resolution in upper layers through large strides and dilation. This setup expands the coverage of the input data effectively. The model employs a filter of width S, with weight dimensions (K, C, S). 1D dilated convolution uses a dilation factor d, applying the filter to every d^{th} element of the input's width, enhancing the network's receptive field without increasing weight parameters (see Eq. (12)):

$$\text{Out}(k, q) = \sum_{c} \sum_{w+d\cdot s=q} \text{In}(c, w) * \text{Weight}(k, c, s) \tag{12}$$

This method effectively captures both granular and extensive contextual information from speech by extending the filter's reach, crucial for detecting long-range emotional cues. Additionally, the architecture accommodates variable-length sequences, essential for processing diverse emotional expressions in speech.

Post-convolution, our model employs instance normalization (IN) to stabilize training and improve generalization across various recording conditions, normalizing activations within 1D CNN layers to diminish sensitivity to input variations. We further utilize the Leaky ReLU activation function to address the "dying ReLU" issue by allowing a non-zero gradient for negative inputs.

Subsequently, a 1D spatial attention mechanism is applied, which recalibrates focus by differentially weighting significant features of the input data. This starts by merging an input feature map l with a global context g, followed by generating a raw attention map c through a 1D convolution with a kernel size of 1 as expressed in Eqs. (13), (14), (15), (16), and (17):

$$L_{combined} = l + g \quad (13) \quad , \quad c = \text{Conv1D}_{\text{kernel\_size}=1}(L_{combined}) \quad (14)$$

$$a = \text{Softmax}(c) \quad (15) \quad , \quad g = a \odot l \quad (16) \quad , \quad G_{aggregated} = \sum_{\text{width}} g \quad (17)$$

Softmax normalizes c to a, distributing attention scores evenly, which allows the model to dynamically focus on pertinent features. This results in an attention-weighted feature map g that accentuates crucial emotional indicators, thus enhancing the model's accuracy in emotion recognition.

Bi-GRU and Temporal Attention: The third component of our architecture is the Bi-GRU Neural Network, positioned after the 1D dilated CNN and spatial attention layer, as illustrated in Fig. 3. GRU operations are expressed as follows:

$$r_t = \sigma(W_{ir}x_t + b_{ir} + W_{hr}h_{(t-1)} + b_{hr}) \quad (18) \quad , \quad z_t = \sigma(W_{iz} + b_{iz} + W_{hz}h_{(t-1)} + b_{hz}) \quad (19)$$

$$n_t = \tanh(W_{in}x_t + b_{in} + r_t * (W_{hn}h_{(t-1)} + b_{hn})) \quad (20) \quad , \quad h_t = (1 - z_t) * n_t + z_t * h_{(t-1)} \quad (21)$$

These equations describe the GRU's reset r_t, update z_t, and new n_t gates, with h_t as the current hidden state and x_t as the input at time t. In our multi-layer setup, higher layers use hidden states from previous layers, adjusted for dropout. The Bi-GRU processes data bidirectionally, integrating both past and

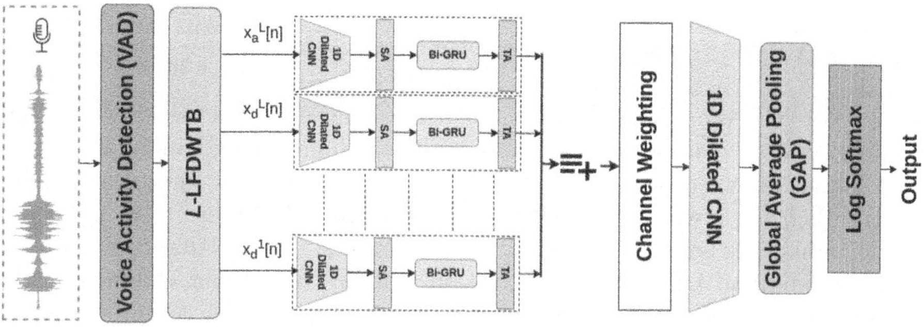

Fig. 4. L-LFDWTB SigWavNet General Architecture

future information, where $\overrightarrow{h_t}$ and $\overleftarrow{h_t}$ are the forward and backward hidden states, respectively, and $\oplus$ represents their concatenation (see Eqs. (22), (23), and (24)):

$$\overrightarrow{h_t} = GRU_{fwd}\left(x_t, \overrightarrow{h_{t-1}}\right), \quad (22) \quad \overleftarrow{h_t} = GRU_{bwd}\left(x_t, \overleftarrow{h_{t+1}}\right), \quad (23) \quad h_t = \overrightarrow{h_t} \oplus \overleftarrow{h_t} \quad (24)$$

After the Bi-GRU, a temporal attention layer enhances emotion recognition by focusing on key temporal features. It processes the sequence of hidden states $H \in \mathbb{R}^{N \times T \times D}$ through a fully connected layer, FC1, evaluating their relevance against the final hidden state h_t, as shown in Eq. (25) and scores are normalized via softmax to produce attention weights A, as per Eq. (26):

$$Score = \text{FC1}(H) \cdot h_t^T \quad (25) \quad , \quad A = \text{Softmax}(Score) \quad (26)$$

These weights create a context vector C through a weighted summation of hidden states, detailed in Eq. (27), which is then concatenated with h_t and transformed via FC2 and tanh to produce the attention vector V (see Eq. (28)):

$$C = \sum_{t=1}^{T} A_t \cdot H_t \quad (27) \quad , \quad V = \tanh(\text{FC2}([C; h_t])) \quad (28)$$

This mechanism dynamically emphasizes crucial features for emotion detection, enhancing sensitivity to emotional cues and refining temporal information aggregation. The resulting attention vectors, derived from each FDWT decomposition level, represent essential information from the respective frequency bands.

Channel Weighting and Global Average Pooling: In the SigWavNet architecture (Fig. 4), attention vectors are concatenated channel-wise to form a composite representation with dimensions $(batch\_size, levels+1, length)$. Each channel represents an attention vector from a specific frequency band, merging critical information across the speech's frequency spectrum into a rich feature set. This composite is then processed by a Channel Weighting layer that dynamically adjusts the importance of each frequency band, as shown in Eq. (29):

$$X' = X \odot W \quad (29) \quad , \quad \log \text{Softmax}\,(x_i) = \log \left(\frac{\exp(x_i)}{\sum_j \exp(x_j)} \right) \quad (30)$$

where X is the input representation, W represents learnable weights for each channel, and $\odot$ denotes element-wise multiplication. The Channel Weighting layer initializes all channel weights to one, indicating equal importance. During training, these weights are refined to prioritize more informative frequency bands, enhancing emotion detection. Following this, a 1D dilated convolutional layer with IN and Leaky ReLU activation matches the output channels to the number of emotion classes, extracting features and producing feature maps for each class. For classification, the network uses a Log Softmax layer, defined by Eq. (30), which enhances numerical stability and computational efficiency, mitigating issues with large input values and simplifying gradient calculations during backpropagation. The architecture also integrates a Global Average Pooling (GAP) layer that averages activations across the convolution output, reducing dimensionality while retaining critical information.

4 Experiments and Results

4.1 Datasets

The NSPL-CRISE Dataset: For our research, we created the NSPL-CRISE [21] dataset by analyzing call recordings from frequent callers to the National Suicide Prevention Lifeline (NSPL) over one month. We identified individuals who called at least 30 times in the previous month. With ethical approval, we collected audio recordings, informing callers of the recording at the start of each call. Recordings were anonymized and screened based on call duration and caller demographics. Trained psychology researchers assessed the emotional content of the first and last call from each individual, noting any changes over the month. Emotional segments were annotated on a confidence scale from 1 (dubious) to 5 (indisputable), with 738 segments as 'angry', 435 as 'fearful/concerned/worried' (FCW), 753 as 'happy', 738 as 'sad', and 909 as 'neutral'.

4.2 Experimental Setup

We standardize all audio signals to a 16 KHz sampling rate and mono-channel format for uniformity. The dataset is split into 90% for training/validation and 10% for testing, using 10-fold cross-validation with stratified random sampling for balanced emotion class representation. For optimal tuning, we use grid search and the Asynchronous Successive Halving Algorithm (ASHA) for efficient hyperparameter optimization. To address class imbalance and overfitting, we implement the focal loss function with L2-regularization, focusing on hard-to-classify examples. The focal loss is defined as in Eq. (31), where p_t is the class probability, α_t adjusts class importance, and γ modulates the focus on difficult examples.

Combined with L2-regularization (Ridge), where λ is the regularization strength, the loss function is expressed as in Eq. (32).

$$FL(p_t) = -\alpha_t(1-p_t)^\gamma \log(p_t) \quad (31) \quad , \quad J_{\text{regularized}}(\omega) = FL(p_t) + \lambda \sum_{i=1}^{n} \omega_i^2 \quad (32)$$

Using the Adam optimizer, we minimize this loss to enhance model generalization across diverse emotional speech states and evaluate performance on the dataset through repeated tests with various seeds. A Voice Activity Detection (VAD) layer is used to improve the processing of the dataset, isolating relevant speech segments for better emotion recognition. Our optimal model architecture includes an 8-layer FDWT block with a kernel size of 20, simulating Daubechies-10 wavelets, followed by a 3-layer dilated CNN and a 6-layer Bi-GRU.

4.3 Results and Comparison with State-of-the-Art (SOTA) Methods

NSPL-CRSE: Table 1 highlights SigWavNet's performance on the NSPL-CRISE dataset. The model excels in detecting 'Happy' emotions with 90.3% precision, though its 73.7% recall indicates room for improvement. It also captures 'Sad' emotions well, with a 73.3% recall but a lower precision of 66.3%, showing sensitivity to broader emotional expressions. 'Neutral' emotions are identified with 68% precision and 76.9% recall, confirming the model's capability to recognize less expressive states. Both F1-score macro and weighted averages are around 72%, reflecting consistent performance across the dataset. An overall accuracy of 72.2% underscores SigWavNet's robust SER abilities.

Table 2 compares SigWavNet with state-of-the-art SER methods on the NSPL-CRISE dataset. Our model outperformed significant SOTA approaches. [19] used a deep RNN with local attention, achieving 51.3% accuracy and 52.1% F1-score. [15] employed a self-attention mechanism on spectrograms and gender classification, scoring 68.7% accuracy and 69.3% F1-score. [7] developed a three-dimensional attention-based CRNN with handcrafted features, achieving 59.6% accuracy and 60.2% F1-score. [30] utilized a 2D CNN-LSTM network to analyze emotional features from log-mel spectrograms, resulting in 67.2% accuracy and 67.9% F1-score. SigWavNet achieved the highest accuracy and F1-score of 72.2% and 72.4%, respectively, demonstrating its effectiveness.

Figure 5 provides a detailed view of SigWavNet's performance, showcasing its ability to recognize various emotions. The model accurately identifies 'Happy' emotions 73.68% of the time, demonstrating effective detection of positive cues. It performs well in recognizing 'Sad' with a 73.33% precision and detects 'Neutral' emotions with a 76.92% accuracy rate, indicating strength in handling both pronounced and subtle emotional states. The confusion matrix shows some misclassifications, such as 'Angry' being confused with 'Sad' or 'Neutral', but these occur at lower percentages, suggesting a good overall grasp of emotional nuances. The model's 61.36% accuracy in distinguishing 'FCW' while occasionally confusing it with 'Sad' highlights areas for potential refinement.

Table 1. SigWavNet performance on NSPL-CRISE in (%)

Emotion	Precision	Recall	F1.
Angry	74.3	70.3	72.2
FCW	64.3	61.4	62.8
Happy	90.3	73.7	81.2
Neutral	68	76.9	72.2
Sad	66.3	73.3	69.6
Overall results			
Macro avg	72.6	71.1	71.6
Weighted avg	73.2	72.2	72.4
Accuracy	72.2		

Table 2. Comparison with SOTA on NSPL-CRISE: best results are in bold font

Compared methods in (%)		
NSPL-CRISE	Test Acc.	F1.
Mirsamadi et al. [19]	51.3	52.1
Li et al. [15]	68.7	69.3
Chen et al. [7]	59.6	60.2
Zhao et al. [30]	67.2	67.9
SigWavNet (ours)	**72.2**	**72.4**

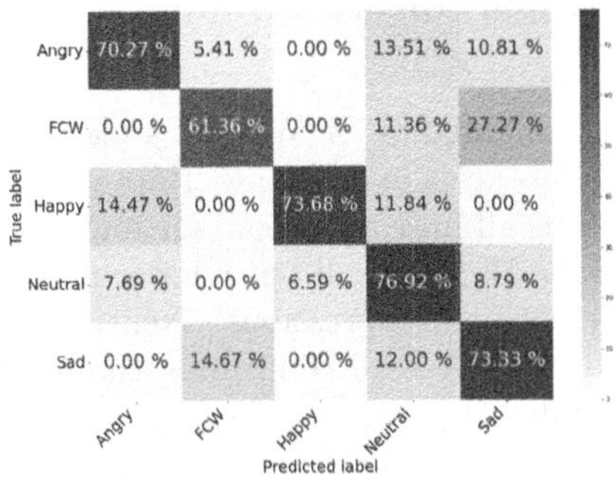

Fig. 5. SigWavNet confusion matrix on NSPL-CRISE

5 Conclusion

In this study, we introduced SigWavNet, a novel SER architecture, and evaluated its performance through extensive experiments. The results demonstrate that SigWavNet significantly outperforms state-of-the-art methods, highlighting its robustness for SER. The architecture leverages FDWT, LAHT, and adaptive kernel learning to capture emotional nuances in speech. This work emphasizes the importance of learnable FDWT and LAHT for enhancing adaptability and performance, while 1D CNN and spatial attention, along with Bi-GRU and temporal attention layers, effectively capture contextual and temporal dynamics. Comparative analyses confirm SigWavNet's superiority over other SER methods, making it reliable for multiple applications. Future research could expand SigWavNet's applicability to more languages and scenarios, increasing its utility.

References

1. Suicide prevention: Centers for disease control and prevention (2024). https://www.cdc.gov/suicide/. Accessed 16 May 2024
2. Abdel-Hamid, L.: Egyptian Arabic speech emotion recognition using prosodic, spectral and wavelet features. Speech Commun. **122**, 19–30 (2020)
3. Andén, J., Lostanlen, V., Mallat, S.: Joint time-frequency scattering. IEEE Trans. Signal Process. **67**(14), 3704–3718 (2019)
4. Andén, J., Mallat, S.: Deep scattering spectrum. IEEE Trans. Signal Process. **62**(16), 4114–4128 (2014)
5. Belouali, A., et al.: Acoustic and language analysis of speech for suicidal ideation among us veterans. BioData Min. **14**(1), 1–17 (2021)
6. Busch, P., Heinonen, T., Lahti, P.: Heisenberg's uncertainty principle. Phys. Rep. **452**(6), 155–176 (2007)
7. Chen, M., He, X., Yang, J., Zhang, H.: 3-D convolutional recurrent neural networks with attention model for speech emotion recognition. IEEE Signal Process. Lett. **25**(10), 1440–1444 (2018)
8. El Ayadi, M., Kamel, M.S., Karray, F.: Survey on speech emotion recognition: features, classification schemes, and databases. Pattern Recogn. **44**(3), 572–587 (2011)
9. Eyben, F., Weninger, F., Wöllmer, M., Shuller, B.: Open-source media interpretation by large feature-space extraction. TU Munchen, MMK (2016)
10. Fahad, M.S., Ranjan, A., Yadav, J., Deepak, A.: A survey of speech emotion recognition in natural environment. Digit. Signal Process. **110**, 102951 (2021)
11. Gao, Y., Li, B., Wang, N., Zhu, T.: Speech emotion recognition using local and global features. In: Zeng, Y., et al. (eds.) BI 2017. LNCS (LNAI), vol. 10654, pp. 3–13. Springer, Cham (2017). https://doi.org/10.1007/978-3-319-70772-3_1
12. Gideon, J., Schatten, H.T., McInnis, M.G., Provost, E.M.: Emotion recognition from natural phone conversations in individuals with and without recent suicidal ideation. In: Interspeech (2019)
13. Huang, L., Shen, X.: Research on speech emotion recognition based on the fractional fourier transform. Electronics **11**(20), 3393 (2022)
14. Huang, Y., Tian, K., Wu, A., Zhang, G.: Feature fusion methods research based on deep belief networks for speech emotion recognition under noise condition. J. Ambient. Intell. Humaniz. Comput. **10**, 1787–1798 (2019)
15. Li, Y., Zhao, T., Kawahara, T.: Improved end-to-end speech emotion recognition using self attention mechanism and multitask learning. In: Interspeech, pp. 2803–2807 (2019)
16. Mallat, S.: A Wavelet Tour of Signal Processing: The Sparse Way. Elsevier Science (2008). https://books.google.ca/books?id=5qzeLJljuLoC
17. Mallat, S.: Group invariant scattering. Commun. Pure Appl. Math. **65**(10), 1331–1398 (2012)
18. Mansoorizadeh, M., Charkari, N.M.: Speech emotion recognition: comparison of speech segmentation approaches. In: Proceedings of IKT, Mashad, Iran (2007)
19. Mirsamadi, S., Barsoum, E., Zhang, C.: Automatic speech emotion recognition using recurrent neural networks with local attention. In: 2017 IEEE International Conference on Acoustics, Speech and Signal Processing (ICASSP), pp. 2227–2231. IEEE (2017)
20. Mustaqeem, Kwon, S.: A CNN-assisted enhanced audio signal processing for speech emotion recognition. Sensors **20**(1), 183 (2019)

21. Nfissi, A., Bouachir, W., Bouguila, N., Mishara, B.: Unlocking the emotional states of high-risk suicide callers through speech analysis. In: 2024 IEEE 18th International Conference on Semantic Computing (ICSC), pp. 33–40. IEEE (2024)
22. Nugroho, H., Fuadiyah, R.N.N.: Development of speech emotion recognition system based on discrete wavelet transform (DWT) and voice segmentation. Int. J. Electr. Eng. Inf. **14**(3), 593–607 (2022)
23. Palo, H.K., Mohanty, M.N.: Wavelet based feature combination for recognition of emotions. Ain Shams Eng. J. **9**(4), 1799–1806 (2018)
24. Silva, J., Narayanan, S.S.: Discriminative wavelet packet filter bank selection for pattern recognition. IEEE Trans. Signal Process. **57**(5), 1796–1810 (2009)
25. Steinbuch, M., Van de Molengraft, M.: Wavelet Theory and Applications: A Literature Study. Eindhoven University Technology Department of Mechanical Engineering Control System Group, Eindhoven (2005)
26. Wang, K., Su, G., Liu, L., Wang, S.: Wavelet packet analysis for speaker-independent emotion recognition. Neurocomputing **398**, 257–264 (2020)
27. Zão, L., Cavalcante, D., Coelho, R.: Time-frequency feature and AMS-GMM mask for acoustic emotion classification. IEEE Signal Process. Lett. **21**(5), 620–624 (2014)
28. Zhang, S., Zhang, S., Huang, T., Gao, W.: Speech emotion recognition using deep convolutional neural network and discriminant temporal pyramid matching. IEEE Trans. Multimedia **20**(6), 1576–1590 (2017)
29. Zhang, W., Li, C., Peng, G., Chen, Y., Zhang, Z.: A deep convolutional neural network with new training methods for bearing fault diagnosis under noisy environment and different working load. Mech. Syst. Signal Process. **100**, 439–453 (2018)
30. Zhao, J., Mao, X., Chen, L.: Speech emotion recognition using deep 1D & 2D CNN LSTM networks. Biomed. Signal Process. Control **47**, 312–323 (2019)
31. Zheng, B.S., Khairunizam, W., Murugappan Murugappan, S., Razlan, Z., Zunaidi, I., Yean, C.: Effectiveness of tuned q-factor wavelet transform in emotion recognition among left-brain damaged stroke patients. Int. J. Simul. Syst. Sci. Technol. **19**(3), 2 (2018)

Dynamic HumTrans: Humming Transcription Using CNNs and Dynamic Programming

Shubham Gupta[1,2](✉), Isaac Neri Gomez-Sarmiento[2], Faez Amjed Mezdari[2], Mirco Ravanelli[1,3], and Cem Subakan[1,2,3]

[1] Mila-Québec AI Institute, Montreal, Canada
Shubham.gupta30@gmail.com
[2] Laval University, Quebec City, Canada
[3] Concordia University, Montreal, Canada

Abstract. We propose a novel approach for humming transcription that combines a CNN-based architecture with a dynamic programming-based post-processing algorithm, utilizing the recently introduced HumTrans dataset. We identify and address inherent problems with the offset and onset ground truth provided by the dataset, offering heuristics to improve these annotations, resulting in a dataset with precise annotations that will aid future research. Additionally, we compare the transcription accuracy of our method against several others, demonstrating state-of-the-art (SOTA) results. All our code and corrected dataset is available at https://github.com/shubham-gupta-30/humming_transcription.

Keywords: Humming · Transcription · Automatic Music Transcription (AMT) · Music Information Retrieval (MIR)

1 Introduction

The field of Automatic Music Transcription (AMT) has made significant progress in developing algorithms that transform acoustic music signals into music notation, positioning it as the musical analogue to Automatic Speech Recognition (ASR). In the piano-roll convention, a musical note is typically characterized by a constant pitch (related to frequency), onset time (the start), and offset time (the end).

One application of AMT is humming transcription, which involves extracting musical notes from a hummed tune. This is a crucial component for melody search engines [6] and automatic music compositions [2]. Such applications enable song identification by mere humming, provide a quick starting point for creating new songs, and democratize music creation for those who may not play an instrument or have disabilities. However, achieving error-free music transcription remains a complex challenge, even for professionals.

In this paper, we explore humming transcription while working with the HUMTRANS dataset [8], a novel dataset that claims to be the largest humming

C. Y. Suen et al. (Eds.): ANNPR 2024, LNAI 15154, pp. 269–280, 2024.
https://doi.org/10.1007/978-3-031-71602-7_23

dataset to date. This dataset is a large collection of clean monophonic humming samples gathered by soliciting help from music students. This provides us with an opportunity for studying transcription in a monophonic setting. Various works in literature explore transcription in more general polyphonic setting like VOCANO [7], Sheet Sage [5], MIR-ST500 [12], and JDC-STP [10]. We propose a novel approach to do an accurate transcription in this monophonic setting and show that we obtain state-of-the-art (SOTA) transcription results. Our contributions are two fold:

1. We identify issues in the ground truth provided by the HUMTRANS dataset and offer heuristics to address them, enabling us to bootstrap the creation of a high-quality subset with more meaningful annotations, which will aid further research in this direction.
2. We introduce a novel approach that combines a CNN-based architecture with a dynamic programming-based post-processing technique, achieving state-of-the-art (SOTA) results.

1.1 Evaluation Metrics

The authors of the HumTrans dataset utilize the library *mir_eval* [9] to evaluate the performance of transcription methods on their dataset. The primary motivation for using this library is to standardize the implementation of metrics for music transcription. More specifically, they employ the method `precision_recall_f1_overlap`, which, according to the documentation, computes the Precision, Recall, and F-measure for reference vs. estimated notes. Correctness is determined based on note onset, pitch, and, optionally, offset, which the authors do not consider.

The authors consider a strict pitch tolerance of ±1 cent (*mir_eval* default value is ±50 cents), or in other words one hundredth part of a semitone, and the default onset tolerance of 50 ms.

Precision, recall and F1-score are metrics that depend on the definition of true positives (TP), false negatives (FN) and false positives (FP) [4].

$$Precision = \frac{TP}{TP + FP} \tag{1}$$

$$Recall = \frac{TP}{TP + FN} \tag{2}$$

$$F1\,Score = 2 * \frac{Precision * Recall}{Precision + Recall} \tag{3}$$

TP: Estimated onset is within the tolerance of the ground truth onset and is within the pitch tolerance.

FP: If N estimated onsets are within the tolerance of the ground truth, only 1 will be considered TP if it's also within the pitch tolerance and the rest N-1 estimated onsets will be considered FP. This means that all ground truth onsets can only be matched once.

FN: No estimated onsets were detected within the tolerance of the ground truth onset.

1.2 Dataset Challenges

A major issue with the HUMTRANS dataset is that the ground truth onsets and offsets are not well aligned. This can be explained because in their methodology they instructed their subjects to synchronize their humming with the rhythm of the played melody, calling this approach as "self-labeling", without any post-processing. We had to overcome this challenge by coming up with semi-supervised ways to correct the provided onsets and offsets.

1.3 Octave Aware vs Octave Invariance

To simplify the problem of pitch estimation, the ground truth pitch given in MIDI file format can be transformed to an octave invariant representation by taking the modulo 12 of the MIDI numbers, which makes the song to be represented only by 12 semitones. In our work we produce results for both octave aware and octave invariant variations of the problem.

2 Transcription Methodology

2.1 Better Ground Truth Annotation

For the purpose of training a neural network, we need precise onsets and offsets for ground truth. We realized that more accurate labeled onsets and offsets result in better-trained models.

To this end, we designed a heuristic-based algorithm to compute improved onsets and offsets. This algorithm calculates a *waveform envelope* and determines onsets and offsets based on when this envelope dips below a specific threshold value. The onsets and offsets obtained in this way can be noisy, so we refine them by eliminating spurious onsets/offsets, enforcing a minimum note length, and maintaining a minimum silence length between notes. This method is supervised by using the number of notes from the ground truth provided by the dataset. We retain only those training, testing, and validation samples where the number of notes detected by our heuristic matches the number provided by the ground truth.

It's important to note that we can only trust the number of notes from the ground truth, as the onsets and offsets cannot be relied upon. By using this approach, we obtained better ground truth onsets and offsets, retaining $6,827$ of the $13,080$ in the training set (52.2%) and 440 of the 769 (i.e., 57%) in the test set. More details on this method are covered in Appendix A.

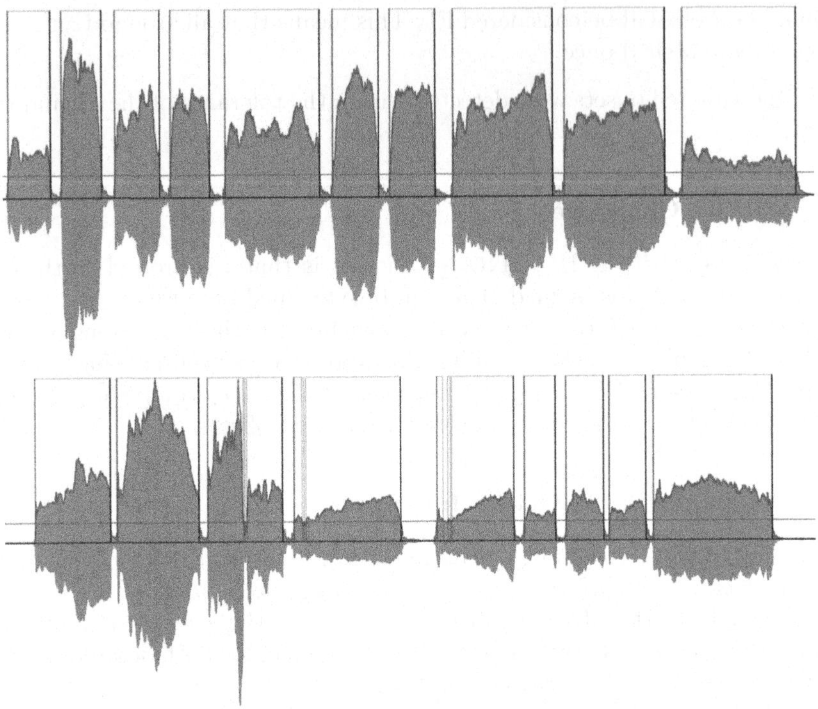

Fig. 1. Examples where the heuristic algorithm for onset and offset detection succeeds (top) and fails (bottom).

2.2 Network Design

Our neural network architecture is a convolution-based network. The model is inspired by the architecture in [3]. We have tailored our network for our use case of a monophonic humming dataset. The following are key elements of this design (Fig. 1):

1. **Input representation**: While *STFT* and Mel spectrograms work well for speech-related tasks, *CQT* representation is much more effective for *MIR* (Music Information Retrieval) as the geometric structure of this transform closely matches the geometric nature of Western classical music.
2. **Harmonic Stacking**: While instruments can produce desired notes precisely, humans aren't very adept at doing so. Human singing/humming normally includes not only the main note but also overtones, which can be spatially far apart from each other in a CQT transform. To address this, [1] introduces Harmonic Stacking, where CQT time frames are shifted by the right amount of offsets to bring overtones closer to each other

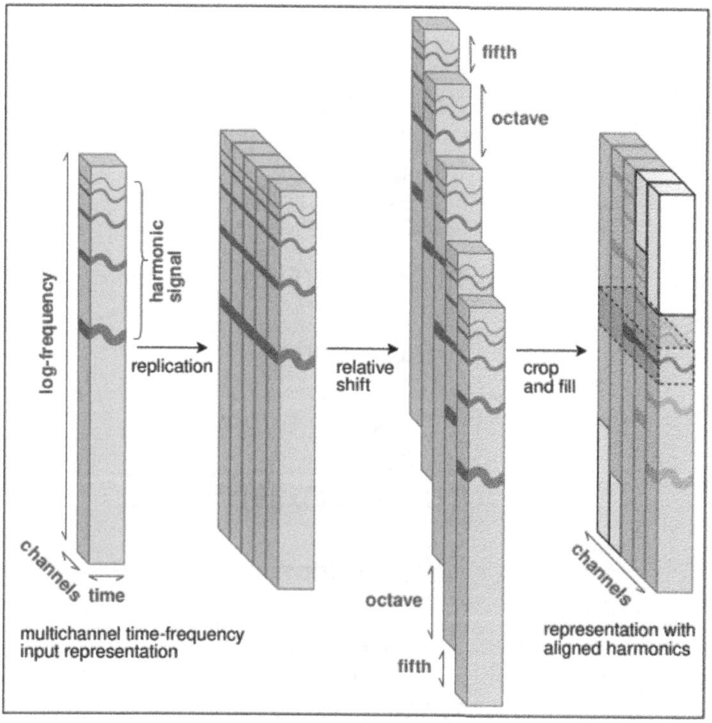

Fig. 2. Harmonic Stacking. Source: [1]

2.3 Training

We train the model with a batch size of 16. For each batch, we randomly select a small sample from the humming recordings by choosing 5 to 10 notes for each batch element. Additionally, we introduce a new dummy note 89 to signify the beginning and end of the recording sample, as well as the silence between notes. Our task for every time frame of the CQT representation is to predict the note that the frame represents. The model is trained using **CrossEntropy** loss and employs **Adam** as the optimizer with a learning rate of 0.001. Unlike other works in this field that first predict onsets and offsets and then condition note prediction on them, as in [2], our method infers onsets and offsets directly from the predicted notes (Fig. 2).

2.4 Inference

Unlike transformers, convolution networks generalize well beyond the training length examples they are trained on. Because of this, we do not have to perform inference on pieces of fixed lengths; instead, we can perform inference on the entire sample as long as available memory allows. During inference, we calculate model logits for each time frame over the possible space of notes. The naive way

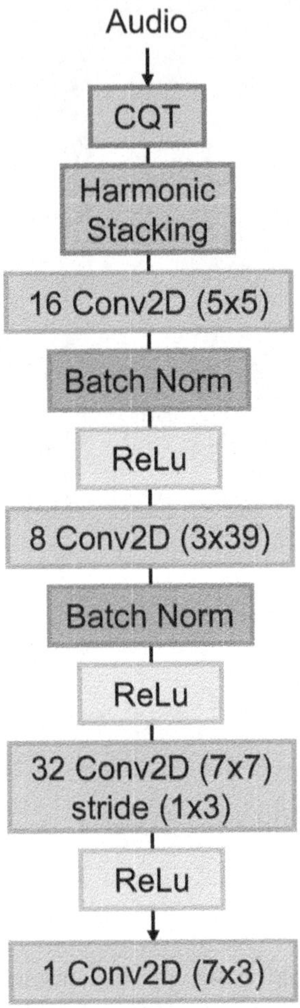

Fig. 3. Our model architecture - A minimal version of Spotify's BasicPitch model.

to convert these logits to actual note predictions would be to take the note at each time frame with the maximum probability. However, this results in noisy note attributions.

We clean up these noisy attributions using a dynamic programming based algorithm inspired by the use of the *Viterbi algorithm* in text-to-speech alignment in speech tasks. Given the affinity matrix denoting the affinity scores of each time frame with the possible notes, we can define a path P through this matrix as valid if it satisfies the following constraints (Note that the dummy note introduced is 89, and T is the length of the segment being inferred) (Fig. 3):

1. P starts at $(0, 89)$ and ends at $(T-1, 89)$.
2. If P is at note $n \neq 89$ at time t, then at time $t+1$, it can be at either n or 89.
3. If P is at note 89 at time t, then it can be at any note at time $t+1$.

These constraints ensure that we do not switch abruptly from one note to another without going through the dummy note, which is a realistic constraint as in all humming samples, the space between two hummed notes is very noticeable. Using a dynamic programming-based method, we can find the path P with the highest probability among all valid paths. We further clean up this path P to enforce minimum note length constraints and report this final cleaned note assignment as our inferred note assignment. We read the onsets and offsets from these note assignments. We reproduce the code for this dynamic programming based postprocessing in Appendix B (Fig. 4).

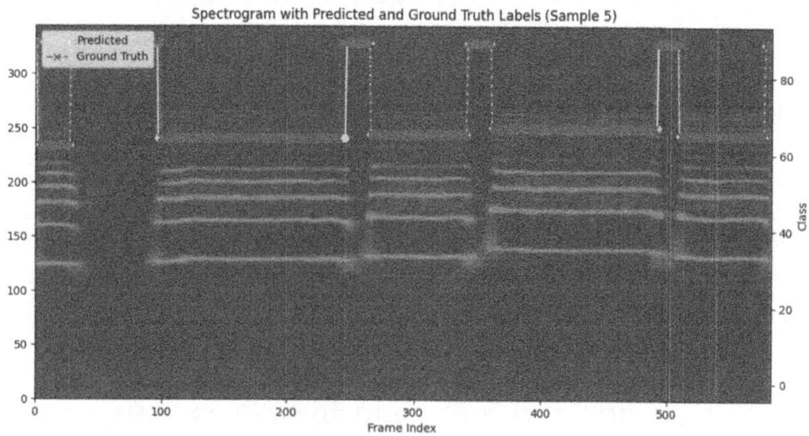

Fig. 4. Example inference: blue represents the ground truth and red the inferred cleaned notes. (Color figure online)

3 Results and Discussion

We compare our methods with various methods discussed in [8]. The authors provide extracted MIDIs for the test set for 4 methods in their GitHub repository [11]. These are - $Vocano$ [7], $MIR-ST500$ [12], $SheetSage$ [5], and $JDC-STP$ [10]. In addition, we compute MIDIs for the test set using Spotify's *basic_pitch* [2], and compare these with the methods proposed in this report.

3.1 Octave Invariant

Following [8], we calculate precision, recall, and F1-score, using the *mir_eval* library, with an onset tolerance of $50ms$ and disregarding offsets. We provide two comparisons here - a comparison with respect to the corrected ground truth we obtain and a comparison where we measure only the note accuracy, disregarding both onsets and offsets. Additionally, these comparisons are octave invariant, i.e. a note is considered to be correctly predicted even if the octave does not match the ground truth exactly. We provide these results on the test set provided by the dataset in Table 1

Table 1. Octave Invariant metrics computed on test set.

Method	Note + Onsets			Notes Only		
	P	R	F1	P	R	F1
Ours	**0.670**	**0.675**	**0.673**	**0.848**	**0.854**	**0.850**
VOCANO	0.568	0.561	0.564	0.729	0.723	0.726
JDC-STP	0.502	0.487	0.490	0.795	0.784	0.783
SheetSage	0.171	0.170	0.170	0.446	0.442	0.444
MIR-ST500	0.601	0.608	0.604	0.808	0.820	0.813
basic_pitch	0.392	0.497	0.434	0.653	0.847	0.729

3.2 Octave Aware

In Table 2, we also provide comparisons of our method with other methods while requiring the models to be octave-aware, i.e., we are looking for an exact note match, including the correct octaves.

Table 2. Octave Aware metrics computed on test set.

Method	Note + Onsets			Notes Only		
	P	R	F1	P	R	F1
Ours	**0.649**	**0.653**	**0.651**	**0.814**	**0.820**	**0.817**
VOCANO	0.344	0.340	0.341	0.446	0.443	0.444
JDC-STP	0.297	0.279	0.286	0.463	0.442	0.450
SheetSage	0.161	0.160	0.161	0.434	0.430	0.444
MIR-ST500	0.360	0.363	0.361	0.486	0.491	0.488
basic_pitch	0.243	0.304	0.268	0.388	0.498	0.432

3.3 Discussion

We observe that our method outperform all tracked methods for humming transcription. We observe that SheetSage performs the worst in all comparisons. Also note that our method performs similarly well in the octave invariant and the octave aware setting, indicating that our architecture is able to learn very robust note representations.

3.4 Future Work

In our work, we provide a novel methodology to accurately estimate monophonic humming transcriptions. A natural extension of this work is to transcribe polyphonic humming samples. This work also provides a novel dynamic programming based post processing and we would like to explore the use of this as postprocessing in other transcription problems. It is also possible to use this postprocessing as a part of the loss function during training thus enabling better transcriptions from the get go.

A Heuristic Algorithm for Better Ground Truth Annotations

We reproduce the python code used to create the waveform envelopes as mentioned in Sect. 2.1. We want our waveform envelops to be as close to the original general waveform shape as possible and thus we utilize a simple heuristic algorithm that locally computes the maximum of the waveform preceding a point of interest and the maximum of the waveform following a point of interest. We found we get a very tight hugging envelope when we take a min of these two values.

```
def get_waveform_envelope(signal):

    # Padding for the signal
    padding = 100
    padded_signal = torch.nn.functional.pad(
        signal,
        (padding, padding),
        mode='constant',
        value=0)

    # Unfold to get sliding windows
    windows_before = padded_signal[:-padding].unfold(0, padding, 1)
    windows_after = padded_signal[padding:].unfold(0, padding, 1)

    # Compute maximums
    max_before = windows_before.max(dim=1).values
    max_after = windows_after.max(dim=1).values

    # Compute minimum of the two maximums
    modified_signal = torch.min(max_before, max_after)

    return modified_signal
```

Listing 1.1. Calculate waveform envelope

We found that the envelope calculated using the above method could still be improved if we calculated the envelope of the envelope again. Having now obtained a tight envelope of the waveform, we now calculate the threshold to use for this waveform to measure the onset and offset boundaries. We further clean these onsets and offsets by disregarding any silences that are too small (the method `adjust_onsets_offsets` in the code below). Finally, we check if only consider this waveform for training or testing purposes if we get the right number of notes through this heuristic, otherwise we disregard this sample altogether. The code to do this is reproduced below.

```
def process(i):
    signal =
torch.abs(torch.tensor(dataset[i]["wav_data"]).float())
    envelope = get_waveform_envelope(signal)[None]
    envelope = get_waveform_envelope(envelope)[None]
    mw_min = torch.min(envelope)
    mw_max = torch.max(envelope)
    thresholds_for_i = mw_min + thresholds * (mw_max - mw_min)
    above_threshold = envelope > thresholds_for_i

    num_notes_known = dataset[i]["midi_notes"].shape[0]


    for t_idx in range(len(threshold_values)):
        current_above_threshold = above_threshold[t_idx]
        onsets = (current_above_threshold[:-1] <
current_above_threshold[1:]).nonzero(as_tuple=True)[0]
        offsets = (current_above_threshold[:-1] >
current_above_threshold[1:]).nonzero(as_tuple=True)[0] + 1

        onsets, offsets = adjust_onsets_offsets(onsets, offsets,
envelope.shape[0])
        num_notes_discovered = len(offsets)
        if num_notes_discovered == num_notes_known:
            file_path = filtered_folder /
f"{dataset[i]['file_name']}_onsets_offsets.txt"
            with file_path.open('w') as f:
                for onset, offset in zip(onsets, offsets):
                    f.write(f"{onset} {offset}\n")
            plot_waveforms(
                signal.numpy(),
                envelope.numpy(),
                thresholds_for_i[0].item(),
                current_above_threshold.numpy(), onsets,
offsets, f"{i}_{dataset[i]['file_name']}.png")
            return True

    return False
```

Listing 1.2. Process a single waveform

B Dynamic Programming Postprocessing

The code for computing a path using dynamic programming as detailed in Sect. 2.4 is reproduced below. Note that the method `clean_path` simply performs a heuristic cleaning on the paths discovered by the dynamic programming solution so that they are more meaningful and make sense.

```
def build_log_path_prob_matrix_with_path(log_affinity):
    L, T = log_affinity.shape
    prob_table = np.full((L, T), float('-inf'))
    path_matrix = np.zeros((L, T, 2), dtype=int)

    # Base case
    prob_table[L-1, 0] = 0

    # Iterating over columns
    for j in range(1, T):
        # Vectorized computation for non-last rows
        non_last_rows = np.arange(L-1)
        max_vals = np.maximum(prob_table[non_last_rows, j-1],
prob_table[L-1, j-1])
        prob_table[non_last_rows, j] = max_vals +
log_affinity[non_last_rows, j]
        path_matrix[non_last_rows, j] =
np.vstack([non_last_rows, np.full(L-1, j-1)]).T
        path_matrix[non_last_rows, j, 0] =
np.where(prob_table[non_last_rows, j-1] == max_vals,
non_last_rows, L-1)

        # Computation for last row, still iterative
        max_index = np.argmax(prob_table[:, j-1])
        prob_table[L-1, j] = prob_table[max_index, j-1] +
log_affinity[L-1, j]
        path_matrix[L-1, j] = (max_index, j-1)

    # Retrace path
    path = []
    current_pos = (L-1, T-1)
    while current_pos[1] != 0:
        path.append(current_pos)
        current_pos = tuple(path_matrix[current_pos])

    path.append(current_pos)
    path = path[::-1]
    path = clean_path(path, log_affinity, L-1)
    return prob_table, path
```

Listing 1.3. Dynammic Programming Postprocessing

References

1. Balhar, J., Hajic, J., Jr.: Melody extraction using a harmonic harmonic convolutional neural network. Dimensions **10**(16x10x540), 16x10x360
2. Bittner, R.M., Bosch, J.J., Rubinstein, D., Meseguer-Brocal, G., Ewert, S.: A lightweight instrument-agnostic model for polyphonic note transcription and mul-

tipitch estimation. In: Proceedings of the IEEE International Conference on Acoustics, Speech, and Signal Processing (ICASSP), Singapore (2022)
3. Bittner, R.M., Bosch, J.J., Rubinstein, D., Meseguer-Brocal, G., Ewert, S.: A lightweight instrument-agnostic model for polyphonic note transcription and multipitch estimation. In: ICASSP 2022-2022 IEEE International Conference on Acoustics, Speech and Signal Processing (ICASSP), pp. 781–785. IEEE (2022)
4. Böck, S., Krebs, F., Schedl, M.: Evaluating the online capabilities of onset detection methods. In: ISMIR, pp. 49–54 (2012)
5. Donahue, C., Thickstun, J., Liang, P.: Melody transcription via generative pretraining. In: ISMIR (2022)
6. Google: Song stuck in your head? just hum to search. https://blog.google/products/search/hum-to-search/. Accessed 12 Dec 2023
7. Hsu, J.Y., Su, L.: VOCANO: a note transcription framework for singing voice in polyphonic music. In: Proceedings of the International Society of Music Information Retrieval Conference (ISMIR) (2021)
8. Liu, S., Li, X., Li, D., Shan, Y.: HumTrans: a novel open-source dataset for humming melody transcription and beyond. arXiv preprint arXiv:2309.09623 (2023)
9. Raffel, C., et al.: Mir_eval: a transparent implementation of common MIR metrics. In: ISMIR, vol. 10, p. 2014 (2014)
10. Kum, S., Lee, J., Kim, K.L., Kim, T., Nam, J.: Pseudo-label transfer from frame-level to note-level in a teacher-student framework for singing transcription from polyphonic music. In: Proceedings of the IEEE International Conference on Acoustics, Speech and Signal Processing (ICASSP) (2022)
11. Shansongliu: Humtrans. https://github.com/shansongliu/HumTrans. Accessed 12 Dec 2023
12. Wang, J.Y., Jang, J.S.R.: On the preparation and validation of a large-scale dataset of singing transcription. In: ICASSP 2021-2021 IEEE International Conference on Acoustics, Speech and Signal Processing (ICASSP), pp. 276–280. IEEE (2021)

Applications in Environmental and Biological Sciences

Leveraging LSTM Embeddings for River Water Temperature Modeling

Benjamin Fankhauser[1(✉)], Vidushi Bigler[2], and Kaspar Riesen[1]

[1] Institute of Computer Science, University of Bern, Bern, Switzerland
{benjamin.fankhauser,kaspar.riesen}@unibe.ch
[2] Institute for Optimisation and Data Analysis, Bern University of Applied Sciences, Biel, Switzerland
vidushi.bigler@bfh.ch

Abstract. River water temperature modeling is a major task in climate research. State-of-the-art methods for water temperature modeling deploy a transductive design, which makes it difficult to generalize to unseen water stations during test time. In the present paper, we isolate one of the common building blocks – a central LSTM, trained for each water station – and propose an embedding scheme in order to increase both the prediction accuracy and the amount of shared parameters and thus the generalization. The proposed embeddings are learned during training time. In an empirical evaluation we show that our method is able to reduce the RMSE by about five percentage points compared to the state-of-the-art reference method while decreasing the tuneable parameters by several orders of magnitude. We also provide a sample analysis of the embedding space of the catchment area of one specific river. Looking at the results of this qualitative analysis, we come to the conclusion that deploying an embedding in water temperature models is not only convincing to decrease the RMSE of water temperature predictions, but also enables better explainable deep learning models. Moreover, the proposed embedding technique opens up various unexplored applications in water temperature research.

Keywords: Water Temperature · Embedding · LSTM · Recurrent Neural Network

1 Introduction

Rivers and their tributaries can be found everywhere in habitable areas. River water temperature has a large impact on various biological processes in our ecosystems [1,2], and thus on flora and fauna, agriculture, or our drinking water. Therefore, water temperature in rivers, and in particular its control and prediction, plays an important role in future climate change.

In general, water temperature is influenced by solar radiation heating the river bed, atmospheric exchange at the surface, and friction. Other important factors are snow melt, rain, ground water inflow or waste water inflows of large

C. Y. Suen et al. (Eds.): ANNPR 2024, LNAI 15154, pp. 283–294, 2024.
https://doi.org/10.1007/978-3-031-71602-7_24

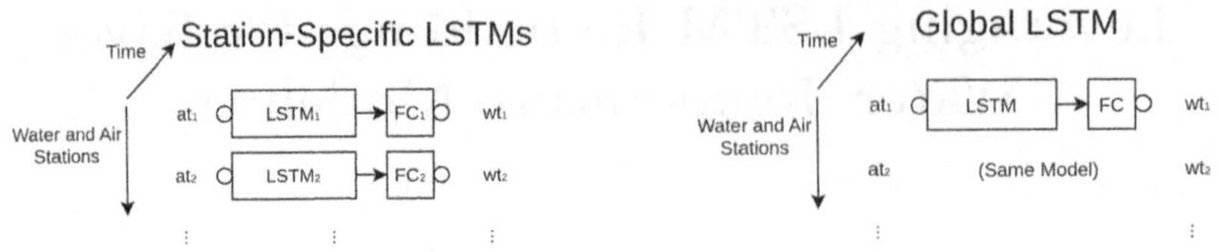

(a) (b)

Fig. 1. (a) Transductive station-specific LSTMs, where every water station has its own tuned neural network, which implies that the water station has to be present during training. (b) Inductive global LSTM where all network weights are shared and generalization to unseen water stations is possible. (Note that at_k and wt_k denote the air and water temperature time series from the k-th water station and its corresponding weather station, respectively.)

cities. Air temperature and – if available – also discharge are usually used as exogenous variables to model water temperature [3].

In order to monitor water temperature increase due to climate change, the *Federal Office for the Environment of Switzerland* (FOEN) is running a water temperature monitoring for more than 40 years (with a doubling of the number of water stations since 2010). The water stations are equipped with a water temperature sensor and they monitor the discharge as well. The *Swiss Meteorological Institute* (MeteoSwiss) provides air temperature and other meteorological measurements from weather stations in the vicinity of such water stations. Recently, the data stemming from both water and weather stations was augmented with data from a *geographic information system* (GIS) in order to model the river segments between such stations as a graph and published under the name *Swiss River Network* [4].

For modeling water temperature based on air temperature, various methods have been proposed in the literature. Physically inspired methods derive models with few parameters, which are in turn tuned using statistical methods [5]. Other recent methods are based on deep learning such as, for instance, the *long short-term memory* (LSTM). The LSTM is a *recurrent neural network* (RNN) which has been successfully applied to many time series problems [6] (including the task of water temperature modeling [7–9]). More sophisticated neural network architectures also take into account neighboring water stations [10–12], modeling the river network as a graph, with nodes representing the water stations and edges representing the river sections.

The above mentioned deep learning methods perform well and provide accurate predictions of the water temperature for practical purposes[1]. However, most of these models are constructed in a way, where water temperature is only considered at single points in the river. In graph learning terminology, they are transductive learning methods, where all nodes have to be available at training

[1] In practice, an RMSE under one degree Celsius is considered a "good model".

time, providing no generalization to unseen nodes at test time [13] (see left side of Fig. 1). This limitation is problematic, as local characteristics of a water station can indeed be learned, but the model does not necessarily generalize to other river segments (with different topological or meteorological characteristics).

As a straightforward way to implement a deep learning method in an inductive way, one could apply one single LSTM to all water stations (see right side of Fig. 1). This method is designed to learn a model which can estimate the water temperature in the entire Swiss River Network, based on the air temperature in the water station's vicinity. However, in preliminary experiments we observe that this global model can not reach the same accuracy as the transductive methods, which train one station-specific LSTM for every single water station. In the present work we bridge the performance gap between the transductive and inductive methods by proposing an embedding based method for water temperature prediction. For the proposed method, we use a single global LSTM for the whole Swiss River Network and move the transductive property into a low dimensional embedding. Due to the small size of the embedding space, the global network is required to generalize to different river sections.

The remainder of this paper is structured the following way. In Sect. 2, we describe the problem of water temperature modeling more formally and describe the proposed method in detail. In Sect. 3, we thoroughly assess the proposed method with an empirical evaluation. As an additional outcome, we discuss the emerged embedding space as well. Finally, we draw conclusions and propose future research activities in Sect. 4.

2 Concatenation Embeddings in an LSTM

2.1 Problem Definition

The problem of water temperature modeling based on air temperature is defined as follows. For a given time series of an air temperature $a_k^{(1)}, ..., a_k^{(T)}$, the task is to find a model f_k which predicts the water temperature $\hat{y}_k^{(t)}$ at time step t (with $1 \leq t \leq T$) at the water station k. Formally, we seek a function f_k

$$f_k(a_k^{(1)}, ..., a_k^{(t)}) = \hat{y}_k^{(t)}, \quad \forall t : 1 \leq t \leq T \tag{1}$$

which can predict the actual water temperature $y_k^{(t)}$ of the k-th station at time t as accurately as possible.

2.2 Concatenation Based LSTM Embedding

The proposed method deploys a station-specific n-dimensional embedding $e_k \in \mathbb{R}^n$ in an LSTM. The embedding $e_k \in \mathbb{R}^n$ denotes the embedding for the k-th water station and can be represented as a point in the embedding space $\mathbb{R}^n$. The embeddings for each station are learned using a gradient descent technique during the training phase and represent the only transductive part of our model. In Fig. 2 the proposed architecture is illustrated.

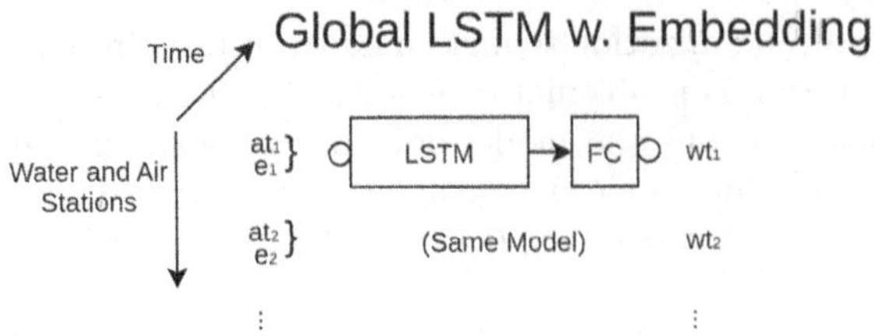

Fig. 2. The proposed LSTM with concatenation based embeddings. The LSTM is shared among all water stations but can encode station-specific characteristics in the embedding e_k.

In order to deploy the embedding to an LSTM, we concatenate the embedding e_k to the input $a_k^{(t)}$ at any time step t. Formally, the input $x^{(t)}$ to the LSTM is defined by

$$x^{(t)} := a_k^{(t)} || e_k, \tag{2}$$

where $||$ denotes concatenation.

At first glance, this behavior might appear cumbersome, as e_k is the same static value at any time step. Yet, it turns out that this method allows the embedding to influence each gate of the LSTM. By factoring out the multiplications with the embedding, this method corresponds to the following LSTM instance:

$$\begin{aligned}
\text{concat_lstm}(a_k^{(t)}, e_k) := LSTM(a_k^{(t)}||e_k) :& \\
f^{(t)} &= \sigma(\boldsymbol{W}_f a_k^{(t)} + \boldsymbol{V}_f e_k + \boldsymbol{U}_f h^{(t-1)} + \boldsymbol{b}_f) \\
i^{(t)} &= \sigma(\boldsymbol{W}_i a_k^{(t)} + \boldsymbol{V}_i e_k + \boldsymbol{U}_i h^{(t-1)} + \boldsymbol{b}_i) \\
o^{(t)} &= \sigma(\boldsymbol{W}_o a_k^{(t)} + \boldsymbol{V}_o e_k + \boldsymbol{U}_o h^{(t-1)} + \boldsymbol{b}_o) \\
\tilde{c}^{(t)} &= \theta(\boldsymbol{W}_c a_k^{(t)} + \boldsymbol{V}_c e_k + \boldsymbol{U}_c h^{(t-1)} + \boldsymbol{b}_c) \\
c^{(t)} &= f^{(t)} \odot c^{(t-1)} + i^{(t)} \odot \tilde{c}^{(t)} \\
h^{(t)} &= o^{(t)} \odot \theta(c^{(t)}) \\
\hat{y}^{(t)} &= MLP(h^{(t)})
\end{aligned} \tag{3}$$

where $\sigma(z)$ denotes the sigmoid activation function and $\theta(z)$ the tanh function. The variables h and c are propagated through time. Bold symbols denote the learnable network weights. The concatenation embedding introduces four matrix multiplications with $\boldsymbol{V}_f$, $\boldsymbol{V}_i$, $\boldsymbol{V}_o$, $\boldsymbol{V}_c$. All matrices are globally trained weights and only e_k depends on the water station k. This static product is then added to the corresponding gate and can (theoretically) influence each gate in each hidden dimension.

The last line of Expression 3 is the final step to project the hidden state to a one dimensional water temperature prediction $\hat{y}^{(t)}$ at time step t. In our model we use a multi layer perceptron (MLP) to accomplish this final prediction.

3 Experimental Evaluation

3.1 Experimental Setup

For our experiment we use the Swiss River Network dataset G_{2010} [4]. As there are fewer weather stations than water stations, we select one water station per weather station (to mitigate an unfair disadvantage in the global reference system, where the same air temperature time series would be mapped to different water temperatures). In total the resulting dataset consists of 42 water stations and corresponding weather stations acquired during the years 2010 to 2021. The temporal resolution are daily averages.

Data from years 2010 to 2018 is used as training set, with a 90/10% validation split. The validation data is only used for model selection. Data from the two years 2019 and 2020 is used to define the hold-out test set and we only report numbers on these two years.

To measure and compare our predictions, we compute the widely used *Root Mean Squared Error* (RMSE), and we also report the *Mean Average Error* (MAE) and the *Nash-Sutcliffe model Efficiency Coefficient* (NSE) as defined in Table 1. While values closer to 0 for the two errors indicate good prediction quality, values of the NSE closer to 1 indicate a better model.

Table 1. The metrics used for evaluation. $y^{(t)}$ is the ground truth value and $\hat{y}^{(t)}$ our prediction.

RMSE	MAE	NSE
$\sqrt{\frac{1}{T}\sum_{t=1}^{T}(y^{(t)} - \hat{y}^{(t)})^2}$	$\frac{1}{T}\sum_{t=1}^{T}\lvert y^{(t)} - \hat{y}^{(t)}\rvert$	$1 - \frac{\sum_{t=1}^{T}(y^{(t)} - \hat{y}^{(t)})^2}{\sum_{t=1}^{T}(y^{(t)} - \bar{y})^2}$

3.2 Training and Hyperparameter Tuning

In order to find the best performing embedding, we compute the gradients of the embedding e_k with respect to the training loss and simultaneously adjust the embedding to the weights of the LSTM during the training phase. We use a grid search on the hidden dimension of the LSTM, the amount of stacked LSTMs, the learning rate for the Adam optimizer [14], as well as the amount of an L2 regularization. We then freeze the network parameters and the embedding and select the model with the lowest validation loss. Note that we fix the dimension of the embedding space to $n = 5$ for our evaluations (in Sect. 3.6 we analyze the embedding space in more detail).

3.3 Reference Systems

Current state-of-the-art methods build on top of a central LSTM and add neighboring water stations to their model [4,10]. Our goal is to improve this central

LSTM. Therefore, we compare our novel model against the two following LSTM based reference methods without the use of neighboring water stations[2] (outlined in Fig. 1).

1. **Station-Specific LSTMs:** The station-specific LSTM reference system deploys a full LSTM_k for each available water station k (as introduced in [8]). We deploy a separate grid search for each trained LSTM_k.
2. **Global LSTM:** The global LSTM reference system is one single LSTM, as described in Sect. 2.2 without the introduced embedding. The same parameters are used and tuned for all available water stations simultaneously and thus we refer to it as global LSTM.

3.4 Growth of Tuneable Parameters

The global LSTM is a *Vanilla LSTM* and has

$$P_g = 4d(hi + hh + h)$$

tuneable parameters, where i refers to the input size, d is the number of stacked LSTMs and h refers to the size of the hidden dimension for the four gates. Using K station-specific LSTMs leads to

$$P_s = K * 4d(hi + hh + h) \tag{4}$$

tuneable parameters.

The proposed method introduces additional parameters to the *Vanilla LSTM*, and uses tuneable parameters to store the embeddings. Therefore our method leads to an LSTM with

$$P_{our} = 4d(hi + hn + hh + h) + Kn \tag{5}$$

tuneable parameters.

We analyze the growth of the tuneable parameters by computing the partial derivative with respect to the variable in question. With respect to the embedding size n, the following increase in tuneable parameters of our model results:

$$\frac{P_{our}}{\partial n} = 4dh + K.$$

This means that one additional dimension in the embedding space adds an additional column to the four weight matrices in the LSTM gates (d-times) plus one tuneable parameter for each of the K water stations. As our embedding space n is usually small, this growth is almost negligible.

[2] In future work, more sophisticated methods can then replace their central LSTM with our proposed method.

Table 2. The average metrics reported on the hold-out test set over 42 water stations.

	Method	Metric		
		RMSE	MAE	NSE
Reference	Station-Specific LSTMs	0.80	0.62	0.95
	Global LSTM	1.54	1.29	0.79
Ours	Global LSTM w. Embedding	**0.76**	**0.59**	**0.96**

The number of water stations K is given by the application and is not subject to our control. The partial derivatives with respect to K of our method P_{our} and the station-specific reference system P_s are

$$\frac{P_{our}}{\partial K} = n \quad \text{and} \tag{6}$$

$$\frac{P_s}{\partial K} = 4dh(i + h + 1), \quad \text{respectively.} \tag{7}$$

Hence, we conclude that when the embedding size n is small, our method provides a parameter growth (Eq. 6) that is orders of magnitude smaller compared to the station-specific parameter growth (Eq. 7).

3.5 Empirical Results

The test results are obtained after an exhaustive grid search and controlled model convergence. We report the metrics on the hold-out test set, after the validation set was used for model selection. Table 2 shows the averages of the reported metrics (RMSE, MAE, and NSE) measured over all 42 test stations.

We observe that the station-specific LSTMs – as expected and already known [4,8] – provide accurate predictions with an RMSE well below 1.0. At the same time, it is clear that the naïve approach of training a global LSTM for all stations does not work sufficiently well and provides values that cannot be used in practice. The proposed method not only closes the performance gap between the station-specific and the global reference systems, but decreases the RMSE even further. That is, compared to the station-specific reference system our model improves the RMSE by about 5% points (improvements on the previous state of the art are observed in the other two metrics as well).

Figure 3 shows the distribution of the RMSEs of all 42 stations stemming from the hold-out test set as box plots. This illustration shows quite nicely that there are large differences in the RMSE among the stations (regardless of which model is used). For example, we observe that in all three systems there are stations that have an RMSE above 1. However, for the station-specific and our novel model using an embedding, the RMSE values for the clear majority of stations are below this 1° threshold (while the global LSTM dramatically loses performance compared to the station-specific methods).

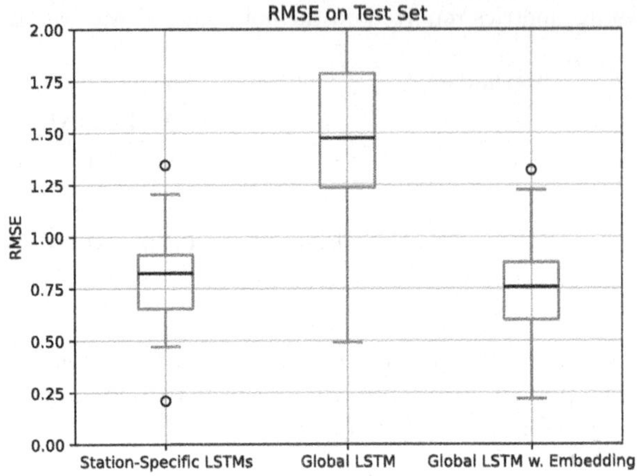

Fig. 3. RMSE on the test set, illustrated with boxplots of the same 42 water stations. Some stations are more difficult to model than others.

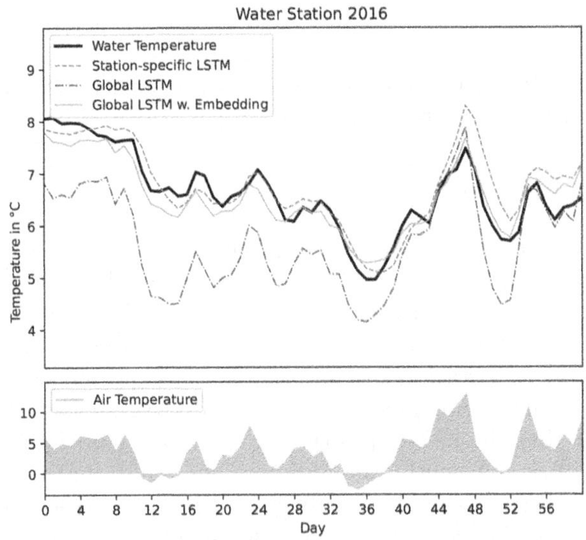

Fig. 4. Sample result at one specific water station of the Swiss River Network over a randomly selected test set period of 60 days. The three thin lines of the plot refer to the predicted water temperatures of the three methods. The thick black line refers to the ground truth water temperature. The lower graph, with the same x-axis, shows the air temperature over the same time period.

Table 3. The number of tuneable parameters of the selected models in the empirical evaluation to serve 42 water stations.

	Method	Parameters
Reference	Station-Specific LSTMs	723,394
	Global LSTM	1,513
Ours	Global LSTM w. Embedding	**1,307**

The lack of performance of the global LSTM is also visible in the illustration of Fig. 4. In this qualitative analysis we randomly select a test period of 60 days at one specific water station. We choose a station in the flatlands, where the river already accumulated to a bigger stream, and thus, the water temperature is less sensitive to small changes in the air temperature. In Fig. 4 it is clearly visible that the station-specific and embedding based methods can adapt to this behavior. The global LSTM, on the other hand, tends to exaggerate influences of air temperature. For example, around days 12, 36, and 52, the global LSTM exaggerates the effect of a drop in air temperature. Moreover, it is also visible that the global LSTM is generally further away from the ground truth water temperature than the other two models.

In Sect. 3.4, we discuss that our method has a more beneficial parameter growth in comparison to the station-specific reference system when deploying it to multiple water stations. In Table 3, we sum up the amount of tuneable parameters of the best performing models. While station-specific models require a total of more than 700K parameters to be optimized, we only observe about 1,300 parameters in our novel approach (including the embedding).

3.6 Analysis of the Embedding Space

One follow up question to our method is the size of the embedding space n. In an additional experiment, we re-run our grid search and evaluate various embedding sizes ($n \in \{1, 2, 3, 5, 10, \ldots, 30\}$). Figure 5 shows the RMSE on the y-axis with respect to the embedding size n (on the x-axis). The RMSE plateaus at an embedding dimension of $n = 10$, but even a very low dimensional embedding space with $n = 1$ or $n = 2$ is beneficial compared to the station-specific baseline model.

We also deploy a two dimensional embedding space and train a new model on all water stations of the Rhine and its tributaries in Switzerland (the Rhine is selected for this analysis as it has the largest catchment area). This allows us to visually analyze spatial properties of all stations in the complete catchment area in a two-dimensional space (see Fig. 6).

The embedding space is unrestricted and the optimization scheme can place embeddings arbitrary. However, four natural clusters (A, B, C, and D) are visually identifiable in the embedding in Fig. 6:

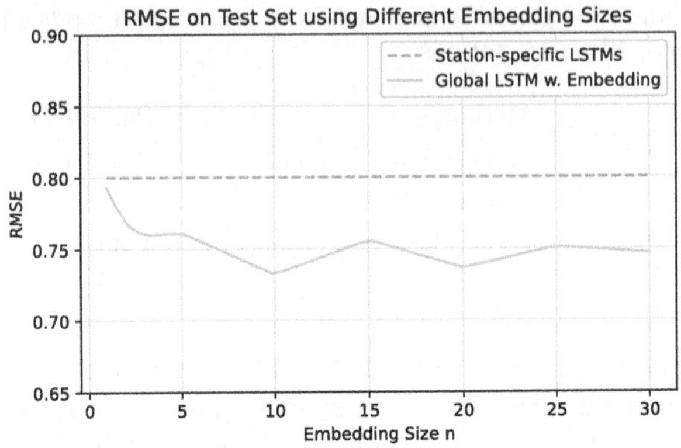

Fig. 5. RMSE with respect to the embedding size n compared to the station-specific LSTMs reference system.

- Cluster A: Stations of this cluster refer to the outflow of the alpine lakes Walensee (2104), Brienzersee (2457) and Thunersee (2030).
- Cluster B: This cluster contains mostly big rivers in the flatlands, like the Rhine in Basel (2091), or the Aare in Brugg (2016).
- Cluster C: This cluster is composed of stations near the outflow of the flatland lakes Bielersee (2029) or Bodensee (2288).
- Cluster D: This cluster is a collection of various tributaries in hilly, pre-alpine landscapes relatively far away from each other (like the Gürbe at Belp (2159), Murg at Wängi (2126), or Linth at Mollis (2372)).

Further away from the origin, we observe various alpine water stations (mainly leaf nodes), that do not belong to a distinct cluster.

4 Conclusion and Future Work

Many ecosystems depend on the well-being of rivers. Hence, monitoring river water temperature plays an important role in research of future climate change. After revisiting state-of-the-art methods for water temperature prediction, we observe that a vast majority of these methods (e.g. [4,7,8]) have a transductive design, meaning that all water stations have to be available during training time. These designs do not share parameters and the method has to learn basic water temperature characteristics at every station from scratch (leading to large amounts of tuneable parameters). We propose to address this issue with an embedding scheme. This embedding is designed so that it can be seamlessly integrated into more sophisticated state-of-the-art methods.

In an empirical evaluation, we demonstrate that our method outperforms the station-specific LSTMs while using two orders of magnitude less tuneable parameters. Moreover, by changing the dimensions of the embedding space, our method allows for flexible tuning between shared parameters and station-specific

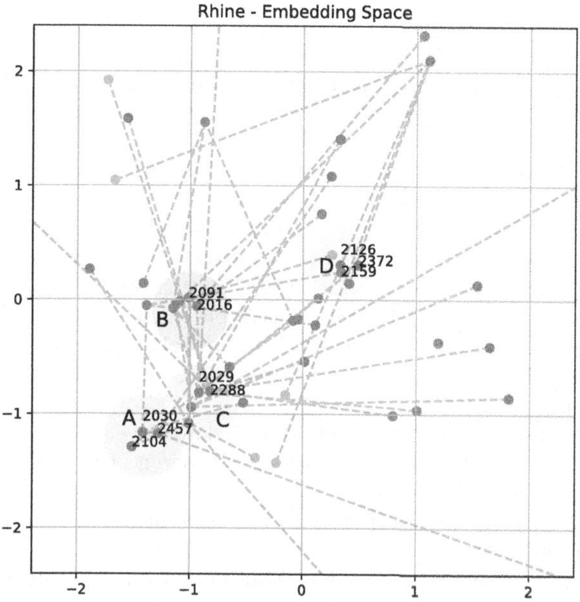

Fig. 6. Embedding space visualization of the Rhine catchment area. Dotted edges represent neighboring stations with respect to the water flow [4]. Thus, leaf nodes denote the first water station after the source of the stream. Blue dots denote water stations in the flatland while red dots are water stations from alpine regions. Four clusters (A, B, C, and D) emerge from the data. (Color figure online)

characteristics. Further, we show that the method is not sensitive to the size of the embedding space, and various configurations lead to good results. Last but not least, we are also able to detect naturally emerging clusters in the embedding space.

We see several possible future research activities. For instance, we can embed even more data or enforce constraints in the embedding space. Another idea is to use embeddings for few shot learning. In this setting, only a few months of water temperature data needs to be available. This could be particularly interesting as we cannot wait until we have measured a decade's worth of water temperatures before developing new models to account for ongoing climate change.

Acknowledgements. This project is supported by the Swiss National Science Foundation (SNSF) Grant Nr. PT00P2 206252. Data are kindly provided by the Federal Office for the Environment and MeteoSwiss. Calculations were performed on UBELIX (https://www.id.unibe.ch/hpc), the HPC cluster at the University of Bern.

References

1. Lessard, J.L., Hayes, D.B.: Effects of elevated water temperature on fish and macroinvertebrate communities below small dams. River Res. Appl. **19**(7), 721–732 (2003)
2. Caissie, D.: The thermal regime of rivers: a review. Freshw. Biol. **51**(8), 1389–1406 (2006)
3. Piccolroaz, S., Calamita, E., Majone, B., Gallice, A., Siviglia, A., Toffolon, M.: Prediction of river water temperature: a comparison between a new family of hybrid models and statistical approaches. Hydrol. Process. **30**(21), 3901–3917 (2016)
4. Fankhauser, B., Bigler, V., Riesen, K.: Graph-based deep learning on the swiss river network. In: Vento, M., Foggia, P., Conte, D., Carletti, V. (eds.) GbRPR 2023. LNCS, vol. 14121, pp. 172–181. Springer, Cham (2023). https://doi.org/10.1007/978-3-031-42795-4_16
5. Toffolon, M., Piccolroaz, S.: A hybrid model for river water temperature as a function of air temperature and discharge. Environ. Res. Lett. **10**(11), 114011 (2015)
6. Hochreiter, S., Schmidhuber, J.: Long short-term memory. Neural Comput. **9**(8), 1735–1780 (1997)
7. Piotrowski, A.P., Napiorkowski, M.J., Napiorkowski, J.J., Osuch, M.: Comparing various artificial neural network types for water temperature prediction in rivers. J. Hydrol. **529**, 302–315 (2015)
8. Qiu, R., et al.: River water temperature forecasting using a deep learning method. J. Hydrol. **595**, 126016 (2021)
9. Moshe, Z., Metzger, A., Elidan, G., Kratzert, F., Nevo, S., El-Yaniv, R.: Hydronets: leveraging river structure for hydrologic modeling. arXiv preprint arXiv:2007.00595 (2020)
10. Jia, X., et al.: Physics-guided recurrent graph model for predicting flow and temperature in river networks. In: Proceedings of the 2021 SIAM International Conference on Data Mining (SDM), pp. 612–620. SIAM (2021)
11. Chen, S., Zwart, J.A., Jia, X.: Physics-guided graph meta learning for predicting water temperature and streamflow in stream networks. In: Proceedings of the 28th ACM SIGKDD Conference on Knowledge Discovery and Data Mining, pp. 2752–2761 (2022)
12. Fankhauser, B., Bigler, V., Riesen, K.: Impute water temperature in the swiss river network using LSTMs. In: International Conference on Pattern Recognition Applications and Methods, pp. 732–738. Scitepress (2024)
13. Hamilton, W.L.: Graph Representation Learning, chap. 5.1.1. Morgan & Claypool Publishers (2020)
14. Kingma, D.P., Ba, J.: Adam: a method for stochastic optimization. arXiv preprint arXiv:1412.6980 (2014)

Research on the Identification of Common Economic Shellfish in Jiangsu Based on Fused-ResNet Network

Zhiyuan Liu[1], Beining Zhang[1], Yangping Wu[2,3], Zhidong Zhang[2,3], Junjie Chen[1], and Hongjun Li[1(✉)]

[1] School of Information Science and Technology, Nantong University, Nantong 226019, Jiangsu, China
lihongjun@ntu.edu.cn

[2] Marine Fisheries Research Institute of Jiangsu Province, Nantong 226007, China

[3] Marine Economic Shellfish Research and Development Center of Jiangsu Province, Nantong, China

Abstract. Shellfish hold significant economic value and are one of the major economic crops in coastal areas of China. To identify a rapid and effective method for distinguishing different Shellfish species in Jiangsu based on morphological differences, this paper proposes a recognition algorithm for common economic Shellfish in Jiangsu based on the Fused-ResNet network. Firstly, the basic structure of the ResNet network is introduced. Then, to address the issues of similarity in shape, color, and texture among Shellfish, a Channel Spatial Attention (CSA) module is incorporated. Additionally, to more effectively differentiate the diversity of Shellfish, a transfer learning training strategy is adopted. Finally, the optimized model is applied to Shellfish recognition tasks. The proposed recognition method demonstrates universality under different targets and achieves a recognition accuracy of 98.8%. Moreover, the algorithm's low computational cost makes it more suitable for deployment on mobile devices, providing effective technical support for the rapid identification of economic Shellfish in Jiangsu.

Keywords: Shellfish recognition · Deep learning · Fused-ResNet Network

1 Introduction

Scallops are an important cultivated shellfish, and currently, the sorting by quality is necessary in both production and sales processes [17]. At present, sorting work is primarily done manually, which presents issues such as harsh working conditions, high labor intensity, and low sorting efficiency [1]. Moreover, in the consumer market, instances of passing off inferior goods and mislabeling are common. For example, some unscrupulous vendors label various other shellfish as clams. These shellfish can vary significantly in market price, ranging from a few

C. Y. Suen et al. (Eds.): ANNPR 2024, LNAI 15154, pp. 295–307, 2024.
https://doi.org/10.1007/978-3-031-71602-7_25

yuan to over twenty yuan. Due to a lack of knowledge about shellfish, consumers often end up paying the higher price for cheaper shellfish. Some enterprises use drum sieve sorting equipment, but this often results in scallop mortality due to shell interlocking, causing significant losses. Utilizing robot-assisted sorting with computer vision offers advantages such as minimal loss and high efficiency, making it a feasible direction for achieving automated scallop sorting [2]. Machine vision technology can exclude debris interference, achieve precise grading and positioning of scallops, and lay the foundation for computer vision robots to sort scallops.

The study of shellfish identification methods has garnered attention from many scholars. However, in China, research on shellfish identification methods is still in its infancy. Due to the numerous species of shellfish, similar morphology, and limitations of sample data [18], the challenges faced by shellfish identification methods are significant. Currently, there are several approaches: Zhao Qing et al. [3] classify shellfish based on the traditional method of DNA barcode COI gene; Fu Rong [4] classifies freshwater shellfish using both pictorial and DNA barcodes; Li Yaru [5] identifies shellfish species through infrared spectral images; Li Fan et al. [6] proposed an image processing-based method, but it requires image feature collection before classification; Yang Mei et al. [7] used a simple neural network for preliminary shellfish classification, showing poor performance in classifying shellfish with similar morphological features and limited generalization ability, with high computational complexity. Zhao Qizheng [8] used a multi-scale perception DenseNet network for shellfish classification, but both the number of categories and the total number were small. Abroad, complex neural networks have gradually been applied to shellfish identification. Li et al. [9] proposed a study on shellfish detection using an end-to-end network; Feng et al. [10] proposed a shellfish identification method based on Gabor transform and principal component analysis; Feng et al. [11] proposed a shellfish detection algorithm based on Faster R-CNN. These studies demonstrate the application of machine vision-based shellfish identification on smooth shell surfaces and well-known shellfish. However, identifying shellfish with similar shapes or colors and processing large-scale shellfish image data face challenges such as high computational complexity and imbalanced sample distribution. In these cases, further algorithm and model improvements may be needed to address these challenges. For instance, incorporating more features such as texture and shape to differentiate similar shellfish, and using more complex machine learning algorithms like deep learning to handle imbalanced sample distribution. These improvements will help enhance the accuracy and robustness of shellfish identification systems.

This paper adopts the ResNet network framework for the recognition of shellfish images. This network not only effectively extracts key features from shellfish images, significantly enhancing recognition accuracy, but its lightweight design and efficient computing capability also enable rapid and precise shellfish identification in environments with limited computational resources. Therefore, not only does it achieve high accuracy in identification, but it also operates at a faster processing speed, making our work in shellfish identification more efficient.

2 Related Work

In the field of shellfish identification, scholars both domestically and internationally have conducted numerous research explorations. For example, Yang Jingyao et al. attempted to classify and recognize shellfish using extreme learning machines, but this algorithm cannot achieve rapid and large-scale detection. Guo Changyou et al. [2] applied image processing technology to estimate and locate the size of scallops based on the area occupied by scallops. However, this method is not suitable for the identification, grading, and positioning of scallops in complex situations with debris interference. Zhao Qizheng [8] used a DenseNet network for shellfish image classification, recognizing 9 categories of shellfish, demonstrating certain potential. In the international research field, complex neural networks have gradually been applied to the field of shellfish identification. Li et al. [9] developed a shellfish detection model integrating residual attention mechanisms; Feng et al. [10] proposed a shellfish identification method based on Gabor transform; similarly, Feng et al. [11] also introduced an improved shellfish detection algorithm based on Faster R-CNN. These studies demonstrate the potential of machine vision technology in identifying shellfish with smooth surfaces and limited types. However, for shellfish with similar shapes, colors, or uneven sample distributions, detection and recognition remain challenging.

In recent years, with the convolutional neural network (CNN) [12] models becoming increasingly deeper, there has been an increase in model complexity and considerable CPU and GPU hardware resource consumption. However, in certain real-world application scenarios such as mobile or embedded devices, such large and complex models are difficult to apply. Firstly, the models are too large, leading to memory shortages, and secondly, these scenarios require low latency. Therefore, researching small and efficient CNN models is crucial in these scenarios. To address these issues, researchers have proposed a series of methods to replace heavyweight networks with lightweight networks, such as the ResNet series networks, and the MobileNet series networks. These deep learning methods optimize the network model structure and enhance the autonomous learning ability of complex features. This study aims to improve the ResNet model and compare different deep learning models, combined with accuracy evaluation metrics, to select the most suitable identification model, providing technical reference for shellfish identification research.

3 Research on Shellfish Recognition Based on Improved ResNet Network

3.1 Overview

This chapter focuses on exploring and addressing two main challenges encountered in the process of single-object shellfish recognition. Firstly, there is a high degree of similarity in features among shellfish images, such as shape, color, and texture. These similarities often reduce the accuracy of traditional recognition methods. Secondly, the extreme imbalance in the number of samples for

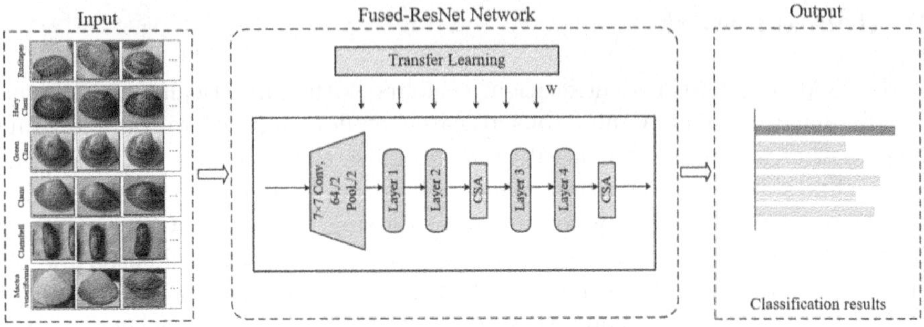

Fig. 1. Shellfish Recognition Technical Route

different shellfish species poses a challenge to recognition accuracy. To overcome these difficulties, this study employs an improved ResNet network to perform single-object shellfish recognition tasks. Through the application of deep learning technology, we aim to extract more discriminative shellfish features to enhance recognition accuracy. The technical roadmap presented in this chapter (see Fig. 1) provides a clear guiding framework for the research.

Figure 1 illustrates the global framework of our shellfish recognition approach. For various input shellfish images, the first step is preprocessing. In this step, the data structure is standardized, followed by a random flip operation with a 50% probability applied to the images. Regardless of whether the image undergoes flipping, normalization is performed to ensure that the data meets the requirements of model training. The normalized image data is then fed into the Fused-ResNet model for training. During the training process, the classifier identifies the shellfish based on the highest probability, ultimately producing the prediction results. This constitutes the overall process of our shellfish recognition.

3.2 ResNet34 Network

ResNet34 is a convolutional neural network structure with skip connections aimed at addressing the problem of increasing errors during training in deep neural networks. In traditional CNNs, as the network layers increase, training errors gradually rise. However, by introducing residual modules, ResNet can directly pass the output of shallow layers to deep layers, effectively avoiding the problem of vanishing gradients. The main highlights of this network model can be summarized as follows: 1) Ultra-deep network structure: ResNet34 breaks through the depth limitation of previous neural networks, with a network depth of 34 layers, which was quite deep for neural networks at that time. This depth of network structure enables better learning and extraction of complex image features, thereby improving the model's prediction accuracy. 2) Introduction of Residual modules: To address the issues of vanishing and exploding gradients in deep neural networks, ResNet34 introduces residual modules. The design of residual modules allows the model to train better by skipping some layers, overcoming the training difficulties of deep networks. 3) Use of Batch Normalization: ResNet34 utilizes Batch Normalization (BN) technique after each convolutional layer. BN can accelerate network training while improving the model's generalization ability. Overall, ResNet34 successfully addresses some of the issues in training deep neural networks through its ultra-deep network structure, introduction of residual modules, and use of Batch Normalization technique, thereby achieving efficient learning and extraction of image features. This makes ResNet34 perform excellently in tasks such as image recognition and classification.

3.3 Channel Spatial Attention Module

In conventional image classification, downsampling operations between convolutional layers are critical for integrating input semantic information. However, this process can inadvertently lead to the loss of detailed information, which is particularly evident in scenarios with high similarity. For instance, in bivalve classification, due to the minor differences between categories and significant intra-category variability, traditional multi-layer convolutional patterns struggle to capture subtle inter-category differences and common intra-category features. Therefore, in this section, we adopt an attention module to enhance the model's ability to capture detailed information from samples. Specifically, we propose the Channel Spatial Attention module based on the original residual network structure, leveraging the concept of the attention mechanism (Fig. 2).

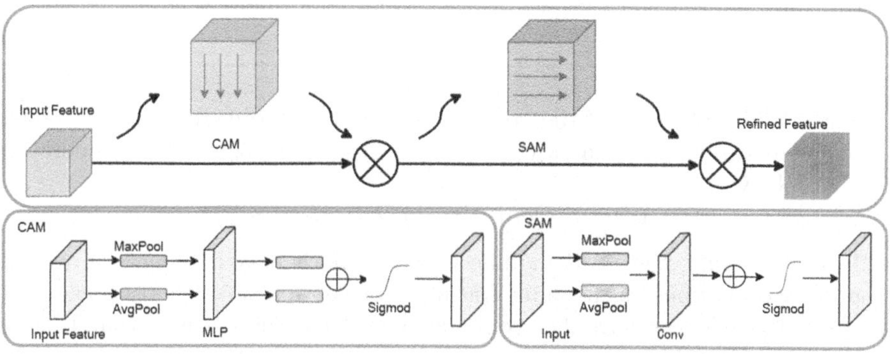

Fig. 2. Channel-Spatial Attention Module Diagram

CSA is an attention mechanism module designed to improve the network's feature representation capability by focusing on critical information in the input feature maps. CSA primarily consists of two sub-modules: the Channel Attention Module (CAM) and the Spatial Attention Module (SAM). These two are serially connected to reinforce the model's attention. According to the schematic in the figure, its working principle is explained as follows:

1) Channel Attention: The channel attention mechanism highlights or suppresses the importance of features in each channel. As shown in the figure, this module

first inputs the features F_{in} into the CAM channel weight generation module to generate corresponding weight coefficients W for each channel, and then uses dot product operations to perform weighted operations on each channel. For terminal features F_c, the channel attention operation can be formalized as follows:

$$F_c = W_c \cdot F_{in} = CAM(F_{in}) \cdot F_{in} \tag{1}$$

where $W_c = CAM(F_{in})$ is the weight vector corresponding to each channel, obtained through internal operations of the CAM module. Specifically, the internal operation can be described as:

$$W_c = \sigma[MLP(Max(F_{in}) + MLP(Avg(F_{in}))] \tag{2}$$

In this formula, the CAM module generates two weight coefficients for the input features F_{in} using a multi-layer perceptron (MLP), then sums them to obtain the final weight vector. The operations Max and Avg represent max pooling and average pooling, respectively, with the outer symbols σ indicating Sigmoid normalization.

2) Spatial Attention: The spatial attention mechanism highlights or suppresses key regions in the feature map, emphasizing important positional features. The latter part of the figure illustrates its working principle. Unlike the channel weighting mechanism, the SAM module performs weighting operations based on spatial positions in the feature map F_c. For terminal output features F_s, the spatial attention operation can be formalized as:

$$F_s = W_s \cdot F_c = SAM(F_c) \cdot F_s \tag{3}$$

where $W_s = SAM(F_c)$ is the weight matrix corresponding to all feature points in the feature space, generated by the SAM module. The internal operation is as follows:

$$W_s = \sigma(Conv^{7*7}([Avg(F_c), Max(F_c)])) \tag{4}$$

In this formula, the SAM module performs both max pooling and average pooling operations on the input features F_c to obtain two types of pooled features. These features are then combined using channel concatenation. Finally, a 7*7 convolution kernel is used to generate the weight matrix for the corresponding spatial positions. Overall, by integrating the CSA attention module into the ResNet34 network, the model can more accurately focus on bivalve information while ignoring background noise, which is crucial for capturing detailed features of bivalves.

3.4 Transfer Learning

In the practical application of machine learning, training a model from scratch for a specific task often presents various challenges, such as high training costs and limited sample data. These factors can significantly impede the effectiveness of model training, especially in specialized fields like shellfish recognition where sample data is scarce [16]. At present, shellfish recognition research also faces the challenge of lacking a standardized dataset with sufficient sample size and broad coverage, which further complicates the task of building effective models from scratch. Transfer learning, as an effective solution, offers a pathway to address the scarcity of samples and reduce training costs. The core idea of transfer learning is to leverage pre-trained models from related or different domains and apply them to new target tasks. Specifically, we utilize models pre-trained on large-scale visual databases such as ImageNet as a starting point. ImageNet, being a diverse image dataset containing millions of images across thousands of categories, pre-trained models have learned to recognize various image features. When applying such a pre-trained model to shellfish recognition tasks, we can retain most of the model's structure and parameters, only fine-tuning or retraining the last few layers to adapt to the specific requirements of shellfish recognition.

In this way, we can effectively utilize the training achievements obtained by the pre-trained model on ImageNet, using part of the model's structure as a powerful feature extractor to capture and learn the key features for the task of bivalve recognition. This method not only fully utilizes the rich knowledge and advanced features learned by the pre-trained model on a large-scale general dataset but also ensures through fine-tuning that the model can efficiently adapt to new and specific classification tasks, thereby achieving more accurate and effective recognition results in the field of bivalve recognition.

4 Experiment and Analysis

In the experiments, two NVIDIA GeForce RTX-1080ti GPUs are utilized, each equipped with 11GB RAM. The system has a memory size of 32GB. The experimental software environment includes CUDA version 11.6, PyTorch version 1.12.0, Python version 3.7 and the operating system is Windows.

4.1 Dataset

Image Preprocessing. This dataset covers six different types of bivalves, including Ruditapes philippinarum, Scapharca subcrenata, Cyclina sinensis, Meretrix meretrix, Sinonovacula constricta, and Mactra veneriformis, totaling 199 images. During the preprocessing stage, since each image contains multiple bivalve samples, we employed a dataset splitting method to ultimately obtain 701 images of different bivalves. Specifically, we have 74 images of Ruditapes philippinarum, 163 images of Scapharca subcrenata, 158 images of Cyclina sinensis, 206 images of Meretrix meretrix, 47 images of Sinonovacula constricta, and 53 images of Mactra veneriformis. To train and validate the model, we divided these images into training and validation sets in a 1:1 ratio. The following Table 1 shows the number of images before and after dataset segmentation.

Table 1. Distribution of Data After Segmentation.

Before Segmentation		After Segmentation	
Category Names	Total Number of Images	Category Names	Total Number of Images
Ruditapes	17	Ruditapes	74
Scapharca	43	Scapharca	163
Cyclina	48	Cyclina	158
Meretrix	44	Meretrix	206
Sinonovacula	22	Sinonovacula	47
Mactra	25	Mactra	53
Total	199	Total	701

As shown in Table 1, we meticulously segmented the bivalve images. Initially, our dataset contained 199 images of bivalves, with some images containing multiple bivalve samples [15]. To enable the model to better learn and understand the features of each type of bivalve, we segmented these images so that each resulting image contains only one bivalve sample. After this segmentation, the number of images increased to 701, which not only enriched our dataset but also helped improve the model's recognition accuracy and generalization capability.

Image Augmentation. Image enhancement techniques aim to improve the visual appearance of images for more accurate analysis and processing by computers [14]. One common method of image enhancement involves the following steps: first, the original image is cropped according to preset scaling factors; then, the cropped image is resized to the desired dimensions using interpolation methods; next, the input image is converted from the format of H × W × C to torch. Tensor format, with the shape of C × H × W. Additionally, we can accept a probability value p (default value is 0.5) as a parameter and randomly decide whether to horizontally flip the input image based on this probability value. Finally, the image can be normalized channel-wise, so that the mean of each channel becomes 0 and the standard deviation becomes 1. The integration of these operations constitutes an effective data augmentation method, which can accelerate the convergence speed of the model during deep learning model training [13], thereby improving the training effectiveness of the model.

4.2 Implementation Details

Training: During the training process, data augmentation techniques such as random cropping and random flipping were employed. The Adam optimizer was chosen for network optimization with an initial learning rate set to 0.0001. The cross-entropy loss function was used to compute the loss. In the transfer learning phase, to improve the model's convergence speed and achieve better training results, the pre-trained weights from ImageNet were loaded into the first half of the network, while the second half of the network was trained without loading any pre-trained weights. The network was trained for a total of 200 epochs with a batch size set to 16 images.

4.3 Experiments

In this section, we performed various predictions for each type of clam, and the accuracy rates of different clams recognized by the Fused-ResNet network are shown in Table 2. The accuracy rates for recognizing Cyclina and Sinonovacula both reached 100%, indicating the model's excellent learning capability for these two types of clams. The prediction accuracy for other categories also exceeded 90%. This result not only demonstrates the effectiveness of the model design but also shows the model's strong performance on this dataset.

We conducted a series of ablation experiments on the shellfish recognition dataset. For these experiments, we used the ResNet34 network as the baseline model.

As shown in Table 3, the baseline model achieved an accuracy of 95.7% in recognition precision. When the Channel Spatial Attention module was added to the baseline model, the recognition precision improved to 96.5%. Furthermore, incorporating the CSA module and using a transfer learning training strategy resulted in a recognition precision of 98.8%, representing a 3.1% improvement over the baseline model.

We also conducted experiments on popular networks to compare the effectiveness of different networks for the shellfish recognition task. The results are shown in Table 4.

Table 2. Fused-ResNet Recognition Accuracy for Each Class.

Class	Precision	Recall	Specificity
Ruditapes	0.974	1.0	0.997
Scapharca	0.988	1.0	0.996
Cyclina	1.0	0.987	1.0
Meretrix	0.98	0.971	0.992
Sinonovacula	1.0	0.957	1.0
Mactra	0.926	0.962	0.994

Table 3. Ablation Experiment

Method	CSA	Transfer Learning	Val_accuracy
Baseline	–	–	0.957
Fused-ResNetv1	✓	–	0.965
Fused-ResNetv1	✓	✓	0.988

Table 4. Comparison with Other Methods

Method	Val_accuracy
MobileNet_V3_small	0.530
MobileNet_V2	0.539
Vision_transformer	0.651
AlexNet	0.919
ResNet34	0.957
Ours	0.988

Table 4 presents the experimental results of different networks on the shellfish recognition task dataset. By analyzing the results, we can observe that the proposed method achieved a precision of 98.8%, which is a 3.1% improvement over the ResNet34 network. This result demonstrates the strong performance and accuracy of our proposed network model in the shellfish recognition task.

4.4 Data Visualization Results and Analysis

Single-Target Data Visualization Results. To demonstrate the generalization ability of the model, the study selected samples from six different shellfish species for recognition and prediction. The prediction results are shown below.

As illustrated in Fig. 3, the model used in this study demonstrates good recognition ability and achieved satisfactory accuracy in the experiments. This highlights the model's excellence in handling complex shellfish recognition tasks. Figures (a)-(g) show that the classification accuracy for six types of shellfish is exceptionally high. It was difficult to find a misclassified sample during the experiments. As shown in Figure (h), due to the similarity between the two types of shellfish, the model mistakenly classified the Mactra as the Meretrix.

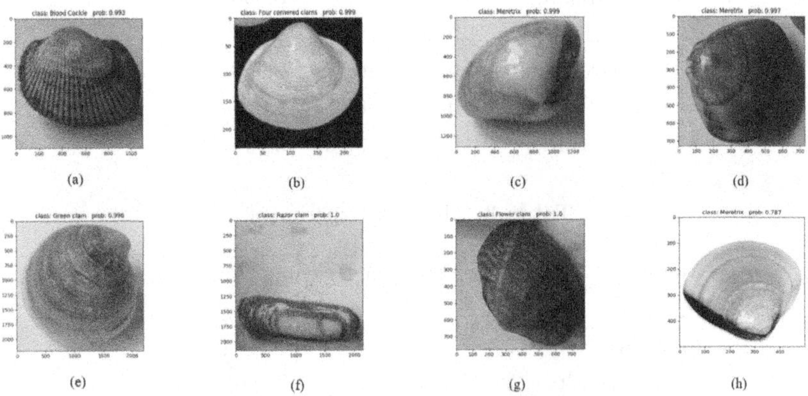

Fig. 3. Recognition Results on Multiple Images

Multi-target Data Visualization Results. For the recognition of multiple shellfish targets, the results are shown in Fig. 4.

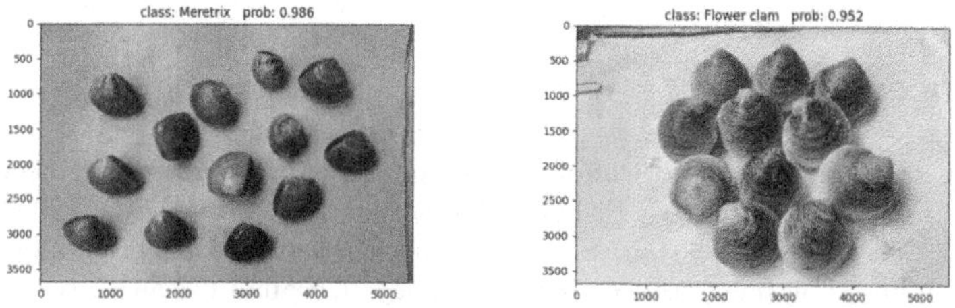

Fig. 4. Recognition Results of Multiple Shellfish by Fused-ResNet Network

To validate the generalization capability of the proposed network, this study utilized Fused-ResNet to recognize multiple shellfish targets. As shown in the figure above, in multi-target scenarios, the model constructed achieved an accuracy of 98.6% for the Meretrix shellfish type, demonstrating its superior recognition performance. Although the accuracy for the Ruditapes type slightly decreased to 95.2%, this result still indicates the model's good recognition effect. This outcome fully proves the robustness and effectiveness of the Fused-ResNet network in multi-target shellfish recognition tasks.

5 Conclusion

In the realm of shellfish image data processing, deep learning technology has become a crucial method for effective recognition and feature extraction. This paper focuses on enhancing the recognition efficiency of scallop sorting and promoting the application of computer vision technology in practical scallop sorting production. Through in-depth research on the ResNet network, we proposed a shellfish recognition algorithm named Fused-ResNet, specifically for common economic shellfish in Jiangsu. Firstly, we introduced the Channel Spatial Attention module into the network to address the difficulties in recognizing shellfish due to similarities in shape, color, and texture. Additionally, we employed transfer learning training strategies and applied the improved model to both single-target and multi-target shellfish recognition tasks. The improved ResNet network demonstrated robust performance in recognizing complex scenes with multiple objects and was able to discern subtle differences among them. In the future, we plan to expand the sample size and diversity, optimize strategies for handling multi-target recognition scenarios, and carefully consider environmental variables such as background and lighting conditions that affect recognition accuracy. We believe that as these research efforts continue to advance, the application of computer vision technology in the field of shellfish sorting will become more mature and widespread.

Acknowledgments. This work is supported in part by Nantong Science and Technology Program JC2023076 and in part by Postgraduate Research and Practice Innovation Program of Jiangsu Province KYCX24_3643 and in part by The "JBGS" Project of Seed Industry Revitalization in Jiangsu Province JBGS (2021) 142.

References

1. Xu, H., Zhang, J.H.: Report on the development of aquaculture engineering discipline in China (2007–2008). Fisheries Modernization **2009**(3), 1–6 (2009)
2. Guo, C.Y., Cao, G.B., Han, S.C.: Automatic identification of scallop size and position based on image processing technology. J. Dalian Ocean Univ. **27**(6), 578–582 (2016)
3. Zhao, Q., Wu, B., Liu, Z.H.: Analysis of the phylogenetic relationships of important shellfish in China using DNA barcoding COI gene. J. Fishery Sci. China **25**(4), 847–857 (2018)
4. Fu, R.: Study on DNA Barcodes and Atlas of Common Freshwater Mollusk Species in Poyang Lake and Parts of the Yangtze River Delta. Nanjing Agricultural University (2019)
5. Li, Y.R.: Rapid Identification and Classification Method of Major Allergens Tropomyosin in Crustaceans and Mollusks Based on Infrared Spectroscopy. Shanghai Ocean University (2019)
6. Li, F., Han, S.J., Wang, Y.Y.: Application of support vector machine in mollusk disease diagnosis. Comput. Simul. **30**(3), 319–322 (2013)
7. Yang, M., Wei, H.L., Hua, S.G.: A scallop image recognition method based on neural networks. J. Dalian Ocean Univ. **29**(1), 70–74 (2016)
8. Zhao, Q.Z.: Clam image classification based on multi-scale perception DenseNet. Ind. Control Comput. **33**(06), 64–65 (2020)
9. Li, G., Li, Z., Zhang, C., Li, Y., Yue, J.: Shellfish detection based on fusion attention mechanism in end-to-end network. In: Lin, Z., et al. (eds.) PRCV 2019. LNCS, vol. 11859, pp. 516–527. Springer, Cham (2019). https://doi.org/10.1007/978-3-030-31726-3_44
10. Yi-Ran, F., Xue-Heng, T., Eung-Joo, L.: Shellfish recognition based on Gabor transformation and principal component analysis. In: 2020 8th International Conference on Orange Technology (ICOT), Daegu, Korea (South), pp. 1–3 (2020)
11. Feng, Y., Tao, X., Lee, E.J.: Classification of shellfish recognition based on improved faster R-CNN framework of deep learning. Math. Probl. Eng. **2021**, 1–10 (2021)
12. Hao, H.M., Liang, Y.G., Wu, H.B.: Symmetric point pattern-deep convolutional neural network infrared spectroscopy recognition method. Spectrosc. Spectral Anal. **41**(3), 782–788 (2021)
13. Graves, A., Mohamed, A., Hinton, G.: Speech recognition with deep recurrent neural networks. In: 2013 IEEE International Conference on Acoustics, Speech and Signal Processing, pp. 6645–6649. IEEE (2013)
14. Muñoz-Benavent, P., Andreu-García, G., Valiente-González, J.M.: Enhanced fish bending model for automatic tuna sizing using computer vision. Comput. Electron. Agric. **150**, 52–61 (2018)
15. Yue, J., Yang, H., Jia, S.: A multi-scale features-based method to detect Oplegnathus. Inf. Process. Agric. **8**(3), 437–445 (2021)

16. Jose, J.A., Kumar, C.S., Sureshkumar, S.: Tuna classification using super learner ensemble of region-based CNN-grouped 2D-LBP models. Inf. Process. Agric. **9**(1), 68–79 (2022)
17. Li, H.J., Tao, X.H., Yu, X.Q.: Application of computer vision technology on quality evaluation of seafood. J. Food Mach. **28**(4), 154–156 (2012)
18. Li, B., Yue, J., Jia, S., et al.: Recognition of abnormal body surface characteristics of oplegnathus punctatus. Inf. Process. Agric. **9**(4), 575–585 (2022)

Generative Plant Growth Simulation from Sequence-Informed Environmental Conditions

Mohamed Debbagh[1(✉)], Yixue Liu[2], Zhouzhou Zheng[2], Xintong Jiang[1], Shangpeng Sun[1], and Mark Lefsrud[1]

[1] McGill University, Montréal, QC H3A 0G4, Canada
{mohamed.debbagh,xintong.jiang}@mail.mcgill.ca,
{shangpeng.sun,mark.lefsrud}@mcgill.ca

[2] College of Mechanical and Electronic Engineering, Northwest A&F University, Yangling 712100, Shaanxi, China
{sunnyliu,zhengzz}@nwafu.edu.cn

Abstract. A plant growth simulation can be characterized as a reconstructed visual representation of a plant or plant system. The phenotypic characteristics and plant structures are controlled by the scene environment and other contextual attributes. Considering the temporal dependencies and compounding effects of various factors on growth trajectories, we formulate a probabilistic approach to the simulation task by solving a frame synthesis and pattern recognition problem. We introduce a sequence-informed plant growth simulation framework (SI-PGS) that employs a conditional generative model to implicitly learn a distribution of possible plant representations within a dynamic scene from a fusion of low-dimensional temporal sensor and context data. Methods such as controlled latent sampling and recurrent output connections are used to improve coherence in the plant structures between frames of prediction. In this work, we demonstrate that SI-PGS is able to capture temporal dependencies and continuously generate realistic frames of plant growth.

Keywords: Plant Growth Simulation · Continuous Frame Synthesis · Conditional Generative Model

1 Introduction

The phenotypic expression of a plant is significantly influenced by the dynamic growing conditions of its environment. In controlled environment agriculture (CEA), the goal is to control these conditions to produce desirable plant traits. As such, computer vision (CV) and sensor fusion-based phenomic models are commonly employed in CEA to assess crop quality and quantify biophysical properties throughout various stages of its growth cycle [7,13,19]. More recently

Supplementary Information The online version contains supplementary material available at https://doi.org/10.1007/978-3-031-71602-7_26.

C. Y. Suen et al. (Eds.): ANNPR 2024, LNAI 15154, pp. 308–319, 2024.
https://doi.org/10.1007/978-3-031-71602-7_26

there has been an effort towards producing plant representation models that simulate plant structural development and predict biophysical attributes [1,6, 15]. These simulations are often realized as parameterized functional-structural plant models (FSPM) that implement strict rules on physiological processes in the form of deterministic computational algorithms [18]. While FSPMs employ mechanisms that mimic plant processes and structural development patterns, they do not capture the true stochastic nature of plant growth and rely on fixed rendered representations of plant structures. In contrast, we propose a probabilistic approach to plant simulation through frame synthesis, a class of generative CV that aims to synthesize novel images of a scene captured by a distribution of observed data. In this paper, we outline a simulation framework that does not rely on rule-based processes to generate structures but rather a model that reflects a distribution of possible outcomes guided by an observation dataset in which the underlying physiological processes are implicitly learned. We summarize the contributions made in this work as follows:

- We develop a sequence-informed generative model that performs a continuous mapping between low-dimensional context or "environment" data with high dimensional synthesized frames.
- We demonstrate that the stochastic elements that contribute to ambiguities in deterministic approaches can be controlled by learning a latent space to isolate variability in representations and maintain spatiotemporal consistency in plant structure development.
- We improve model performance by employing a recurrent connection from the output layers to capture dependencies from prior states.
- We evaluate the framework on perceptual quality and frame-by-frame coherence metrics along with qualitative observations.

2 Related Work

Recent works have approached the task of plant growth modeling by representing structural elements through images and treatment data. Two prominent neural network (NN)-based approaches have been employed to predict future stages of plants' structural development from previous states found within a temporal scene.

Next Frame Prediction. In this approach, a frame-by-frame analysis using a convolutional recurrent neural network (CRNN) has been suggested as a method to extract features from a sequence of spatial image data [8]. Early works have taken a deterministic approach to the plant growth prediction problem by suggesting that the next sequence of a plant's structural development can be predicted by feeding a sequence of prior images through some implementations of a CRNN [16,22]. These methods use encoder-decoder networks to produce subsequent frames of the temporal scene. However, due to the ill-posed nature of the prediction task, there exist many possible outcomes for the plant's trajectory such that any deterministic generative process will result in ambiguities in outputs over large datasets [9]. These ambiguities are further exacerbated by unguided predictions that provide no influence from scene conditions.

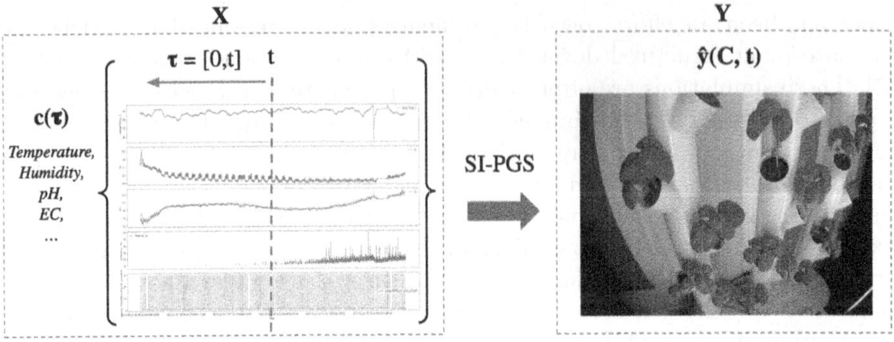

Fig. 1. Conceptual frame synthesis framework relating the X sequence condition domain to the Y structured output domain.

Conditional Generative Forecasting. Generative models have typically been employed as a data augmentation technique for training on sparse agricultural datasets [11,12]. There have been a few developments in plant growth forecasting which employ conditional generative models (CGM) for discrete predictions in which potential outcomes are generated at an arbitrary point in the future [3,10,14]. These predictions are often controlled by fixed treatment attributes passed as conditional inputs. Stochastic elements are introduced in the form of random noise to capture variability. However, these conditions are represented as discrete treatment embeddings and do not incorporate the dynamic and compounding environmental conditions that play a significant role in plant phenotype expression. Discrepancies in temporal consistencies in plant structures are observed when continuous predictions are made, as frames are not coherently coupled.

3 Methods

The task of plant simulation is formulated as a structured representation problem where we generate the frames of a scene, y, from a low-dimensional discrete or continuous random variable, c (Fig. 1). The generated outputs, $\hat{y}$, of the model represent a simulated visualization of the plant's structural development and phenotypic attributes at a given point in time, t, with respect to its growth cycle. We construct a CGM framework, particularly a conditional variational auto-encoder (CVAE), with the objective of learning a prior distribution, $p_\theta(z|x)$. The latent representation of the scene, z, is directed by a set of encoded sequence vectors, $X = \{x_i \mid i \in \{1, 2, \ldots, N\}\}$, and is determined by the dataset of environmental conditions, denoted as $C = \{c_i(\tau) \mid i \in \{1, 2, \ldots, N\}, \tau \in [0, t]\}$ with N i.i.d. samples.

In this section, we outline the various components of the directed graph models and the implementation of our sequence-informed plant growth simulation (SI-PGS)(Fig. 2–3).

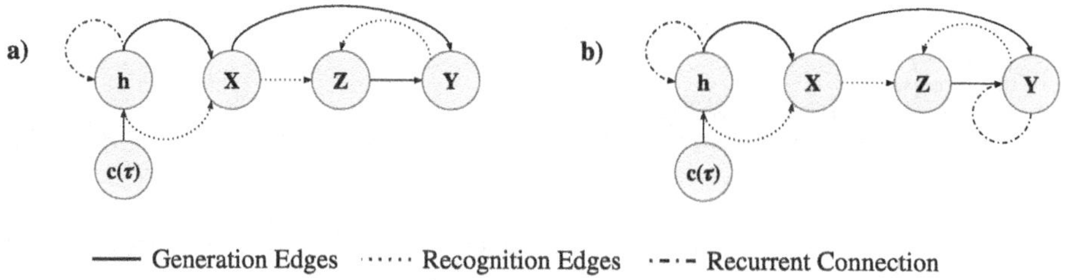

Fig. 2. Directed graph models. **(a)** SI-PGS. This model incorporates generation edges that represent the forward generative process and recognition edges that depict the encoding processes of the latent space, Z. The conditional vector, X, is influenced by the last layer hidden outputs, h, of an RNN that processes compounding sequential context data, $c(\tau)$, and is subsequently used to derive structured outputs, Y. **(b)** SI-PGS with recurrent output connections (SI-PGS-R). Recurrent connections are introduced to the model to utilize feedback from the generated output Y to refine X and Z.

3.1 Sequence-Informed Conditioning

c is defined as a configuration of random variables that describe temporal observations contributing to the development and phenotypic expression of the plant within the scene. Typically, this is in the form of data observed from external sensors, intrinsic scene context, or discrete attributes. Accounting for the temporal aspect of our simulation, c incorporates two time attributes with respect to the scene: The time elapsed since the inception of the plant's growth trajectory, defined at the first frame, and the time-interval between frames. Integrating these attributes allows us to relax our definition of t to represent the sequence indices of each frame with variable time intervals and allows for the implicit interpolation of missing data between these periods. A sequence encoder, $S_\theta : c(\tau) \rightarrow x$, is employed to capture the accumulated effects of these observations on plant development at t.

$$h_t = f_{\mathrm{rnn}}(c(\tau), h_{t-1}; \theta) \quad \tau \in \{1, 2, \ldots, t\} \tag{1}$$

$$x = g_\theta(h_t) \tag{2}$$

S_θ introduces an RNN, particularly a LSTM, denoted as $f_{\mathrm{rnn}}(.)$, with parameters θ, whose inputs represent compounding sequential information. The hidden states of the RNN, h, are mapped to a MLP forward process, $g_\theta(.)$ (Eq. 1–2).

3.2 Latent Representation and Sample Generation

A prior distribution, $p_\theta(z|x)$, is learned so that in turn a distribution of generated scene outcomes, $\hat{y} \sim p_\theta(y|z, x)$, can be derived. The simulated outcome space, Y, is obtained by sampling the latent space, Z, such that $z \sim p_\theta(z|x)$ (Fig. 2a).

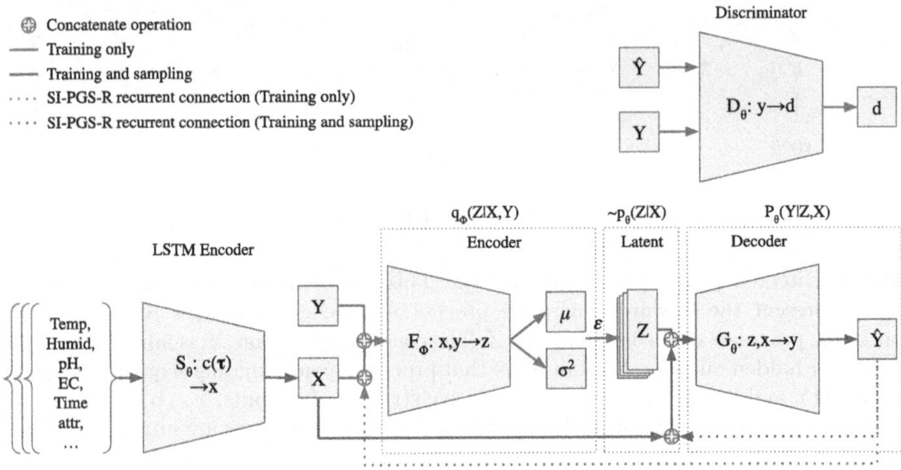

Fig. 3. SI-PGS network architecture. Input, $c(\tau)$, is composed of a sequence of scene attributes (i.e. temperature, humidity, pH, EC and temporal attributes) that are encoded by the LSTM Encoder S_θ. The encoded output x and the ground truth Y are concatenated and fed into the encoder F_ϕ, which produces the latent distribution variables μ and σ^2. latent variable z is sampled and passed through the decoder G_θ to generate the output $\hat{y}$. The discriminator D_θ distinguishes between real and generated outputs. Implementation of the recurrent connection is denoted as SI-PGS-R.

Recognition Model. The recognition model is given by the variational distribution, $q_\Phi(z|x, y)$. It is obtained implicitly by learning the distribution parameters, Φ, through an encoding process, $F_\phi : x, y \rightarrow z$.

$$\mu, \sigma^2 = f_{\text{enc}}(x, y; \Phi) \tag{3}$$

$$z = \mu(x, y) + \sigma(x, y) \odot \epsilon \tag{4}$$

We define $q_\Phi(z|x, y)$ as a Gaussian distribution with Φ consisting of mean, μ, and standard deviation, σ^2, such that $\mathcal{N}(\mu(x, y), \sigma^2(x, y))$. $f_{\text{enc}}(.)$ describes the convolutional encoding network which outputs a vector representation of the distribution parameters in Eq. 3. Z can then be sampled by defining a deterministic functional parameterized by $\Phi(x, y)$. In order to perform backpropagation a stochastic auxiliary variable, $\epsilon \sim \mathcal{N}(0, I)$, is introduced to the functional and is defined as the reparameterization trick in Eq. 4 [9].

Generation Model. The generation model is outlined by, $G_\theta : z, x \rightarrow y$, the process in which a frame of the scene is synthesized such that $y \sim p_\theta(y|x, z)$. In our implementation of SI-PGS we employ a condition-dependent parameterization to achieve a sampling process that describes $z \sim p_\theta(z|x)$. First we obtain z as a sample from a fixed Gaussian distribution $\mathcal{N}(0, I)$. Then, we introduce x as the conditioning term that will modulate the parameters during generation.

Both terms are passed as input to a convolutional decoder network, $f_{\text{gen}}(.)$, in Eq. 5 to produce a constrained output.

$$\hat{y} = f_{\text{gen}}(z, x; \theta) \mid z \sim \mathcal{N}(0, I) \tag{5}$$

The recognition and generation models are characterized by an objective function which aims to optimize the variational lower bound, $\mathcal{L}(\theta, \Phi)$ [17]. The lower bound can be expressed by two terms; the first term, $D_{\text{KL}}(q_\phi \| p_\theta)$, is the Kullback-Leibler (D_{KL}) divergence between the approximate posterior and the prior distributions. The second term, $\mathbb{E}_{q_\phi(z|x,y)}[p_\theta(y|x,z)]$, is the expected value of the log-likelihood. These terms act to balance the model between accurate reconstruction of the output, $\hat{y}$ with y, and a regularization on the learned latent space, such that the $q_\Phi(z|x, y)$ remains close to the prior $p_\theta(z|x)$. In previous work [2], we demonstrate that the parameterization of $\mathcal{L}(\theta, \Phi)$ can be done to prioritize model learning between reconstruction and regularization. We employ a weight, β, in Eq. 6 which introduces a tunable parameter that balances the importance placed by each term during training. $\beta > 0.5$ signifies more importance placed towards the regularization term, while $\beta < 0.5$ places more significance towards the reconstruction term.

$$\mathcal{L}_{\text{cvae}}(\theta, \Phi) = -(\beta) D_{\text{KL}}(q_\phi(z|x, y) \| p_\theta(z|x)) + (1 - \beta)\mathbb{E}_{q_\phi(z|x,y)}[p_\theta(y|x, z)] \tag{6}$$

Discriminator. $G_\theta(.)$ can be interpreted as the generator of a generative adversarial network (GAN) which employs a predefined random noise distribution, ϵ, to guide image generation [4]. As such, we can implement a discriminator, $D_\theta : y \rightarrow d$, whose output, d, represents the probability that input, y, is real or fake. SI-PGS employs a convolutional encoder network, $f_{\text{discr}}(.)$, whose output features are passed through a sigmoid function to quantify d as shown in Eq. 7. This GAN mechanism is employed to refine the realistic generation of the synthesized frames by optimizing the value function, V, of a min-max game between the generator and discriminator with the objective function described in Eq. 8. The combined SI-PGS objective function is given by Eq. 9.

$$d = \text{SIGMOID}(f_{\text{discr}}(y; \theta)) \tag{7}$$

$$\min_G \max_D V(D, G) = \mathbb{E}_{y \sim p(y)}[\log D(y|x)] + \mathbb{E}_{z \sim q_\phi(z|x,y)}[\log(1 - D(G(z|x)))] \tag{8}$$

$$\mathcal{L}_{\text{SI-PGS}} = \mathcal{L}_{\text{CVAE}}(\theta, \phi) + \min_G \max_D V(D, G) \tag{9}$$

3.3 Recurrent Connection

A recurrent connection is incorporated into the architecture such that output, $\hat{y}_t$, is conditioned on the previous output state, $\hat{y}_{t-1}$ (Fig. 2b). The aim of this recurrent step is to introduce a known spatial state that will add continuity over time, resulting in improved frame generation. Thus the generation model is updated to $G_\theta : z, x, y_{t-1} \rightarrow y_t$, where $f_{\text{gen-R}}(.)$ represents the modified convolutional decoder network in Eq. 10. The implementation of the recurrent connections is shown in Fig. 3.

$$\hat{y}_t = f_{\text{gen-R}}(z, x, \hat{y}_{t-1}; \theta) \mid z \sim \mathcal{N}(0, I) \tag{10}$$

4 Experiments

In this section, we evaluate the various components of the simulation model. The objective of these experiments is to demonstrate that SI-PGS is able to synthesize continuous frames that capture the realistic development of a plant given a sequence of environmental data and outperforms the existing baseline model.

4.1 Gardyn Dataset

The Gardyn dataset (GD) is comprised of images and sensor data that observe Genovese basil (*Ocimum basilicum*) grown in a controlled vertical hydroponics system. The data was collected over four months, covering two growth cycles and is used as a preliminary dataset for model testing. It contains time-series data which include 40,000 640 by 480 pixel RGB images, captured at two angles of the scene, each with corresponding sensor data taken at a frequency of 10 min. The sensor data include the following information: ambient temperature, ambient humidity, reservoir level, reservoir temperature, nutrient pH, and electrical conductivity (EC). GD includes metadata such as lighting schedule, irrigation schedule, timestamp information and image metadata.

4.2 Quantitative Model Evaluation

Frame Generation. One major criterion for evaluating the generative component of SI-PGS is the realistic synthesis of scene frames. As such, two metrics are selected to quantitatively assess the generated outputs: reconstruction loss and Fréchet inception distance (FID) score.

A reconstruction loss, particularly average mean squared error (MSE), is employed to quantify the pixel-wise discrepancies between the generated frames and their corresponding real images from GD. This metric measures the overall visual fidelity of the target scene frame. The FID score evaluates the distance between feature vectors calculated for real and generated images, which are extracted using an Inception-v3 model pre-trained on ImageNet [20]. This metric considers both the feature mean and covariance, capturing not just the visual appearance but the statistical properties of the image distributions. A lower FID score indicates that the generated images are not only visually similar to the actual images but are statistically comparable, reflecting better perceptual quality in the synthesized frames [5].

Coherence Between Frames. Temporal coherence, within the bounds of our simulation, is defined by the smooth and proportional development of plant structures between frames with respect to time. In other words, objects within a scene should not drastically change from one frame to the next if the time increments are small. These temporal inconsistencies are referred to as "jitters". We employ two coherence metrics: Time-weighted MSE and structural similarity index (SSIM). These measures are used to compare differences in sequential frames of predictions.

Since MSE emphasize larger errors and is sensitive to outliers and abrupt changes between frames, it is employed to assess the stability of frame transitions. SSIM, measures the similarity between two images by assessing the visual impact of three characteristics of an image: luminance, contrast, and structure. This metric provides a more comprehensive measure of similarity by considering how changes in the structural information and luminance affect the perceived quality of video sequences [21]. SSIM is measured between 0 and 1, with more similarities between images as the value approaches 1. Since the frames synthesized by SI-PGS are not necessarily generated at fixed time intervals, we extend these metrics by applying a weighted time parameter, w_t, which accounts for the time interval between generated frames, normalized across all weights.

4.3 Experimental Setup

We evaluate the performance of the SI-PGS framework compared with a baseline discrete embedding conditional GAN (cGAN), conceptually used in preceding work for predictive plant growth modeling [14]. The cGAN model is deployed with a comparable network architecture and depth, which is obtained by detaching the generator and discriminator from the latent encoding process of SI-PGS and using the objective function given by Eq. 8. A modified SI-PGS with recurrent output connections, outlined in Sect. 3.3, is tested alongside the original model and is denoted as SI-PGS-R. We experiment with various β values and observe the effect of reconstruction versus regularization on the representation of plant structures during simulation. The β values selected for the experiments are: 0.10, 0.25, 0.50, 0.75 and 0.90. We further test a controlled and fixed latent space sampling process, denoted as controlled sampling (CS), to isolate the stochastic elements that lead to variability in consecutive frames. The models are trained on a subset of GD containing 5,892 images and corresponding sequence data. Moreover, all models are trained for 70 epochs, with early stopping to avoid overfitting once the model converges on the validation set. The experiments in this study were conducted using a workstation equipped with an NVIDIA GeForce RTX 4080 GPU. The full implementation and supplemental materials from these experiments are available on GitHub.[1]

[1] Code and supplementary material can be found at: https://github.com/mohas95/Sequence-Informed-Plant-Growth-Simulation.

Table 1. Comparative analysis of model performance evaluating generation quality and frame coherence metrics. Time-weight (tw) applied to frame coherence metrics follows an exponential decay, $w_t = e^{-\lambda \cdot \Delta t}$, with a decay factor, $\lambda = 1.9 \times 10^{-4}$, and is normalized over $\sum_{t=2}^{n-1} w_t$.

	Generation Quality		Frame Coherence			
	FID	MSE	MSE_{tw}	$SSIM_{tw}$	$CS\text{-}MSE_{tw}$	$CS\text{-}SSIM_{tw}$
SI-PGS ($\beta = 0.25$)	142.78	8915	2942	0.844	1.08	≥ 0.999
SI-PGS ($\beta = 0.50$)	88.86	6359	2081	0.904	1.16	≥ 0.999
SI-PGS ($\beta = 0.75$)	62.25	3389	1279	0.937	1.04	≥ 0.999
SI-PGS ($\beta = 0.90$)	47.04	**2441**	**1024**	**0.951**	**0.89**	≥ 0.999
SI-PGS-R	**45.09**	6858	6144	0.889	5.98	≥ 0.999
cGAN	218.33	14002	3365	0.930	≤ 0.001	≥ 0.999

5 Results and Discussion

The results from the quantitative experiments are shown in Table 1 and demonstrate that SI-PGS outperforms cGAN in generating sequence-consistent plant structures. FID scores remain consistently lower for SI-PGS based models, with the lowest obtained from SI-PGS-R, indicating more apparent similarities in identifiable structures between generated and reference images. When fixing the stochastic elements of each model (z for SI-PGS and ϵ for cGAN) we observe a significant decrease in CS-MSE_{tw} while CS-$SSIM_{tw}$ approaches 1 across all models suggesting that jitters between frames are suppressed. This is interpreted as the effective isolation of variability that leads to ambiguities in deterministic models. Thus, each latent variable can be used to produce a unique instance of a scene outcome with spatiotemporal consistent structural development. Finally, we notice a consistent decrease in FID score and MSE with increasing β values. This suggests that the outputs of the models that favor the regularization term over the reconstruction in the objective function produce more realistic structures. This is more apparent with the qualitative assessment of model outputs.

5.1 Qualitative Assessment

Outputs of the experiment are shown in Fig. 4 and illustrate the sequential frames of a scene generated from low-dimensional scene attributes. SI-PGS models are able to capture plant structures across all references, while the cGAN model fails to capture early stages of the plant. Moreover, we suggest that cGAN requires a longer training phase to generate realistic plant structures when compared to SI-PGS, as later-stage frames fail to capture finer plant details.

Regularization vs Reconstruction. We observe that as the β value increases, the outputs of the model become more realistic when favoring the regularization

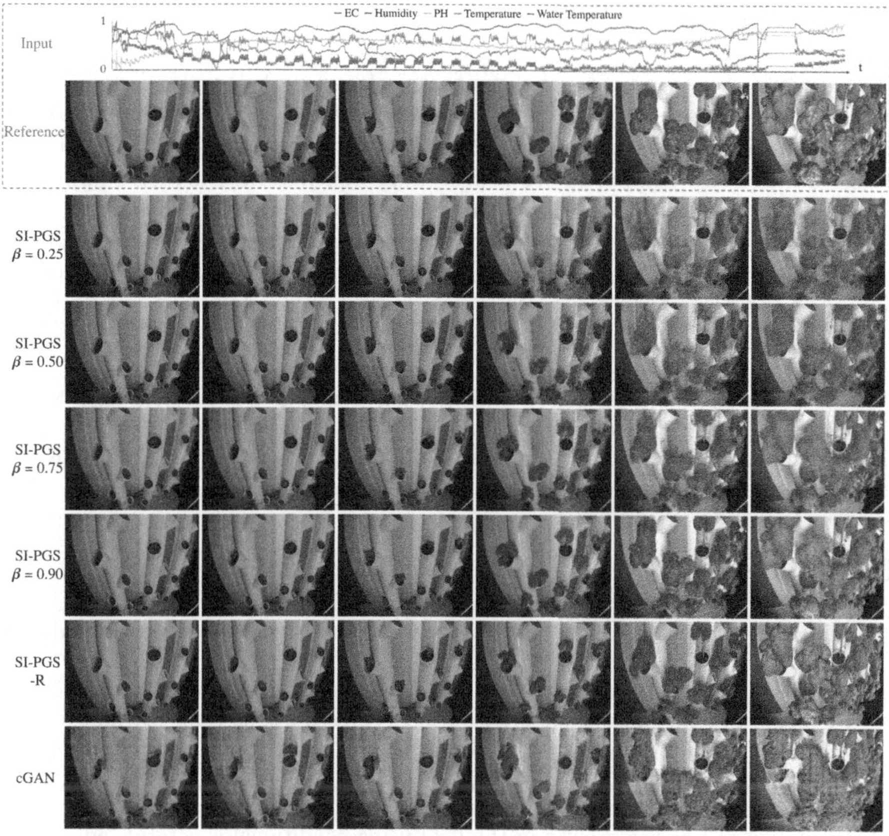

Fig. 4. Comparative generated model outputs. Top two rows show the input signals over time and corresponding real reference frames. Subsequent rows show the outputs of corresponding generative models.

term. While intuitively the reconstruction term aims to increase the model's pixel-wise fidelity, in reality, it results in an averaging effect across large image datasets. In effect, the reconstruction term favors a deterministic outcome, leading to the ambiguous structures in generated outputs.

Effects of Recurrent Connection. While the effects of the SI-PGS-R are not captured well in Fig. 4, it becomes apparent when observing all frames compiled into video format. The addition of a recurrent connection from the output layers of the model results in a feedback mechanism that enhances the model's ability to capture dependencies and improve generative performance. The generated frames capture movements of the plant structure during growth more realistically over the standard SI-PGS model.

6 Conclusion

In this work, we propose a novel generative framework for plant growth simulation that captures continuous temporal dependencies. We demonstrate its effectiveness in reconstructing plant structural development using a typical CEA image dataset and compare it to a preceding benchmark generative approach. Furthermore, we introduced variations to the baseline model to improve realistic and coherent scene synthesis. Additional research should be done to link phenotypic information to synthesized frames. Such efforts should be directed towards building larger datasets that capture the temporal dependencies in plant structural development and phenotypic attributes.

Acknowledgements. This study was partially funded by Gardyn and Mitacs (IT16220). We thank the Gardyn team for providing the vertical growth systems essential for building our datasets. Special thanks to Anita Parmar, Ollivier Dyens, and the Building 21 members for their engagement in creative discussions.

References

1. Cieslak, M., et al.: L-system models for image-based phenomics: case studies of maize and canola. in Silico Plants **4**(1), diab039 (2021). https://doi.org/10.1093/insilicoplants/diab039
2. Debbagh, M.: Learning structured output representations from attributes using deep conditional generative models. arXiv preprint arXiv:2305.00980 (2023). https://doi.org/10.48550/arXiv.2305.00980
3. Drees, L., Junker-Frohn, L.V., Kierdorf, J., Roscher, R.: Temporal prediction and evaluation of brassica growth in the field using conditional generative adversarial networks. Comput. Electron. Agric. **190**, 106415 (2021). https://doi.org/10.1016/j.compag.2021.106415
4. Goodfellow, I., et al.: Generative adversarial nets. In: Advances in Neural Information Processing Systems, vol. 27 (2014). https://doi.org/10.48550/arXiv.1406.2661
5. Heusel, M., Ramsauer, H., Unterthiner, T., Nessler, B., Hochreiter, S.: Gans trained by a two time-scale update rule converge to a local nash equilibrium. In: Advances in Neural Information Processing Systems. NIPS 2017, vol. 30, pp. 6629-6640 (2017). https://doi.org/10.5555/3295222.3295408
6. Hitti, Y., Buzatu, I., Del Verme, M., Lefsrud, M., Golemo, F., Durand, A.: Growspace: a reinforcement learning environment for plant architecture. Comput. Electron. Agric. **217**, 108613 (2024). https://doi.org/10.1016/j.compag.2024.108613
7. Jiang, Y., Li, C., Paterson, A.H., Sun, S., Xu, R., Robertson, J.: Quantitative analysis of cotton canopy size in field conditions using a consumer-grade RGB-D camera. Front. Plant Sci. **8**, 2233 (2018). https://doi.org/10.3389/fpls.2017.02233
8. Keren, G., Schuller, B.: Convolutional RNN: an enhanced model for extracting features from sequential data. In: 2016 International Joint Conference on Neural Networks (IJCNN), pp. 3412–3419. IEEE (2016). https://doi.org/10.1109/IJCNN.2016.7727636
9. Kingma, D.P., Welling, M.: Auto-encoding variational bayes. arXiv preprint arXiv:1312.6114 (2013). https://doi.org/10.48550/arXiv.1312.6114

10. Leonhardt, J., Drees, L., Jung, P., Roscher, R.: Probabilistic biomass estimation with conditional generative adversarial networks. In: Andres, B., Bernard, F., Cremers, D., Frintrop, S., Goldlücke, B., Ihrke, I. (eds.) DAGM GCPR 2022. LNCS, vol. 13485, pp. 479–494. Springer, Cham (2022). https://doi.org/10.1007/978-3-031-16788-1_29
11. Li, Y., et al.: Self-supervised plant phenotyping by combining domain adaptation with 3D plant model simulations: application to wheat leaf counting at seedling stage. Plant Phenomics **5**, 0041 (2023). https://doi.org/10.34133/plantphenomics.0041
12. Lu, Y., Chen, D., Olaniyi, E., Huang, Y.: Generative adversarial networks (GANs) for image augmentation in agriculture: a systematic review. Comput. Electron. Agric. **200**, 107208 (2022)
13. Luo, L., et al.: Eff-3DPSeg: 3D organ-level plant shoot segmentation using annotation-efficient deep learning. Plant Phenomics **5**, 0080 (2023). https://doi.org/10.34133/plantphenomics.0080
14. Miranda, M., Drees, L., Roscher, R.: Controlled multi-modal image generation for plant growth modeling. In: 2022 26th International Conference on Pattern Recognition (ICPR), pp. 5118–5124 (2022). https://doi.org/10.1109/ICPR56361.2022.9956115
15. Prusinkiewicz, P., Cieslak, M., Ferraro, P., Hanan, J.: Modeling plant development with L-systems. In: Morris, R.J. (ed.) Mathematical Modelling in Plant Biology, pp. 139–169. Springer, Cham (2018). https://doi.org/10.1007/978-3-319-99070-5_8
16. Sakurai., S., Uchiyama., H., Shimada., A., Taniguchi., R.: Plant growth prediction using convolutional LSTM. In: Proceedings of the 14th International Joint Conference on Computer Vision, Imaging and Computer Graphics Theory and Applications (5: VISAPP), pp. 105–113. INSTICC (2019). https://doi.org/10.5220/0007404901050113
17. Sohn, K., Lee, H., Yan, X.: Learning structured output representation using deep conditional generative models. In: Advances in Neural Information Processing Systems, vol. 28 (2015). https://doi.org/10.5555/2969442.2969628
18. Soualiou, S., et al.: Functional-structural plant models mission in advancing crop science: opportunities and prospects. Front. Plant Sci. **12**, 747142 (2021). https://doi.org/10.3389/fpls.2021.747142
19. Sun, S., et al.: In-field high throughput phenotyping and cotton plant growth analysis using lidar. Front. Plant Sci. **9** (2018). https://doi.org/10.3389/fpls.2018.00016
20. Szegedy, C., Vanhoucke, V., Ioffe, S., Shlens, J., Wojna, Z.: Rethinking the inception architecture for computer vision. In: 2016 IEEE Conference on Computer Vision and Pattern Recognition (CVPR), pp. 2818–2826 (2016). https://doi.org/10.1109/CVPR.2016.308
21. Wang, Z., Lu, L., Bovik, A.C.: Video quality assessment based on structural distortion measurement. Signal Process. Image Commun. **19**(2), 121–132 (2004). https://doi.org/10.1016/S0923-5965(03)00076-6
22. Yasrab, R., Zhang, J., Smyth, P., Pound, M.P.: Predicting plant growth from time-series data using deep learning. Remote Sens. **13**(3) (2021). https://doi.org/10.3390/rs13030331

A Simulation Study on Energy Optimization in Building Control with Reinforcement Learning

Peter Bolt[1(✉)], Volker Ziebart[1], Christian Jaeger[1], Nicolas Schmid[1], Thilo Stadelmann[1,2], and Rudolf M. Füchslin[1,2]

[1] Zurich University of Applied Sciences Winterthur, Winterthur, Switzerland
bolt@zhaw.ch

[2] European Centre for Living Technology (ECLT), Ca' Bottacin, Venice, Italy

Abstract. We propose and evaluate a deep reinforcement learning control paradigm for building energy systems. In comparison to other advanced control techniques, namely Model Predictive Control, the reinforcement learning paradigm avoids the costs and uncertainties associated with the requirement for a control-oriented model. We apply a mixed agent for the Proximal Policy Optimization algorithm, similar to the algorithm proposed in [7] as well as a non-discounted finite horizon optimization scheme.

We investigate the capabilities of the proposed reinforcement learning controller regarding energy efficiency, comparing it against the most widely used rule-based control paradigm as a baseline controller. We verify our proposed paradigm in a simulation study with building models implemented in Dymola.

Keywords: Smart building · Building control · Reinforcement learning

1 Introduction

The transition from fossil fuels to renewable energy sources entails increased complexity in energy management, due to varying availability and the difficulties of storing electrical and thermal energy. Building energy systems (BES), representing an important share of overall energy consumption, offer significant potential in reducing the carbon footprint, provided that an optimized energy management adapted to both availability and usage patterns can be realized.

Prior work has established that advanced control techniques such as Model Predictive Control (MPC) can yield substantial energy savings in BES [3,5,12]. However, despite intensive research, especially in MPC for BES, most buildings still rely on simple rule-based controllers (RBC). A possible reason for this is the high computational complexity associated with online MPC, along with the need for an accurate, control-oriented model to describe the physical system in

Funded by Innosuisse 31326.1 IP-EE.

C. Y. Suen et al. (Eds.): ANNPR 2024, LNAI 15154, pp. 320–331, 2024.
https://doi.org/10.1007/978-3-031-71602-7_27

a way amenable to real-time optimization. For many applications, this restricts the model to being essentially linear. Creating this control-oriented model is often the most time-consuming part of MPC design and can result in a compromise between prediction accuracy and computational complexity that leads to suboptimal controller performance.

This work is motivated by the idea that Deep Machine Learning (ML) holds promise in solving the energy management problem. ML, as a data-driven technique, has gained traction as a powerful paradigm across various application domains, including building control [11,15]. Particularly in an engineering context [1], ML can complement the model-driven methodologies. We explore the potential advantages of a control system based on deep reinforcement learning (Reinforcement Learning Control, RLC) in the context of BES. We investigate various RLC variants and compare them to conventional controllers. Notably, the investigated RLC schemes do not necessitate a control-oriented model, potentially reducing development time and enhancing controller performance. However, they do require a numerical model of the BES to gather training data—often spanning several simulated years.

We apply a mixed agent for the Proximal Policy Optimization algorithm (PPO) [9], capable of handling the mixed continuous and discrete action space inherent to our problem (see Sect. 3.2) and a finite horizon optimization scheme to replace the traditional infinite horizon discounting scheme (see Sect. 3.3). This change is driven by the observation that the discounting scheme made delaying the import of energy appear cheaper to the controller, whereas in reality, it led to increased costs due to energy import becoming an urgent need at an ill-suited moment.

2 Building Energy System (BES)

The investigated system is a simplified version of the one used by the authors in [3], which is based on a real application scenario. It integrates a small-scale, grid-connected photovoltaic collector (PV), one heat pump (HP) charging one thermal storage and supplying energy to a low-temperature surface heating. The consumer is a single-family house modeled as a second-order Lumped Parameter Model, consisting of two resistors and two capacitors. Figure 1a shows the schematic of the installation which integrates

- Ice storage (IceSt) 84 m^3
- Room heating storage (RhSt) 500 liter
- Photovoltaic (PV) collector, 30 m^2, south-facing
- Brine to water HP with nominal heat output 3.93 kW at B0/W35 with a Coefficient of Performance (COP) of 4.65

The heat pump has a variable speed compressor, allowing it to adjust its operating electrical power (power modulation). Ice storage is a low-temperature storage with a very long time constant and a high capacity. The room heating storage is a high-temperature storage with low capacity.

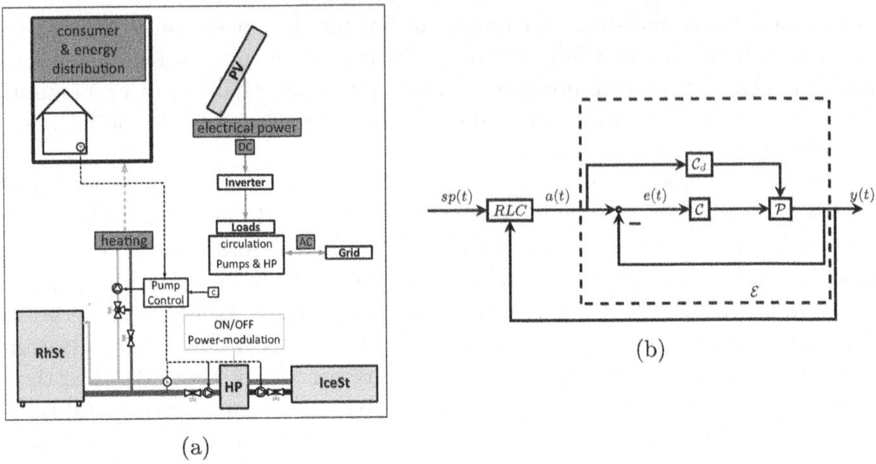

Fig. 1. a) Scheme of the HVAC system simulated with Dymola. Heating storage tank (RhSt), heat pump (HP), ice storage (IceSt), and photovoltaics (PV). **b)** Hierarchical control architecture. $\mathcal{P}$: Plant to be controlled, $\mathcal{C}$:continuous (PI) controllers, $\mathcal{C}_d$: discrete HP controller of the inner loop, and $\mathcal{E}$: Environment, constituting the inner closed-loop system. The RL module establishes setpoints for room temperature and HP electrical power, and handles the on/off operation of the HP.

3 Control Architecture

We employ the hierarchical architecture depicted in Fig. 1b. Proportional-Integral (PI) controllers are used in the inner loop to achieve system stabilization for the fast dynamics. The outer loop is a reference governor (RG) [2] for the inner loop. The RG's objective is to minimize the energy consumption of the inner loop while adhering to setpoints and system limitations.

3.1 The Reinforcement Learning Controller

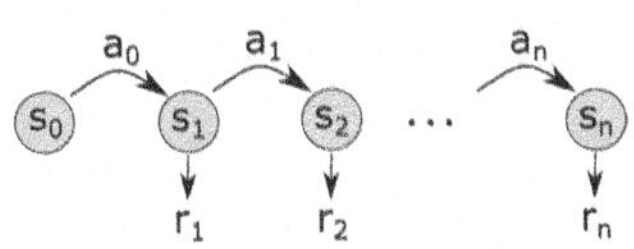

Fig. 2. Markov Decision Process with states s_t, actions a_t and rewards r_t

We formulate the control problem as a Markov Decision Process as depicted in Fig. 2 within the environment $\mathcal{E}$ [13]. The control episode starts a state s_0.

The controller then successively chooses actions a_t from a set of possible actions $\mathcal{A}$ according to the control policy $\pi(a, s)$ which is the probability of choosing the action a in the state s. Due to the applied actions and the dynamics of the environment, the states s_t will evolve over time. The performance of the controller is measured by the reward r_t which is a function of s_t and the action a_t.

In most RL applications the return R_t is defined as the discounted sum of all future rewards in the control episode of length n

$$R_t = \sum_{l=0}^{n-t} \gamma^l r_{t+l} \tag{1}$$

where $\gamma \in (0, 1]$ is the discount factor, which must be tuned during the training process. A value of $\gamma < 1$ means that rewards count less, the further they lie in the future, reflecting the fact that increased temporal distance introduces more uncertainties. Consequently, smaller yet more assured rewards in the near future are preferred over potentially larger but riskier rewards further from the present.

The return can be used to define a quality measure of the chosen actions, namely the Action-Value Function (Q-function):

$$Q^{\pi,\gamma}(s_t, a_t) := \mathbb{E}[R_{t+1} \mid s_t, a_t] \tag{2}$$

The Q-function represents the expected return when starting from state s_t, taking action a_t, following policy π thereafter, and using the discount factor γ. Similarly, the State-Value Function (V-function) is defined as the expected return when starting from state s_t, following policy π thereafter.

$$V^{\pi,\gamma}(s_t) := \mathbb{E}[R_{t+1} \mid s_t] \tag{3}$$

The function $A(s_t, a_t)^{\pi,\gamma}$ is called advantage and specifies how much better or worse the value of the action a_t is compared to the π-weighed average of all possible actions in the state s_t. It is given by

$$A(s_t, a_t)^{\pi,\gamma} = Q^{\pi,\gamma}(s_t, a_t) - V^{\pi,\gamma}(s_t) \tag{4}$$

In our research, we predominantly utilized the Proximal Policy Optimization (PPO) algorithm as outlined in [9]. PPO is a state-of-the-art reinforcement learning algorithm that combines policy gradient techniques with a clever clipping mechanism. This clipping approach is employed to mitigate the risk of excessively aggressive policy updates that could lead to instability in the learning process. Another advantage of PPO is its ability to engage with multiple environments in parallel, significantly expediting the training process. As described in [10], we approximately maximize the following objective during PPO training:

$$L = \hat{\mathbb{E}}[L_A - c_V L_V + c_S L_S] \tag{5}$$

Here, c_V and c_S are non-negative weighting factors, and L_S contributes to entropy loss for enhanced exploration. Despite the potential benefits of promoting exploration through the entropy term (e.g. [10]), our experimentation did not reveal any notable improvements. Therefore, we set the weighting factor c_S to zero. L_V corresponds to the aggregation of squared deviations between the learned value function V_θ-represented by the neural network's parameters θ-and the estimated returns $\hat{V} = R_{t+1}$ derived from interactions with the environments.

The first and most important term is given by

$$\hat{\mathbb{E}}[L_A] = \sum_{k=1}^{m} \sum_{t=0}^{n} A_{\theta_0}(s_t^k, a_t^k) \frac{\pi_\theta(a_t^k, s_t^k)}{\pi_{\theta_0}(a_t^k, s_t^k)} \tag{6}$$

where θ_0 denotes the parameter set of the neural network that was used when interacting with the environment and generating the data set $\{s_t^k, a_t^k, r_t^k\}_{k=1,\dots,m}$ for the m parallel episodes. When optimizing the loss (Eq. 6), we employ the following estimate $\hat{A}$ for A [9]:

$$\hat{A}(s_t, a_t) = r_t + \gamma\, V_{\theta_0}(s_{t+1}) - V_{\theta_0}(s_t) \tag{7}$$

The optimization of L_A can be understood as follows: For $\theta = \theta_0$, the quotient $\lambda = \pi_\theta / \pi_{\theta_0}$ is equal to 1. Starting from $\theta = \theta_0$, the parameters of the neural network are optimized such that λ will be increased for positive advantages (actions with better-than-average action-value) and decreased for actions with negative advantages. Therefore, actions with positive advantages will have a higher probability of being chosen in the next iteration.

In our application, the environment comprises the sub-controllers $\mathcal{C}$ and the Building Energy System $\mathcal{P}$ (see Fig. 1b). The possible actions applied every hour are as follows:

(i) Heat pump on/off
(ii) Heat pump power modulation 60–100%
(iii) Indoor temperature setpoint 20–23 °C

From a BES perspective, if necessary, we can easily convert (ii) and (iii) into discrete actions, provided that the discretization has a fine enough granularity. Concerning (i), the operation of the HP can be approached either discretely by switching the HP on or off every hour or continuously by turning it on or off for a duration between 0 s and 3600 s.

The state information comprises the current supply and return flow temperatures, the flow rates of the heating circuit, the room temperature of the building, as well as the room heating storage temperature and the power produced by the photovoltaic collector. Additionally, the state includes a 12-h forecast of outdoor temperature and solar irradiation, updated every hour.

The reward consists of two terms: (i) a deviation of the room temperature from the tolerance band 20–23 °C results in a linearly increasing negative reward (a penalty), and (ii) an energy penalty proportional to the energy drawn from the grid. The use of electricity from the PV collector incurs no penalties.

3.2 Mixed Agent

In practice, even actions that are naturally continuous, such as temperature set points, are often discretized and handled by a discrete agent. However, if high resolution is required, the number of discretized actions can become large and it's uncertain whether the same performance can be achieved as with a continuous agent. The mixed Agent is analogous to the one proposed in [7] for the Maximum a Posteriori Policy Optimization (MPO) agent.

In the case of **continuous action** spaces, the Gaussian policy is commonly used. Here, actions $a_c = \{a_{c,i}\}$ are drawn from a parameterized uncorrelated normal distribution represented by the equation:

$$\pi_c(a_c \mid s) = \prod_i \frac{e^{-\frac{(a_{c,i}-\mu_{\theta,i}(s,a_{c,i}))^2}{2\sigma^2_{\theta,i}}}}{\sqrt{2\pi}\sigma_{\theta,i}} \tag{8}$$

The functions $\mu_{\theta,i}(s,a)$ and $\sigma_{\theta,i}(s,a)$ are approximated by a neural network.

For **discrete action** spaces with actions a_d , the softmax policy is used. It is defined by the equation:

$$\pi_d(a_d \mid s) = \frac{e^{\phi_\theta(s,a_d)}}{\sum_{a'_d \in \mathcal{A}} e^{\phi_\theta(s,a'_d)}} \tag{9}$$

Here, the function $\phi_\theta(s,a_d)$ is represented by a neural network, and $\pi_d(a_d \mid s)$ calculates the probability of taking the action a_d given state s. The softmax policy ensures that the probabilities of all possible actions sum up to 1.

For the **mixed agent** with actions $a = \{a_c, a_d\}$ we used the combined distribution given by the product

$$\pi(a,s) = \pi_c(a_c \mid s)\ \pi_d(a_d \mid s) \tag{10}$$

The functions $\mu_{\theta,i}(s,a)$, $\sigma_{\theta,i}(s,a)$, and $\phi_\theta(s,a_d)$ as well as $V_\theta(s)$ are all approximated by the same neural network with parameters θ, which are found by optimizing the loss given by Eq. 5.

Exploration in the case of a continuous action space is accomplished by sampling actions according to the latest version of the policy network μ_θ and then adding Gaussian noise. This results in the action $a_c = \mu_\theta + \mathcal{N}(0, \sigma_\theta)$, which is subsequently applied within the environment.

For the scenario of a discrete action space, we generate samples from the categorical distribution $\pi_d(a_d \mid s)$. In this case, the most recent iteration of the policy network ϕ_θ is perturbed according to: $a_d = \text{argmax}(\phi_\theta - \log(-\log(u)))$, with $u \sim \text{Uniform}(0,1)$, and the resulting action is applied to the environment.

3.3 Finite Horizon Optimal Control Problem

Discounting future rewards (see Eq. 1) suffers from three disadvantages in our application.

The first one arises from the fact that future energy draws from the grid are discounted which means that the HP seems to need less energy the more the action "HP on" lies in the future. This is briefly illustrated by the following example: Let us assume that we have a sunny winter afternoon and in order to maintain the required room temperature we need to run the HP for one hour in the coming 6 h. Currently, the PV system provides about 20% of the power the HP needs and therefore 80% must be taken from the grid. Thus, it seems to make sense to immediately switch on the HP and make use of the free PV power. However, if $\gamma = 0.95$ the return, which is always negative in our application, is higher if the power-up of the HP is delayed for the maximum of 5 hours because $\gamma^5 = 0.77$. But running the HP after sunset would certainly cost more grid energy.

This underweighting of future grid energy draws can be reduced by choosing a discount factor very close to one. But then the second disadvantage becomes apparent. Due to the very slow decay of the weighing factor γ^l if $\gamma \approx 1$ rewards that will be generated far in the future with almost no causality to the current action will be taken into account. Furthermore, the long optimization horizon increases the complexity exponentially, resulting in longer comutation time and higher variance in training performance.

The authors of [14] discuss how different values of γ would influence the behavior of an RL controller for a heat pump. They compare results with $\gamma = 0.25$ and $\gamma = 0.75$ (both strongly discounted) and observe the following: (i) higher values assign greater importance to achieve long-term rewards and accordingly lead to more frequent operation of the heat pump when the outdoor temperature is high (higher coefficient of performance (COP)). Pre-heating during periods with a high COP consumes more energy at the current time step while saving energy for upcoming time steps. If γ is small, the discounted future savings could not justify the current costs.

Because of these shortcomings we introduced a finite non-discounted optimization window for PPO which favors achieving long-term rewards over a reasonable time horizon without increasing the complexity of the optimization problem too much. The return is given by

$$R_t = \sum_{l=0}^{N_h} r_{t+l} + V_F \tag{11}$$

The prediction horizon N_h replaces the discounting factor γ. V_F accounts for the return value of the final state, reached after N_h steps. We obtained our results by simply setting $V_F = 0$. This can be justified by the fact that the planning horizon is sufficiently long in comparison with the time constants of the system that the final cost term is negligible for the determination of a_t. Additional work may be needed to fully understand the behaviour with respect to this term. A reasonable value for N_h can be estimated based on the time constants present in the system. In our case, the time constants of two thermal heat storages, RhSt and building capacity, are relevant. Equations 2 to 10 remain applicable without modification to this new return formulation.

In the following chapter, we present and compare results for various configurations of the PPO algorithm using both discounted and finite non-discounted reward windows. In addition, we also present results we obtained using the Soft Actor-Critic (SAC) [6] algorithm to find an optimal control strategy. Table 1 lists the investigated configurations.

Table 1. Evaluated parameter sets and discrete or continuous actions of PPO and SAC. Actions: (1) Heatpump turned on/off every hour or in case of the SAC the controller can choose a variable runtime ("var.rt.") between 0 s and 3600 s every hour, (2) Heatpump power modulation in percentage of maximum power, (3) Indoor temperature setpoint.

#	algorithm	action 1	action 2	action 3	γ	N_h
1	$\text{PPO}_{d:\gamma=0.95}$	on/off	$\{60, 65, ..., 100\}\%$	$\{20, 21, 22, 23\}°\text{C}$	0.95	-
2	$\text{PPO}_{d:\gamma=0.99}$	on/off	$\{60, 65, ..., 100\}\%$	$\{20, 21, 22, 23\}°\text{C}$	0.99	-
3	$\text{PPO}_{d:N_h=12}$	on/off	$\{60, 65, ..., 100\}\%$	$\{20, 21, 22, 23\}°\text{C}$	-	12
4	$\text{PPO}_{m:\gamma=0.95}$	on/off	$[60, 100]\%$	$[20, 23]°\text{C}$	0.95	-
5	$\text{PPO}_{m:\gamma=0.99}$	on/off	$[60, 100]\%$	$[20, 23]°\text{C}$	0.99	-
6	$\text{PPO}_{m:N_h=12}$	on/off	$[60, 100]\%$	$[20, 23]°\text{C}$	-	12
7	SAC	rt. var.	$[60, 100]\%$	$[20, 23]°\text{C}$		

4 Simulation Results

The RLCs were trained in co-simulation with a BES modeled in Dymola [4]. For the implementation of the RLCs, we employed the OpenAI framework [8]. Our optimization approach for the RLCs aimed at achieving two primary goals: (i) minimizing electricity consumption from the grid and (ii) maintaining room temperature within the range of 20–23 °C. The range enables the RLC to effectively utilize the building mass as an additional heat storage mechanism (see Fig. 5b). Leveraging the building's mass roughly doubles the short-term storage capacity of the system. In the upper plot of Fig. 5b, it is evident that the RLC capitalizes on this by activating the heat pump during periods of available surplus PV electricity. Throughout the investigation, only a negligible number of small constraint violations were observed. To assess and compare the performance of the RLCs, we have included a baseline RBC and an improved RBC in our evaluation. The baseline RBC uses a heat pump with fixed operating power (no power modulation) and a constant setpoint for the room temperature of 20 °C. Furthermore, the heat pump is turned on and off according to a constant threshold temperature in the RhSt. In the improved RBC, both the heatpump power and the room heating storage setpoint were decreased until there was just enough energy left to cover the requirement. Moreover, a time-dependent threshold temperature in the RhSt was introduced, which increases if PV electricity is available. This adjustment results in significantly higher self-consumption of PV electricity. It is only possible with prior knowledge of the energy requirement and availability of PV-energy, therefore it demonstrates the quasi-upper-performance limit achievable with the used RBC structure. Figure 3 shows, that this dramatically reduces the RBC's grid electricity consumption.

4.1 Performance Comparison and Training Speed

Figure 3 illustrates the electricity export versus import from the grid. The controllers minimize electricity import, while export is neither optimized nor constrained. The RLCs demonstrate remarkable energy efficiency results. Among the top-performing RLCs, PPO_d with a prediction horizon of $N_h = 12$ hours consumed an average of 940 $\frac{kWh}{a}$ of energy from the electricity grid. A improved RBC, also utilizing the building as a heat storage, required 1280 $\frac{kWh}{a}$, while a simple RBC required 2060 $\frac{kWh}{a}$. Furthermore, PPO_d exported more electrical energy to the grid than the improved RBC. Overall, RLCs with non-discounted finite windows, specifically PPO_d and PPO_m, exhibit outstanding performance with significantly lower variance compared to RLCs with discounted infinite windows. All simulations were conducted under identical circumstances.

Table 2 shows the number of samples required to achieve 85%, 90%, 95%, and 99% of the maximum reward. The mixed agents achieve high rewards more quickly. For instance, to reach 90% of the maximum performance, $PPO_{m:N_h=12}$ requires 1.8×10^6 samples, while $PPO_{d:N_h=12}$ requires 3.7×10^6 samples.

Table 2. Number of samples in millions required to reach 85%, 90%, 95%, and 99% of the maximum reward of each agent for the first time.

% of max reward	PPO parameters					
	$d{:}N_h{=}12$	$d{:}\gamma{=}0.99$	$d{:}\gamma{=}0.95$	$m{:}N_h{=}12$	$m{:}\gamma{=}0.99$	$m{:}\gamma{=}0.95$
85%	2.0	3.1	2.0	1.0	1.8	1.6
90%	3.7	3.9	2.9	1.8	2.7	1.8
95%	4.3	4.9	3.1	3.7	3.3	5.1
99%	9.8	8.8	9.2	9.2	7.1	6.9
yearly electricity consumption kWh	951	1022	1032	939	957	928
mean ± std	±17	±55	±20	±11	±33	±121

4.2 Discounted Versus Non-discounted Windows

We conducted an empirical investigation to examine the impact of different types of optimization windows. The proposed non-discounted window consistently matches or surpasses the discounted versions in all scenarios, resulting in not only lower mean energy consumption but also reduced energy consumption

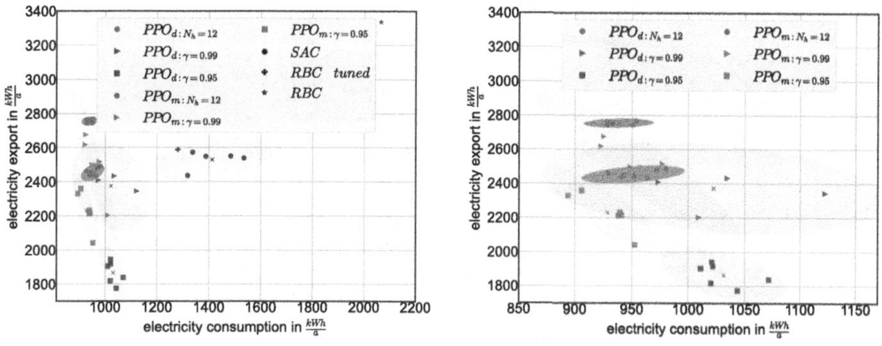

(a) (b)

Fig. 3. a) Energy efficiency comparison among different controllers optimized to minimize grid electricity consumption (kWh) while maintaining room temperature within 20–23 °C. The vertical axis: surplus energy produced by the PV system and supplied to the grid. Subscripts denote specific properties of the PPO Controller: PPO_d indicates discrete actions, PPO_m indicates mixed actions. The parameter γ represents the discount factor used in the optimization window, if a discounted window was employed. All simulations were run under identical (environmental) conditions. The variance over different runs is due to the stochastic nature of the sampling inherent to reinforcement learning. Figure **b)** Zoomed-in View of Fig. 3a.

variance. The outcomes are depicted in Fig. 3, where various RLCs with different windows were tested with identical environments.

Optimization becomes more challenging when there's a limited availability of surplus PV electricity. In Fig. 4, we reduced the PV area from 30 m^2 to 15 m^2. Notably, in this scenario, the non-discounted PPO_d outperforms the discounted $PPO_{d:\gamma=0.95}$ by approximately 13% in grid electricity consumption reduction. Concurrently, the exported electricity increases by over 30%.

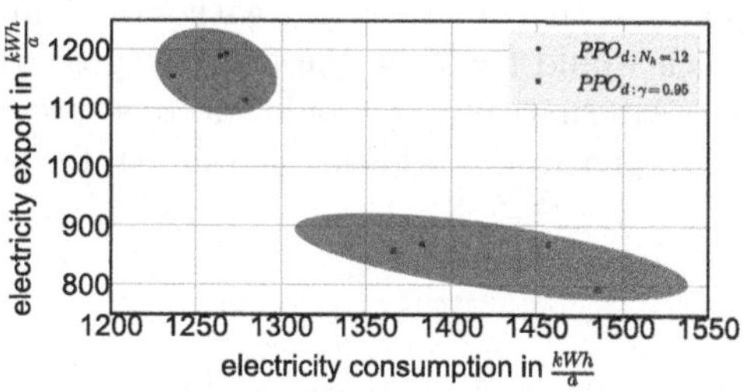

Fig. 4. Simulation with Reduced PV Area. The PV area is 15 m^2, compared to the original 30 m^2.

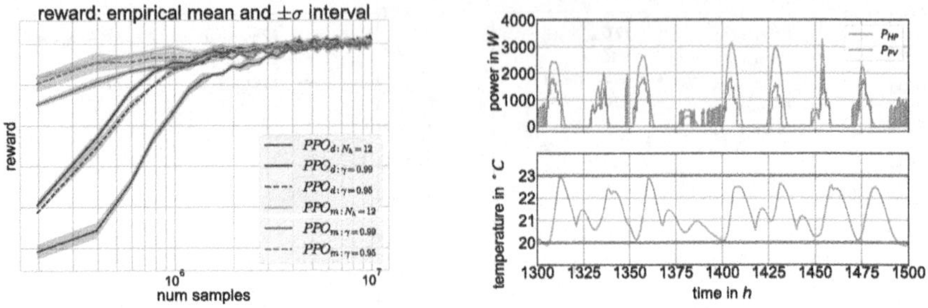

(a) (b)

Fig. 5. a) Evolution of reward during training. Solid lines: mean performance. The shaded regions: lower 10th and upper 90th percentiles over 6 trials. Note the significant variation in the initial phase. **b)** Top: Electricity production with PV and the actively modulated power consumption of the heat pump. Bottom: Room temperature within the target range of 20–23 °C.

5 Conclusion and Outlook

We have shown that an RLC has a great potential for saving primary energy. In contrast to MPC the RLC approach avoids the expenses and compromises associated with developing a control-oriented model.

The use of a mixed agent has significantly accelerated the learning process, achieving performance that is at least comparable to, if not superior to, a standard PPO agent. Nevertheless, the amount of data required remains impractical for online learning applications. In fact, we would need a reduction of several orders of magnitude to eliminate the dependency on building simulations.

While deep RL provides a comprehensive framework for deriving optimal control strategies directly from data, further progress is needed to effectively apply RL to real-world BES. Within the current range of RL methods, two intriguing paths for exploration emerge: One involves initiating the learning process with a RBC as the initial policy, rather than starting with a randomly initialized neural network. This approach would facilitate the learning process with a reasonable performance baseline. Alternatively, we could leverage pre-trained models and adapt them, either with or without additional training, for applications in BES.

References

1. Balali, Y., Chong, A., Busch, A., O'Keefe, S.: Energy modelling and control of building heating and cooling systems with data-driven and hybrid models-a review. Renew. Sustain. Energy Rev. **183**, 113496 (2023)
2. Bemporad, A.: Reference governor for constrained nonlinear systems. IEEE Trans. Autom. Control **43**(3), 415–419 (1998). https://doi.org/10.1109/9.661611
3. Bolt, P., Ziebart, V., Jaeger, C., Ritzmann, R., Meier, O., Füchslin, R.M.: Model predictive control for building automation, pp. 1330–1341. International Solar

Energy Society. https://doi.org/10.21256/zhaw-3296, https://digitalcollection.zhaw.ch/handle/11475/16903
4. Fritzson, P.: Principles of Object-Oriented Modeling and Simulation with Modelica 2.1. Wiley, Hoboken (2004)
5. Fux, S.F., Ashouri, A., Benz, M.J., Guzzella, L.: EKF based self-adaptive thermal model for a passive house **68, Part C**, 811–817. https://doi.org/10.1016/j.enbuild.2012.06.016, http://www.sciencedirect.com/science/article/pii/S0378778812003039
6. Haarnoja, T., Zhou, A., Abbeel, P., Levine, S.: Soft actor-critic: off-policy maximum entropy deep reinforcement learning with a stochastic actor (2018). https://doi.org/10.48550/arXiv.1801.01290, arXiv:1801.01290 [cs, stat]
7. Neunert, M., et al.: Continuous-discrete reinforcement learning for hybrid control in robotics. In: Conference on Robot Learning, pp. 735–751. PMLR (2020)
8. OpenAI: OpenAI. https://openai.com/ (2003–2021). Accessed Sept 2021
9. Schulman, J., Moritz, P., Levine, S., Jordan, M., Abbeel, P.: High-dimensional continuous control using generalized advantage estimation. http://arxiv.org/abs/1506.02438
10. Schulman, J., Wolski, F., Dhariwal, P., Radford, A., Klimov, O.: Proximal policy optimization algorithms (2017)
11. Stadelmann, T., Tolkachev, V., Sick, B., Stampfli, J., Dürr, O.: Beyond imagenet: deep learning in industrial practice. Applied data science: lessons learned for the data-driven business, pp. 205–232 (2019)
12. Sturzenegger, D.: Model predictive building climate control - steps towards practice
13. Sutton, R.S., Barto, A.G.: Reinforcement Learning: An Introduction. Adaptive Computation and Machine Learning, MIT Press
14. Vázquez-Canteli, J., Kämpf, J., Nagy, Z.: Balancing comfort and energy consumption of a heat pump using batch reinforcement learning with fitted q-iteration **122**, 415–420. https://doi.org/10.1016/j.egypro.2017.07.429, https://www.sciencedirect.com/science/article/pii/S1876610217332629
15. Wang, Z., Hong, T.: Reinforcement learning for building controls: the opportunities and challenges. Appl. Energy **269**, 115036 (2020)

[illegible]
10. [illegible]
[illegible]
11. [illegible]
[illegible]
12. Sutton, [illegible] ... [illegible]
[illegible]
13. [illegible] ... [illegible]
[illegible]
14. [illegible] ... [illegible]

Author Index

C. Y. Suen et al. (Eds.): ANNPR 2024, LNAI 15154, pp. 333–334, 2024.
https://doi.org/10.1007/978-3-031-71602-7

The manufacturer's authorised representative in the EU is Springer
Nature Customer Service Centre GmbH, Europaplatz 3, 69115 Heidelberg,
Germany. If you have any concerns regarding our products, please
contact ProductSafety@springernature.com

Printed and bound by CPI Group (UK) Ltd, Croydon, CR0 4YY
28/09/2026
02224076-0012